SUDDEN ORIGINS

Fossils, Genes, and the Emergence of Species

Other books by Jeffrey H. Schwartz

The Red Ape

What the Bones Tell Us

Skeleton Keys: An Introduction to Human Skeletal Morphology, Development, and Analysis

Orang-utan Biology (edited volume)

SUDDEN ORIGINS

Fossils, Genes, and the Emergence of Species

Jeffrey H. Schwartz

John Wiley and Sons, Inc.
New York • Chichester • Weinheim • Brisbane • Singapore • Toronto

This book is printed on acid-free paper. ♾

Copyright © 1999 by Jeffrey H. Schwartz. All rights reserved
Published by John Wiley & Sons, Inc.
Published simultaneously in Canada

Illustrations copyright © 1999 by Jeffrey H. Schwartz. All rights reserved

No part of this publication may be reproduced, stored in a retrieval system or transmitted in any form or by any means, electronic, mechanical, photocopying, recording, scanning or otherwise, except as permitted under Sections 107 or 108 of the 1976 United States Copyright Act, without either the prior written permission of the Publisher, or authorization through payment of the appropriate per-copy fee to the Copyright Clearance Center, 222 Rosewood Drive, Danvers, MA 01923, (978) 750-8400, fax (978) 750-4744. Requests to the Publisher for permission should be addressed to the Permissions Department, John Wiley & Sons, Inc., 605 Third Avenue, New York, NY 10158-0012, (212) 850-6011, fax (212) 850-6008, E-Mail: PERMREQ @ WILEY.COM.

This publication is designed to provide accurate and authoritative information in regard to the subject matter covered. It is sold with the understanding that the publisher is not engaged in rendering professional services. If professional advice or other expert assistance is required, the services of a competent professional person should be sought.

Library of Congress Cataloging-in-Publication Data:

Schwartz, Jeffrey H.
Sudden origins : fossils, genes, and the emergence of species / Jeffrey H. Schwartz.
p. cm.
Includes bibliographical references (p.) and index.
ISBN 0-471-32985-1 (cloth : alk. paper)
1. Evolution (Biology) 2. Homeobox genes. 3. Fossils.
I. Title.
QH366.2.S386 1999
576.8—dc21 98-45724

Printed in the United States of America

10 9 8 7 6 5 4 3 2 1

For Lynn

Contents

Preface

Everyone who will read this book probably has not only a general idea of what evolution is but also a mental outline of the major events in human evolution. The fact that interest in the general theory, and in human evolution specifically, seem naturally to go hand in hand is no extraordinary coincidence. For as long as people exist there will be a yearning to understand our place in nature. In the course of trying to unravel this mystery, one is inevitably forced to consider the broader issues concerning life as we know it on earth. Often, thoughts on human origins set the stage for broader evolutionary speculations. Equally often, however, one sees either a book on human evolution or a book on evolutionary theory. Because these two stories are entwined, I have kept them together.

In the pages that follow, I will explore the historical threads that have brought us to our present appreciation of the reality of fossils and genes and the ways in which these aspects of an organism's existence have enhanced our comprehension of evolution. You will see that there was a diversity of interpretations along the way, many of which differed dramatically from the more popular presentations based on Darwinism. All too often, the past is ignored, as if it no longer had any relevance. The scientific literature rarely contains references to studies that are five, much less ten, years old. But it is imperative that we return to the past as often as we can in order to better understand how we came to be where we are and why we think the way we do. In doing so, we are frequently amazed at how much of what we take for granted emerged from an intellectual battle in which one of an array of alternative theories was the victor. Intellectual victory does not, however, necessarily equate with correctness. In the evolutionary sciences, where we are all struggling to piece together a history that can be perceived only through the fragments of fossils or the living termini of a past that is now lost, it would be foolhardy to cling unreservedly to a particular set of models and hypotheses without at least occasionally questioning their very bases. Unfortunately, however, the urge to defend rather than dissect our intellectual roots is quite strong.

It is my hope that you, the reader, will take my discourse and think about its content in its own right. Regardless of whether you and I come to the same conclusion about how evolution works, I think you will see that the time is ripe for a new look at old issues. Recent discoveries, both of fossils that muddy the presumed clear picture of human evolution and of previously unknown kinds of genes that can control whether an organism develops into a fruit fly or a mouse, serve to take us back to where the early evolutionists began: wrestling with the past through the eyes of a scientist as well as an organism that can contemplate its own existence. Because this contemplation is predicated on our own individual life histories and experiences, it is understandable that alternative worldviews have been, and will continue to be, formulated. This book, then, tells part of that story.

Acknowledgments

The process of writing this book was often a lonely one. Thank goodness for the few good-hearted colleagues who not only accepted my request to send them earlier versions of the manuscript for comment but actually took the time to read some or all of what I delivered to them. My brother and literary agent, Laurens Schwartz, did the painstaking early line editing, and my editor, Emily Loose, brought the project into its final form. I thank Emily for her enthusiasm and encouragement. For the discussion of *Hoxd-13* genes, I thank Bjorn Olsen of the Department of Cell Biology at Harvard Medical School; for suggestions, especially on the chapters on Mendel and Bateson, I thank Robert Olby of the University of Pittsburgh's Department of History and Philosophy of Science; and for a thorough critique of the entire work I am eternally grateful to Thomas Gill III, of the Departments of Genetics and Pathology at the University of Pittsburgh Medical Center. Tom could be complimentary and critical at the same time, and he never let an unclear or incorrect thought get by him. And, as so often happens, a chance conversation with Niles Eldredge of the American Museum of Natural History provoked me to consider in greater depth the implications of some of my earlier conclusions. I thank my friends and colleagues one and all for contributing to the completion of this effort. Once again, I am indebted to Timothy D. Smith for his artistic contributions.

A final note: My peers may object to the format of the bibliography. This was dictated by the publisher, but it does fully reference all sources consulted in the preparation of this manuscript.

Introduction

Thanks to the media's having popularized the discoveries of new fossils and their potential for shedding light on our origins, we have a good idea of many of the important events in human evolution. If we were questioned about some of these fossils, most of us would probably mention the very early *Australopithecus afarensis,* the species that the famous specimen Lucy represents, or much more recent species, such as *Homo erectus* or *Homo neanderthalensis.* Although the earliest members of our specific evolutionary group—the hominids—are more than 4 million years old, the story of human origins actually begins hundreds of millions of years earlier, with the origin of cellular life and the subsequent emergence of a mind-boggling diversity of fabulous and complex organisms. For if these evolutionary events had not occurred there would never have been occasion to investigate the evolution of our own group and to muse about the greatness of being humans. Although the finer historical details of our own evolutionary group may remain shrouded in mystery, many of the clues to the earlier evolutionary events that gave rise to the potential ancestors of our earliest ancestors seem to be lost forever.

We think we know that, somehow, humans and apes are related. And that, perhaps, one of the 4- to 14-million-year-old fossils from Africa or Eurasia was the ancestor from which apes and hominids split and proceeded along their separate evolutionary paths. We also think that one of the fossils from even older deposits in northern Africa, well over 30 million years ago, might have been the ancestor of the group of primates that includes humans, apes, and monkeys. Fortunately, in spite of the normally patchy fossil record of land mammals, the evolutionary history of primates and the evolutionary relationship of primates to other animals is reasonably well understood. But what about the origin of other groups of animals? Of mammals? Of the group that subsumes mammals, the vertebrates? Or of the first life forms that were not invertebrates, the chordates?

You and I are chordates. A chordate is an animal that, for at least some part of its life, has a stiff rod of cartilage along its back. The primary trunk of a chordate's central nervous system runs parallel to this rod. A chordate's mouth is at the front end of its body. Its anus is at the opposite end, near the base of the tail. A chordate is also a bilaterally symmetrical animal, which means that it has a right side and a left side. You and I belong to a subgroup of chordates, Vertebrata. Vertebrates are animals in which the stiffening rod becomes segmented during early development into separate, often bony units called vertebrae. Vertebrates also have a brain that is encased in a projective, usually bony shell. Vertebrates have paired appendages: two behind the head region and two closer to the anus. Most vertebrates have teeth, which may be continually replaced by new teeth throughout their lives.

Nobody knows for certain where chordates came from, who their last common ancestor was. As far as the fossil record goes, for millions of years there were no chordates. Then, suddenly, they appear in the fossil record. Equally as suddenly, after millions of years, vertebrates emerge in the fossil record, replete with lots of vertebrae, paired appendages, and dermal plates that covered not only the head but the entire body. These vertebrates, however, lack jaws. But after additional millions of years, vertebrates with jaws appear, and they are quickly followed and replaced by jawed vertebrates with teeth. And these vertebrates don't merely have some teeth; they have complete sets of teeth, which are not only spread along the full length of their jaws but also last a lifetime, with waves of new teeth forever replacing the old set of teeth.

Where is the trail of intermediates, of missing links leading from invertebrates to chordates to vertebrates? The typical paleontologist's answer has been, simply, that these fossils haven't been found yet. These missing links existed, of course, and should be there in the fossil record waiting to be discovered. But, the explanation goes on, the vicissitudes of preservation and fossilization have—though hopefully only temporarily—continued to thwart paleontologists who have attempted to fill in these annoying gaps. Otherwise, the fossil record would provide us with the full picture of the evolutionary transition of one form into another. This expectation was first articulated most fully more than a century ago by Charles Darwin, who based his speculations on the realization that all life forms are related by virtue of descent from a common ancestor.

When Darwin was formulating his particular theory of evolution, in which natural selection adapted an organism to its environment, he, like his contemporaries and the naturalists of the next few generations, hadn't a clue about inheritance. But anyone could see that there is continuity from one generation to the next, because offspring are similar to their parents in some aspects of anatomy and behavior. If species emerge from other species as individuals do from other individuals, then, Darwin and other evolutionists speculated, the lessons we learned from studying individuals could be applied to species and, ultimately, to evolution.

For Darwin, most aspects of an organism's anatomy or behavior served a purpose and was important for the survival of the individual. Differences between individuals reflected the degree to which they were better or less well adapted in terms of any given feature. Change was brought about by natural selection, which chose the variations that would better serve the individual bearing them, and this, in turn, would be borne out by the better-adapted individuals producing more offspring than those that are less well endowed. Over time, individuals became fitted into, or adapted to, their particular circumstances, and as these circumstances changed, so, too, did individuals, and, ultimately, species.

The only illustration Darwin published in *On the Origin of Species* was a diagram depicting his view of evolution: species descendant from a common ancestor; gradual change of organisms over time; episodes of diversification and extinction of species. Given the simplicity of Darwin's theory of evolution, it was reasonable for paleontologists to believe that they should be able to demonstrate with the hard evidence provided by fossils both the thread of life and the gradual transformation of one species into another. Although paleontologists have, and continue to claim to have, discovered sequences of fossils that do indeed present a picture of gradual change over time, the truth of the matter is that we are still in the dark about the origin of most major groups of organisms. They appear in the fossil record as Athena did from the head of Zeus—full-blown and raring to go, in contradiction to Darwin's depiction of evolution as resulting from the gradual accumulation of countless infinitesimally minute variations, which, in turn, demands that the fossil record preserve an unbroken chain of transitional forms.

Even in Darwin's day there were alternative ideas about how anatomical novelty might arise. The precocious and outspoken English comparative anatomist Thomas Henry Huxley, who did not fully agree with Darwin on gradualism or the role of natural selection but nonetheless championed his intellectual right to advance these concepts, was convinced that new forms came into being not through the modification of the details of their morphology but through the abrupt, large-scale reorganization of entire anatomical systems. For Huxley and his intellectual kin, novelty emerged in a leapfrog, or saltatory, fashion rather than through a tranquil process of gradual evolution. Coincidentally, a version of Huxley's saltational picture of the history of the diversification of life had earlier been demonstrated in the fossil record by one of the most vocal of the opponents of evolution.

During the late eighteenth to the early nineteenth century, the anti-evolutionary, biblically informed yet brilliant French comparative anatomist and father of paleontology, Georges Cuvier, documented that the extinct plants and animals entombed in the limestones of the Paris Basin did not form a gradual succession of species across time. Rather, species in one stratum were usually not found in the next one up in the geologic sequence. Consequently, with the rare case of a species persisting from one layer to the next, older species were completely replaced by suites of younger species. In addition, the strata themselves did not flow smoothly into one another but were delineated by breaks or discontinuities between them. Cuvier believed that the disjunction between adjacent geologic layers and the species they contained could have been brought about only by a catastrophic event, such as the flood described in the Bible. Although there was only one biblical flood, Cuvier invoked a history of floods to explain the disjointed succession of life and rocks preserved in the geologic and paleontological records. After each catastrophic flood, his god repopulated the earth with new species. Cuvier also expanded the biblical reference to a single Garden of Eden to many such gardens, each of which provided a haven or refuge for the species that managed to survive each flood. The biblical interpretations aside, however, Cuvier's observations were real. Species did not grade smoothly one into another over time. As would be proposed by Huxley and other evolutionary saltationists, novelty seemed to appear out of nowhere.

Beginning in the late nineteenth and continuing into the twentieth century, the debate among evolutionists was phrased in terms of continuous versus discontinuous variation. Were the features of individuals and, consequently, of their species, analogous to the colors of the spectrum, which, when viewed collectively, blend imperceptibly one into another? If so, then the individuals within a species, as well as the species themselves, could be arranged to form a continuum of variation. Or, to the contrary, no matter how similar these individuals might seem, were their features discretely different from one another? If this was so, individuals within species, as well as species, did not form an unbroken continuum of slight gradations of difference: Individuals and species were distinguishable because their differences were disjunct, or discontinuous. The side of the debate in favor of discontinuous variation was promoted by two of the leading figures in the emergent field of genetics: the Englishman William Bateson and the Dutchman Hugo de Vries.

At the beginning of their careers, both Bateson and de Vries were supporters of Darwin. In fact, Bateson, who trained in physiology and comparative developmental anatomy, sought to document Darwin's suggestion that organismal adaptation was tied to—indeed, that it tracked—change in the environment. After years of field studies, however, he was forced to admit that he could not find any support for Darwin's suppositions. Upon discovering and then championing the earlier genetic studies of the Austrian monk Gregor Mendel, Bateson became further convinced that Darwin's evolutionary explanations were incorrect.

In Darwin's eyes, variation formed an unbroken continuum from one individual to the next within a population, because he believed that the features of both parents became "blended" together in the creation of their offspring. Mendel, however, demonstrated through controlled experimental breeding of garden peas that physical features are tied to discrete units of inheritance that do not lose their integrity from one generation to the next through a process of blending. The reason that one or more characteristics present in one parent might not be expressed in the offspring is that the underlying units of inheritance are masked by the counterparts inherited from the other parent. Because units of inheritance are discrete entities, they can persist over many generations, and then, when the right combinations come together, produce the morphological features they represent. If, as seemed to be the case, Mendel was correct about units of inheritance being distinctly separate entities, then, Bateson argued, the features these units of inheritance generate would never form, as Darwin maintained, an uninterrupted continuum of variation within a species or between species. Consequently, at no level—the unit of inheritance, the feature, the individual, or the species—was there such a thing as continuous variation. Rather, variability, or, to be more exact, difference, was real and discontinuous.

Hugo de Vries also saw disjunction, rather than a continuum, as the underlying motif of the diversity of life. Like Bateson, he embraced Mendel's ideas. Indeed, de Vries was one of the original rediscoverers of Mendel's work. But de Vries was particularly predisposed to discreteness in morphology because of his own studies on spontaneous mutation in the evening primrose. The sudden occurrence—within the space of a generation—of plants with differently configured leaves or flowers made it impossible for de Vries to accept Darwin's gradualism and its underlying premise of continuous variation. In his mutation theory, de Vries argued that the origin of species is an abrupt phenomenon. New species arise through significant genetic change, not, as would be deduced from Darwin's model, by shifting around the expression of already-present genetic material. De Vries was also skeptical of Darwin's concept of natural selection. Rather than picking and choosing the "best" or "most adaptive" features, natural selection, as de Vries envisioned it, was a force that eliminated the worst features, leaving behind an array of characteristics that were of no harm to the individual. It was only at the level of the individual and the daily lives of the members of a species that de Vries envisioned any possible role for natural selection, weeding its way among minor individual variations.

The rejection of Darwin's theory by an increasing number of population geneticists continued into the twentieth century. Among the most vocal of these opponents of Darwinism was the American Thomas Hunt Morgan. Like many of his colleagues, Morgan initially trained in comparative developmental anatomy. In the early phase of his career at Bryn Mawr College, he was convinced that neither Darwinian nor Mendelian explanations were relevant to an understanding of evolution and the origin of species. Even when he later converted to both doctrines, Morgan refused to embrace natural selection as a force in producing change. As he saw it, new features just happened, or chanced, to appear, and whether they persisted in a species was also a matter of chance. Essentially in keeping with de Vries, Morgan's attitude was: If a new feature didn't kill you, you had it. As for Darwin's concept of continuous variation, Morgan rejected this conceit as well.

But in 1915 Morgan and his Columbia University collaborators in experimental studies on fruit-fly population genetics unearthed what was seemingly the Rosetta stone of genetics and evolution. One day, upon returning to the laboratory, they discovered that, amid their fruit flies, there was now an eyeless mutant. When they bred this mutant with normal individuals, they found that they could, over many generations, gradually increase the number of eyeless individuals. By such manipulative experimental breeding, Morgan and his colleagues could change the dominant character of a population of fruit flies, and they did so many times. They were particularly impressed by their ability to shift a population of fruit flies from one in which the wings were typically a bit longer than the body to one in which the wings were usually a bit shorter than the body. They speculated that were this process to continue indefinitely, their fruit flies would end up being wingless.

In one fell swoop, these discoveries convinced Morgan that both Mendel and Darwin were correct. As Mendel had predicted, morphological characters are indeed tied to discrete genetic units—which, by then, had become identified as genes—that were inherited according to very consistent rules. A gene can appear in one of two states, or alleles: dominant or recessive, with the former always being expressed over the latter. Each parent possesses a pair of alleles (in whatever combination) for each gene and passes on to their offspring one allele from each pair. Even if there is only one dominant allele in the offspring's pair, the offspring will still possess the morphology dictated by that allele. Only when an individual inherits two recessive alleles will it develop the feature these gene states represent. Given the behavior of dominant and recessive alleles, a recessive allele can be passed on for generations, spreading throughout the population, without ever being expressed.

Since Darwin's theory of evolution relied on natural selection to shift the average frequencies of traits and so produce change, the question was: Where did new features come from? De Vries invoked mutation, but mutations on a large scale. Morgan also underscored the role of mutation in introducing novelty into a population. However, because of his experiments on fruit flies—such as gradually changing wing length—Morgan thought that the only evolutionarily meaningful mutations would be minor, not of the magnitude espoused by de Vries. Putting the whole package together, Morgan proposed the following scenario.

As Bateson had demonstrated decades earlier, mutations typically arise in the recessive state. This means that it will take some time before the effect of a mutation is expressed in members of the population. In keeping with Darwin's notion of continuous variation, the mutations and the morphological novelties they produce add only slightly to the variability in the population. By themselves, these novelties are minor. But over time the gradually accumulated effects of these small mutations would be profound. Natural selection had nothing whatsoever to do with provoking a mutation, with introducing novelty. That happened by chance. Only after a new feature became established in a population might it be fine-tuned by natural selection. This was the process that pushed a species along evolutionarily, allowing it to fluctuate smoothly and continuously as its surroundings changed. The case seemed closed: Mendelism and Darwinism were compatible with the notion that low-level variation is the fodder of gradual evolutionary change.

Fired up by the impact of melding Mendelism with Darwinism, R. A. Fisher, a young British geneticist and mathematical wizard, burst onto the scene in 1930 with a theory that represented adaptation in terms of an individual's success in passing on his or her genes. Among Fisher's concerns was the question of how a genetic theory of evolution and adaptation could be applied to the origin of species. Of course, he accepted mutation as introducing novelty. But because a mutation usually arose in the recessive state, and would not be accessible to selection, Fisher proposed that there had to be a mechanism by which this mutation was quickly converted into the dominant state. If the mutation was advantageous, it would spread by way of natural selection. As he saw it, the spread of mutations would be slow and species change would be gradual.

Fisher also tackled the problem of how species multiplied in number, which was a question that Darwin had left unanswered. Fisher suggested that, to begin the process, members of the same species had to become isolated from one another. The simplest way to achieve separation was to have a barrier, probably a geographic barrier, split a species into two groups—in Fisher's mind, two large groups. This barrier would curtail gene flow between the two groups. The two groups having been separated, mutation and natural selection would gradually push them in different directions. Eventually, each group would be recognizable as a different species.

But Fisher was not the only mathematically minded population geneticist to emerge during the 1930s with new ideas about evolution. In Great Britain, there was J. B. S. Haldane, and the United States had Sewall Wright.

Haldane's and Wright's ideas differed from Fisher's in various significant ways. They did not demand that the mutant recessive allele be quickly converted into the dominant state so that selection could act on it. Rather, they preferred to allow the mutation to spread silently throughout the population while still in the recessive state, until it was widely distributed. Eventually, when the mutation was expressed in offspring that inherited a pair of these mutated recessive alleles, there would be a number of individuals with the new feature. Then natural selection could decide what to do with it. At some point, the mutant recessive would be converted into the dominant state. In addition to mutation, Wright proposed that evolutionary novelty results from natural selection's choosing different combinations of adaptive features, which, in turn, result from combining genes that are already present in the population in different ways. Haldane preferred to see the environment as affecting the development and, consequently, the final outcome of an organism.

Haldane and Wright also rejected Fisher's notion that change was slow and that it had to occur in large groups of individuals. The two men, but especially Haldane, took a lesson from successful animal-breeding experiments and maintained that, because a mutation will spread quickly in a relatively small population through inbreeding, new species will arise rapidly. But at this point Wright and Haldane parted ways. Wright's "small population" was not as small as Haldane's, and it was never completely separated from the parent population. It was just far enough on the periphery of its species' range to be subject to different forces of natural selection and small enough for changes to accrue quickly. The novelties that arose in Wright's peripheral population—by mutation, but most likely by new genetic combinations—would be infused into the large parental population, which itself would then experience rapid change. Haldane's model was simply that change would be rapid because the truly small peripheral population would be totally isolated and inbreeding would hasten the spread of the mutation. In addition, being on the periphery implied an element of interaction of the developing individual with its particular environmental circumstances.

By the late 1930s, genetics had taken the foreground in discussions of evolution. Although such naturalists as Charles Darwin had begun the revolution in thinking about evolution, and paleontologists, comparative anatomists, and embryologists had been instrumental in making evolutionary science a viable field of study, these scientists' endeavors were no longer considered central to the mission. In the early 1940s, when representatives of these different disciplines convened to try to put together a uniform evolutionary theory—the Synthetic Theory of Evolution, or, as it became known, the Synthesis—genetics became the language of evolution. Consequently, when naturalists, such as the ornithologist Ernst Mayr, who was a member of this committee, discussed the question of how species might form, they followed Fisher in couching the problem in the context of inheritance: how to disrupt gene flow between subgroups of a species so that mutation and natural selection would push the two groups apart morphologically, behaviorally, and genetically.

Faced with alternative suggestions as to how slowly or quickly speciation could occur, and whether it would involve many or few individuals, Mayr and his colleagues opted for "slowly" and "many." They had no time for the alternative theory proposed by the German geneticist Richard Goldschmidt. Goldschmidt maintained that the genetics of speciation is totally different from the population genetics of everyday life. Rather than resulting from the gradual accumulation of minor mutations, the novelties that distinguish a new species, he argued, derive from an abrupt, almost catastrophic genetic reorganization that occurs within the space of a single generation. Mayr, the geneticist Theodosius Dobzhansky, and the vertebrate paleontologist George Gaylord Simpson all went out of their way to ridicule and discredit Goldschmidt. Unfortunately, these attacks were made even more vicious not only because Goldschmidt had been rather abrasive about his contemporaries' narrow-mindedness but also because he proposed that the unsuspecting bearer of his hypothesized major genetic revolution, given that it was so potentially different from its parents and siblings, should be called a "hopeful monster"—"monster" referring to its novelty and "hopeful" to its need of a similarly mutated mate. To make matters worse, Goldschmidt could not deal adequately with the problem of ensuring that there would be more than one "hopeful monster," of each sex, at any point in time. This was a necessity because, for sexually reproducing organisms, there would have to be at least one male and one female of a new species in order for the species to become established.

Goldschmidt's theory did find favor with the German paleontologist Otto Schindewolf. By the time of the Synthesis, many prominent paleontologists, led especially by the American G. G. Simpson, were convinced that the fossil records of various groups, particularly the horse, demonstrated the cumulative process of evolutionary change. Schindewolf, however, looked at the same fossil records, including that of the horse, and saw otherwise. New species emerged suddenly. If the new species turned out to be the ancestor of a number of species, then the group they formed would have had an abrupt beginning that was heralded by the extraordinary novelties of their ancestor. Although the magnitude of the new features of species within the group may not be as profound as their ancestor's, their origins were, nonetheless, still abrupt. The picture of evolutionary change as portrayed in the fossil record was totally consistent, Schindewolf noted, with the expectations of Goldschmidt's theory of major genetic reorganization. One might have expected Schindewolf's defense of Goldschmidt's ideas to elicit scorn, but while Goldschmidt received brutal attacks, curiously, Schindewolf emerged relatively unscathed. However, Schindewolf's interpretations of the fossil record also remained unappreciated.

As far as the founders of the Synthesis were concerned, there was no question that evolution occurred by the gradual accumulation of minor mutations. A new species could arise from another by the simple process of gradual transformation. A single species could give rise to two species when a barrier to gene flow was imposed between its members, creating groups that then went off in different directions and were then gradually transformed into new species. This, of course, was Fisher's general idea about species, but Mayr's concept of a species, which dominated the Synthesis, went farther. As Mayr saw it, a species is an aggregate of individuals that, even if separated by a barrier, can potentially interbreed and produce offspring that, in turn, can successfully interbreed. For Mayr, it was not just geographic separation but genetic incompatibility that distinguished one species from another. In addition, Mayr argued, speciation could occur only when there was a vacant niche into which the new form could expand and become adapted. As Darwin had theorized, a species was a perfect fit for its environment.

Because it was based on the reproductive potential of living populations, Mayr's species definition could not be applied directly to fossils. Since one could not observe extinct individuals interbreeding, the next best thing was to fall back on Mayr's claim that speciation will occur only when there is a vacant econiche into which a new species can become adapted. Being adapted to a particular econiche meant that there should be something (or somethings) distinctive in the morphology of the members of a species that reflects that adaptation. With any luck, one or more of these adaptive features will be preserved in the fossilized anatomy of extinct organisms.

With this synthesis of Darwinism and Mendelism, the evolutionary sciences, paleontology included, could then go about the business of trying to figure out the details of the past and the present. But the notion of gradual transformational evolution was challenged again in the early 1970s, not by geneticists this time but by paleontologists. Two Columbia University graduate students in invertebrate paleontology, Niles Eldredge and Stephen Jay Gould, proposed the model of punctuated equilibria. Their emphasis was on the fact that the picture of evolution that one gets from the fossil record is not of species gradually changing over time but, rather, of the emergence of species in a very short period of time. Once established, a species actually remains relatively unchanged for the duration of its existence.

Eldredge and Gould's evidence derived from their studies of trilobites and snails, which, being preserved in large numbers, provided a reasonably intact record of their evolutionary past. Because of this, Eldredge and Gould could see that what could have been a gap between related species was actually filled by a fine trail of links between the old and the new species. This was not the picture evolution should present if it occurred, as Mayr and others proclaimed, by the gradual transformation of large groups of individuals. The rather abrupt appearance of a species, together with the preservation of small numbers of individuals connecting an ancestor to its descendant species, suggested to Eldredge and Gould that speciation begins in small peripheral populations—as Wright and especially Haldane had proposed. As viewed through the eyes of punctuated equilibria, the gaps in the fossil record are not inconvenient potholes in the path of a gradual picture of evolutionary change. Instead, these gaps are a reflection of a very rapid process of speciation that involves only a fraction of the original population.

For the most part, neither geneticists nor morphologists found anything acceptable about the model of punctuated equilibria. Even when Eldredge and Gould responded to five years of accumulated criticism, a great wall of resistance to punctuated equilibria remained, especially among population geneticists. The geneticists maintained that there was absolutely no need to invoke an unprovable hypothesis—and one based on supposedly negative evidence, to boot—when simple population genetics could easily explain how a species could be changed over a long period of time by the introduction of minor mutations and the action of natural selection.

Although during the 1980s many paleontologists embraced the idea of rapid speciation, the model of punctuated equilibria continued to run up against one key objection: Natural selection, acting on minute variations, which were introduced by equally minute mutations, was all that was necessary to explain evolutionary change. Moreover, the kind of large-scale genetic alteration that seemed necessary to produce a new species, according to Eldredge and Gould, was simply incompatible with what population geneticists knew at the time. But while these debates were going on, another kind of geneticist—who focus was how organisms develop from embryo to adult form—was busy at work.

Almost from its beginnings, the discipline of biology has had its share of embryologists, scientists who study the growth and development of an organism from fertilized egg to mature adult. When embryology was applied to evolution, it was believed that an organism's evolutionary past was revealed through the course of its development. Various comparative anatomists, such as Thomas Huxley, interpreted the developmental stages shared by different organisms as common evolutionary phases that preceded those during which an organism acquired the characteristics of its species. Ernst Haeckel, the greatly influential German comparative anatomist, embryologist, and all-around evolutionist, wrongly believed, however, that these common developmental phases (such as the gill phase) actually represented real ancestors (such as a fish). In addition, Haeckel proclaimed that each phase (such as the gill phase) represented an adult form (an adult fish). As Haeckel saw it, all one had to do was study the development of an organism from the beginning and a parade of its adult ancestors would be revealed. This, of course, was a ridiculous proposition, but it did impact the study of human evolution from the end of the nineteenth century into the early decades of the twentieth. In turn, increasing interest in human development played a significant role in the formulation of novel evolutionary ideas.

A striking realization was that, compared with most other animals, human adults retain many features that are characteristic of the fetus or child. For instance, human adults have big heads and brains and small faces, and they are essentially hairless. Also, compared with most other animals, humans take a long time to mature physically. In most other animals, the morphological changes in their bodies occur more rapidly and are more substantial. One embryologist, Louis Bolk, went so far as to proclaim that humans are actually reproductively mature fetuses.

Bolk's musing aside, many embryologists became interested in how the timing of sexual versus physical maturation determines the adult state of an organism. The fundamental discovery was that when an animal becomes reproductively mature, it essentially stops changing. In many animals, the developing individual loses its juvenile features and becomes morphologically adult at the same time that it becomes reproductively mature. Humans, however, prolong the rate at which they mature physically relative to the time it takes to become reproductively mature. This is why adult humans look physically childlike. In other animals, accelerating the rate of sexual maturation freezes the individual in the body of a juvenile.

Using this kind of information, in the 1920s Walter Garstang proposed a simple and elegant explanation for the origin of chordates from an invertebrate ancestor. Garstang's invertebrate of choice came from among the tunicates, a group of marine organisms that includes sea squirts. As adults, many species of tunicate attach themselves permanently, as do barnacles, to a rock or some other immobile structure. The larvae of some tunicate species, however, are free-swimming. And some of these free-swimming larvae look like primitive chordates. They have a mouth at one end of the body and an anus at the other. They have a short cartilaginous stiffening rod and a central nerve running along the back. And they have a right and a left side.

To make a chordate, all that needed to happen was for this larval state to persist until the individual reached sexual maturity. A mere change in developmental timing, and the result would be a primitive chordate, whose emergence upon the scene would have occurred without a trail of ancestors. Although Garstang's hypothesis remains a popular explanation for chordate origins, its broader evolutionary implications have not been appreciated by many: Alterations in how development occurs can produce profound morphological differences and lead to new species, essentially in an instant. Despite the promise of this striking insight, developmental biology was all but ignored in the formulation of the Synthesis.

But the field of developmental biology did not curl up and die because of this neglect; and, in the 1980s, the door to incorporating developmental insights into evolutionary theory was finally opened with the discovery of a class of highly conserved regulatory genes, called homeobox genes.

Homeobox genes control an organism's development by means of sending signals from one to another in the form of the proteins they produce. As demonstrated in the fruit fly, the cell that eventually gives rise to the egg cell receives the messages that determine what will be the head and tail and up, down, right, and left sides of a potential offspring by a back-and-forth signaling, carried by proteins, between homeobox genes in this cell and the cells of the ovary around it. The animal that then emerges from the egg that derives from this predetermined, pre–egg cell obtains its specific features through the process of turning on and off certain homeobox genes at different times in different regions of its developing body.

All animals, from unsegmented worms to fruit flies, starfish, tunicates, zebra fish, chickens, mice, and humans, share essentially the same basic homeobox genes. Since all of an organism's genes are contained in each and every one of its cells, the striking morphological difference between animals lies basically in which cells and when during development one or more homeobox genes are active. For instance, a radially symmetrical animal, such as a starfish or a sea urchin, which is arranged like spokes of a wheel around a central spoke, begins life the same way as an animal that has head and tail ends and up, down, right, and left sides. But then three particular homeobox genes are turned on simultaneously in a series of cells that surround a central axis. The result is a starfish or a sea urchin. In a fish, for example, the same three homeobox genes are activated sequentially from head to tail and produce structures of the midline skeleton and nervous system. Tunicates have the homeobox for producing teeth, but this gene is active only in vertebrates. Fruit flies, frogs, mice, and humans have the same homeobox genes for making an eye, yet the eyes of fruit flies have multiple fixed lenses, compared with the single, flexible lens of frogs, mice, and humans, and the protein in the frog lens is different from that of the mammals. Tetrapods, animals with forefeet and hind feet, have the same homeobox genes as fish. In fish, these genes are activated only along the back side of the developing fin bud, whereas in tetrapods they are turned on along the back as well as across the front of the elongating limb bud.

The importance of homeobox genes for understanding evolution has certainly not been lost on the developmental geneticists studying and identifying them, or on the evolutionary biologists who have sought to inform their comparative anatomical studies with insights about growth and development. And for good reason. It is mind-boggling to realize that, for all intents and purposes, many differences between a fruit fly and a human may lie pretty much in where and when certain homeobox genes are activated. To be sure, there are some other differences between a fruit fly and a human at the molecular level. But, fundamentally, the main difference between organisms lies in alterations in development that result from differences in the timing of homeobox gene activity.

When Darwin proposed his theory of evolution by way of evolution, perhaps the most troubling thing for him was that he was going against church doctrine. According to the Bible, a Divine Creator placed all living creatures on this earth. As Darwin saw it, all life was connected by way of a common origin. In retrospect, he was correct, but for the wrong reason. Life is less connected by a trail of transformation from one state to another than by a commonality of homeobox genes. Differences between organisms derive less from the addition of new regulatory genes than from the novel combination of existing homeobox genes. In short, genes do not evolve specifically for the creation of a particular organism, whether it be a human being or a worm. There are no such things as genes for a worm or genes for a human being. Rather, particular combinations of the genes that are already present in all organisms lead to the development of organisms that have specific morphologies. Viewed in this light, it becomes clear that different organisms—new species—would not arise by a gradual accumulation of minor changes. Rather, a mutation affecting the activity of a homeobox gene can have a profound effect—such as turning Garstang's free-swimming larval tunicates into the first chordates.

Clearly, the potential homeobox genes have for enacting what we call evolutionary change would seem to be almost unfathomable. But are we at the same place Hugo de Vries was when he proposed his mutation theory, or Richard Goldschmidt his "hopeful monsters"? Must we discard a model that has enjoyed so much success in explaining evolution because we cannot integrate it with those tantalizing pieces of a puzzle that suggest something punctuational? And must we reject the notion of punctuation because it is not clear how we should incorporate it into popular evolutionary views? The answer to all of these questions is no. Their resolution lies in understanding how a mutation in a homeobox gene—either a mutation that simply turns a gene on or off, or one that duplicates or changes it slightly—fits into the generally accepted Darwinian-Mendelian framework of evolution. And, it turns out, there is a very simple way in which these seemingly incompatible schools of evolutionary thought can come together—without invoking special pleas or unknown causes. In fact, many of the pieces of the answer have been in front of us for decades. But in order to see them we have to start at the beginning, with how people have perceived the world around them and their place in it. This is where our story begins.

1

A Rash of Discoveries

[I]f the historical pathway should forever remain hidden, we can still develop bodies of theory and experiment to show how life might realistically have crystallized, rooted, then covered our globe. Yet the caveat: nobody knows.

—Stuart Kaufman (1995)

The Earliest Human Ancestor

In late 1993, a team of paleoanthropologists, whose members came from Japan, Ethiopia, and the United States, discovered the oldest fossil remains of a potential human ancestor—almost 4½ million years old. These fossils were even older than the famous specimen nicknamed Lucy, which had been referred to as the species *Australopithecus afarensis.* For more than fifteen years, this species had held the exalted position of being *the* earliest human ancestor. The new fossils were found along the Awash River in the northern part of Ethiopia. They came from the same general area in which, some twenty years earlier, Don Johanson and his colleagues had discovered the fossilized remains of Lucy and an array of skeletons of male and female *Australopithecus afarensis* of different ages, which they nicknamed the first family. Once a forested riverine environment, the Awash is now one of the most desolate and inhospitable places on the face of the earth.

The specimens—the bits and pieces—that actually formed the basis of this new, most ancient of potential human ancestors were not very spectacular as far as fossil finds go. There were only fragments of arm bones and skulls and, literally, a handful of isolated teeth, representing different individuals of various ages. But a discovery is a discovery. If these fossils did represent a new and more distant human relative, as the team claimed in the 1994 article that was published in the international journal *Nature,* the rest of the scientific world was going to have to deal with it.

The leader of the group that made this discovery was Tim White, a professor of human paleontology and paleoanthropology at the University of

California at Berkeley. White is a weathered paleontologist who knows the Awash region of Ethiopia well. He has spent years there prospecting for early human fossils, on his own and with the recently retired University of California at Berkeley–based prehistoric archaeologist J. Desmond Clark. It must have been something to see the two of them working together in the field. The much younger Tim White, tall, thin, blond. Desmond Clark, short, wizened, gray, and balding. White, very American. Clark, very British, very proper.

Earlier in his professional life, White had worked with Johanson during the field seasons of the seventies that yielded the Lucy skeleton and other specimens of *Australopithecus afarensis.* White also excavated with Johanson in the eighties at the fossil site of Olduvai Gorge, in Tanzania, East Africa. Decades earlier the British husband-and-wife team Louis and Mary Leakey made this site famous with their discoveries of what were then the oldest hominid fossils. Louis was the paleoanthropologist and Mary the prehistoric archaeologist. White's colleagues and collaborators in the new Ethiopian discovery, Gen Suwa of the University of Tokyo and Berhane Asfaw of the Ethiopian Ministry, were equally well-seasoned paleoanthropologists and fossil collectors, and also had experience hunting for fossils in the Awash region.

In their article in *Nature* announcing this discovery, White, Suwa, and Asfaw gave this most ancient of potential human ancestors a new species name, *ramidus,* to signify that this truly did represent an organism distinct from any other. To a taxonomist, the species is the basic unit of nature and is distinguished by one or more characteristics that are unique to it. As formalized by Carl Linnaeus, the father of modern taxonomy, a species is always subsumed in a genus, which is the next highest category in a biological classification. If two or more species are similar enough in a particular way, the taxonomist may choose to put them in the same genus, which would be defined on the basis of these common features. Following the accepted procedure of assigning a species to a genus, White and his colleagues kept their new species in the already established genus *Australopithecus,* which had become the paleoanthropologist's choice for subsuming species of potential early human ancestors. Consequently, when they announced the discovery of a new and even older potential human ancestor, it was presented as *Australopithecus ramidus.*

White and his colleagues decided on the genus and species *Australopithecus ramidus* to accommodate their fossil for the following reasons. First, they considered that, although their new find could easily be distinguished from the species *Australopithecus afarensis*—primarily because their fossil was more apelike and less generally human or hominid-like—the species *ramidus* could still be comfortably accommodated in the genus *Australopithecus.* As for choosing the species *ramidus,* White, Suwa, and Asfaw explained in their article that they had coined the name in honor of the Afar people of the Awash region. *Ramidus* is the Latinized version of *ramid,* which is the Afar word for "root"—which *Australopithecus ramidus* would certainly be if it truly were the ancestor of all other hominids, the group that includes all potential, non-ape, human relatives.

News of the newest and oldest hominid ancestor spread through the public media as well as the major scientific journals. Since all that was known skeletally about *Australopithecus ramidus* came, essentially, from the waist up, White, Suwa, and Asfaw were anxious to get back into the field to discover more evidence of this tantalizing fossil hominid. However, after their lead article, in another issue of *Nature,* sequestered in its back pages, where corrections and addenda to articles previously published in the journal are placed, was a short note in which White, Suwa, and Asfaw took an unusual action. There, they removed the species *ramidus* from the genus *Australopithecus* and put it into a new genus they had created for it, *Ardipithecus.* According to the news reports, White and his colleagues could not be reached for comment because they were in the field at the time of publication.

The reason White and his colleagues' action was unusual was that no new information had been provided about *ramidus,* either in the form of a description of additional specimens or in the addition of previously overlooked details of anatomy. The note essentially constituted the belatedly expressed opinion that perhaps the features that distinguished the species *ramidus* from the species *afarensis* and other species assigned to *Australopithecus* should more appropriately be recognized at the higher taxonomic level of the genus. *Ardipithecus ramidus* was still supposed to belong to the family group—the hominids—into which taxonomists classified *Homo sapiens* and potential fossil relatives of *H. sapiens.* But because it was supposed to be the most primitive and apelike of any putative hominid, it seemed, in retrospect, that placing it in its own genus was a taxonomically reasonable and logical thing to do.

For those in the profession who are sticklers about the procedures and protocol of naming a new species or genus, White, Suwa, and Asfaw's emendation was not the proper way to go. If, as these critics objected by way of the professional underground, White and his colleagues had taken a bit more time to think about it, they would have published the new genus name, in the appropriate manner of presentation, in the original *Nature* article. I overheard one paleoanthropologist say, however, that he was glad that White and his colleagues had kept the first letter of the new genus name the same as in *Australopithecus.* At least this paleoanthropologist would not have to change his newly redone slides of hominid evolutionary trees, in which he had used the convention of denoting a species' genus by the first letter of the name. In his slides, *A. ramidus* remained *A. ramidus,* in spite of the change in the genus name.

But the paleoanthropological rumor mill was circulating two possible reasons why White, Suwa, and Asfaw might have taken such an untraditional course in naming *Ardipithecus.* One scenario was that they had become nervous about providing enough information on this new fossil hominid to enable some other paleoanthropologist to step in and suggest that their discovery actually deserved to be recognized at the genus level. The potential taxonomic interloper whose name was being bandied about was Walter Ferguson, a zoologist on the faculty of Tel Aviv University.

Ferguson has made a career of assigning new genus and species names to other people's fossil discoveries. This is not to say that the discoverer of a fossil is necessarily the person most qualified to study it and decide whether it represents a new species, or even a new genus. Sometimes this is the last person who should do the definitive study, while someone who has never done any fieldwork but was trained as a systematist—someone schooled in the problems of species identification and the sorting out of evolutionary relationships—would be most qualified for such a task. I have even renamed, at the species and genus levels, specimens that had been collected by others and housed in various museums and institutions under what my studies suggested were the wrong labels. But Ferguson is famous for doing that, and his particular prey are hominid fossils of the ancient vintages collected by the most visible of paleoanthropologists, such as Johanson and the Leakeys. *Ramidus,* so the scuttlebutt went, was a sitting duck for Ferguson.

The other possible reason why White, Suwa, and Asfaw had been so quick to change the genus name of *ramidus,* the rumormongers speculated, was that they had already discovered more parts of the skeleton of their putative hominid ancestor—and these bits and pieces were not anatomically hominid-like at all. By removing *ramidus* from a genus, *Australopithecus,* that everyone in the discipline was convinced was hominid, and that the media and interested public had been told for decades was hominid, White and his colleagues were setting the stage for distancing *ramidus* from the hominid group.

In the first place, the features that had been presented in support of *Ardipithecus ramidus* being a hominid were few: mainly, the elliptical shape of the upper end of its upper arm bone, and, on the base of the skull, the forward position of the region that articulates with the vertebral column. On the other hand, the lower end of *A. ramidus*'s upper arm bone and the elbow joint to which it contributed were distinctly apelike. And so were its teeth. Not only did the teeth of *A. ramidus* differ from the teeth of accepted hominids in shape and relative proportions but *ramidus*'s molar teeth bore only a thin covering of enamel. The teeth of most monkeys and both African apes, the chimpanzee and the gorilla, have thin molar enamel. To paleoanthropologists, thick molar enamel is a hallmark of the hominid group.

But White, Suwa, and Asfaw interpreted the forward position on the skull base of the articulation with the vertebral column to mean that *Ardipithecus ramidus* had been a bipedal primate. After years of haggling among paleoanthropologists about which features distinguish our evolutionary group from all other primates, bipedalism had come to be regarded as one, if not the only, truly diagnostic feature of being hominid. Obviously, in White and his colleagues' opinion, evidence of bipedalism took precedence over all else in declaring *A. ramidus* a hominid.

White and his colleagues had to provide the additional evidence—from the pelvis, the hip, knee, and ankle joints, or from the arrangement of the toes—that would back up their suggestion that *Ardipithecus ramidus* had been a biped and that, as a biped, it qualified evolutionarily as a hominid. They would have to get more information, because some primates that are clearly not bipeds also have a markedly forward position of cranial-vertebral articulation. Maybe, the armchair paleoanthropologists speculated, White, Suwa, and Asfaw had already found pieces of the hips, knees, ankles, or feet of *A. ramidus,* and the shapes of these parts of the skeleton did not conform to those of a biped. Maybe, as would have been a reasonable conclusion if only the elbow joint or the teeth had been all that was known, *A. ramidus* was actually a fossil ape and not a hominid after all.

The issue here is not what motivated White and his colleagues to remove the species *ramidus* from the genus *Australopithecus* and put it into a new genus created especially for it. What is more important is whether their data are accurate—and accuracy is what White, Suwa, and Asfaw, being careful paleontologists and comparative primate anatomists, are known for. What this situation emphasizes is a more fundamental question: What makes a hominid distinct from other primates?

When we consider living primates alone, this seems a relatively easy question to answer. Aside from *Homo sapiens,* there is no other living primate that is a habitual upright biped. Aside from *H. sapiens,* there is no other living primate with such a large brain-to-body ratio. Taxonomists of the seventeenth, eighteenth, and nineteenth centuries emphasized the anatomical gulf between humans and other primates, including the most humanlike or "anthropomorphous" of the apes. In turn, this emphasis made it appear that we humans are so different from the rest of the animal world that whatever we "discover" about ourselves has to be unique to us and us alone. But if you think about it for a moment, you will realize that our situation is a very unnatural one.

Humans, *Homo sapiens,* are the only living species of what we now know had once been a very species-diverse evolutionary group. The three living great apes—the orangutan of Southeast Asia and the two African apes, the chimpanzee and the gorilla—are also the only surviving species of their individual evolutionary groups. Unfortunately, we have not yet recovered any fossils that appear to be specifically related to either of the African apes, but we do have a good fossil record of numerous potential orangutan relatives. Orangutans are like humans: They are the sole surviving species of what had once been a very diverse evolutionary group of species.

When you put the whole picture together, you get a very unreal situation in nature—unreal when compared with virtually any other group of related organisms. Even though the three living great apes are more closely related to us than to any of the other living primates, and one or two of these great apes are probably our closest living relatives, the actual, *closest* relatives of *H. sapiens,* and of each great ape, as well, are now extinct.

The vast evolutionary spaces between the four of us large-bodied primates—the great apes and *H. sapiens*—had at one time been filled by relatives that we can hope to know only through the good fortune of discovering their fossilized remains. So when we humans look out at our closest living relatives, the apes, we see animals that are so different from one another, as well as from us, that, by comparison, almost any fossil hominid looks like kissing kin. In contrast, our everyday experience with other animals, such as squirrels or warblers, demonstrates vividly the more common situation in nature: The closest relatives of most living species are also living. And the differences between these close relatives are trivial compared with the differences that exist between most of the specimens of fossil hominids that have been jammed into our own species.

Many of the difficulties in paleoanthropology arose with the discovery of new, different, and totally unpredicted fossil forms. Foremost among these problems is the matter of coping with the boundary, if you will, between being ape and being human. This is the same problem the young anatomist Raymond Dart had to confront when, in 1925, he described the first specimen of *Australopithecus.* Braininess versus bipedalism was in the forefront of early paleoanthropological debates and deliberations on this issue. And they still are.

The second stumbling block in the pursuit of human origins is the reluctance—perhaps even the inability—of most paleoanthropologists to recognize diversity in the human fossil record. Linnaeus, the Swedish botanist credited with formalizing the approach to taxonomic classification that is still used in biology today, coined the name of our genus, *Homo,* in 1758. In 1925, Raymond Dart took the bold step of creating the genus *Australopithecus* for the first known fossil "man-ape," which was clearly neither fully ape nor fully human. Since 1925, quite a few genus names have been proposed to embrace newly named potential species of fossil hominid. But with the notable exception of the genus *Paranthropus,* which Dart's senior colleague, Robert Broom, named in 1938 to accommodate a robust species of early hominid, *Homo* and *Australopithecus* remain the only two genera (the plural of genus) accepted by the majority of paleoanthropologists.

As the case of *ramidus* reveals, paleoanthropologists have at least become more accepting of species diversity within the earliest hominids. Many species of *Australopithecus* have now been identified. But, as the case of *ramidus* also demonstrates, there is still a reluctance among paleoanthropologists to allow the kind of diversity in human evolution that can be reflected only at levels above the species. In blatant contrast with the narrowness of taxonomic vision in paleoanthropology, it is commonplace in systematic studies of other animals to recognize species diversity. And since *Homo sapiens* is an animal, there is no reason to suppose that the picture of human evolution was any different from that of any other group of animals.

"Little Foot"

For a number of years, a pile of excavated rubble lay near its source at the South African fossil hominid site Sterkfontein. Finally, in 1980, this pile of rubble, dubbed Dump 20, was picked over for bones and potential artifacts. Four foot bones were recovered. Along with other bits and pieces of animal skeleton, these four bones were taken back to Johannesburg, where they lay unnoticed for more than fourteen years in the collections of the University of the Witwatersrand. Perhaps these bones were ignored because they had been placed in separate drawers and containers. Or perhaps they were overlooked because Phillip Tobias, the head of this lab—the Palaeo-Anthropology Research Unit of the Department of Anatomy and Human Biology—was preoccupied with other early hominid fossils from elsewhere in Africa. No matter. It is sufficient that these bones were found in the first place, and subsequently rediscovered.

The person who rediscovered the four foot bones was Ron Clarke, an Englishman who had spent his early adult years as an assistant to Louis and Mary Leakey at Olduvai Gorge, the famous fossil hominid site in Tanzania. In the early 1970s, Clarke took leave of the work at Olduvai to return to England to pursue first his undergraduate and then his doctoral degree. His intellectual bravado, good humor, and sharp wit—but, overall, his kindness toward others—helped sustain me during the year that I was a foreign student in London, at University College. When Clarke finished his formal education, he returned to Africa and eventually went to work with Phillip Tobias in Johannesburg.

Clarke is a genius with fossils. He can locate and excavate them. He is brilliant at describing their anatomy and interpreting them evolutionarily. And he is a genius when it comes to putting the bits and pieces back together to create a specimen that is complete enough to be described and analyzed. He can take what to the outsider, and even to many experts, looks like an undecipherable jigsaw puzzle of bone chips and reassemble them, recreating a skull or a jaw as it had originally been prior to its bearer's death. Clarke's knowledge of fossil hominid skeletal anatomy, which often seems like preknowledge, is so keen, and his memory so viselike, that he can remember pieces of skull or jaw from one museum drawer to the next and knows exactly which pieces should go together. He has demonstrated this phenomenal ability on several occasions.

The first time was when he figured out that a number of cranial fragments that had been scattered among various drawers that contained other fossil fragments actually came from the same skull. In relatively short order, Clarke assembled these far-flung bone fragments into what would emerge as a relatively complete skull of the first definitive cranial specimen of fossil *Homo* from any South African early hominid site. The site was Sterkfontein, which had been known primarily for its treasure trove of *Australopithecus* remains.

Now, Clarke turned his attention to the four fossil foot bones from Dump 20. In collaboration with Phillip Tobias, he published the results of this rediscovery in the journal *Science,* July of 1995. The four Sterkfontein foot bones consisted of a series of bones that actually do articulate one in front of the other. The series began with the ankle bone that sits right under the larger of the two leg bones, the tibia, and goes right through to the bone in the sole of the foot that is part of the big toe. Compared with the size of these bones in a modern human foot, the bones of the fossil foot were much smaller. Clarke and Tobias nicknamed their rearticulated specimen Little Foot.

The discovery of Little Foot was provocative in a number of ways. First, there was every reason to believe that this multiboned specimen constituted the oldest known set of articulated hominid foot bones. In fact, if dated correctly, this set of bones represented the earliest known hominid ever excavated from a South African site.

Although the bones had not been discovered in situ, in a particular stratum of the site, it was known that the material in Dump 20 had been excavated from the deepest levels of Sterkfontein, which are identified as Members 1, 2, and 3. All other hominid fossils had come from Member 4, which is the youngest stratum. As Clarke and Tobias discussed in detail in their article, they ruled out the possibility that Little Foot came from Member 1, which is the oldest level of the site, because it is poor in fossils. Member 3, the youngest of these three levels, has yielded more fossils than Member 1, but the bones of Little Foot did not match in quality of fossilization the bones recovered from Member 3. Clarke and Tobias concluded that the middle level, Member 2, was the likeliest source of Little Foot. This stratum is jam-packed with fossils, and many of the skeletal elements found

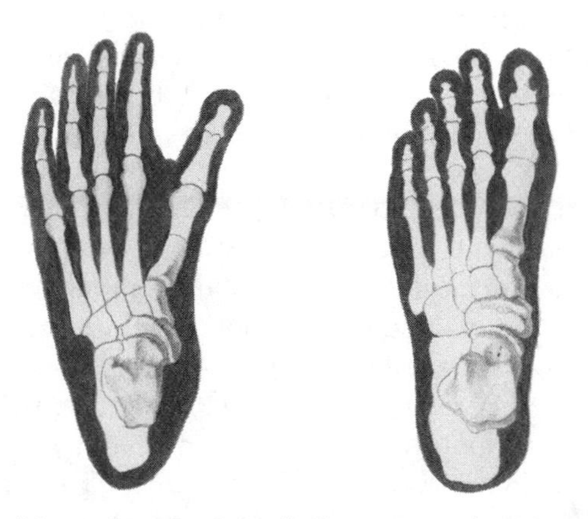

Reconstruction of foot of Little Foot (left) compared with an adult modern human's foot (right). The three complete and fourth partial foot bones of Little Foot are shaded, as are their counterparts in the human foot. (Adapted from Clarke and Tobias)

there are still in articulation. Because of its stratigraphic depth and relation to other dated levels, the kinds of species of non-primate animals recovered from it, and the apparent depositional history of the sediments, Clarke and Tobias concluded that Sterkfontein Member 2 was definitely older than 3 million years. Conservatively, they thought that Member 2 could be at least 3.5 million years old.

Another exciting aspect of the discovery of Little Foot was that most of the fossil hominids from Sterkfontein had come from Member 4, which was considerably younger than Member 2. Being higher up the depositional sequence of the site than Member 2, Member 4 had been dated at roughly 2.5 to 2.6 million years old.

Prior to the discovery of Little Foot, the Leakeys' site at Olduvai Gorge had held the record for divulging the oldest set of articulated hominid foot bones. This earlier fossil foot had been so completely preserved that only the toe bones were missing; every other bone of the foot, which had been cataloged as OH (Olduvai Hominid) 8, had been preserved, either in part or in its entirety. OH 8 came from below a layer of volcanic ash that was dated (using the potassium-argon technique and other methods) to approximately 1.8 million years before the present. But Little Foot was almost twice as old as OH 8. Little Foot was also almost as old as the earliest known footprints left by any hominid.

These fossil footprints—two parallel sets of tracks that went on for many meters—had been excavated during the late 1970s by Mary Leakey, Tim White, and Ron Clarke at a site called Laetoli, which is south of Olduvai Gorge. There, some 3.7 million years ago, two hominids had walked across the savanna together, on a thin layer of volcanic ash that had become muddy with the onset of the rainy season. These hominids were not alone on the landscape. Now-extinct kinds of elephants, horses, baboons, and antelope were making their normal seasonal migrations across this part of the country. And these animals also left their foot- or hoofprints in the moistened ash. The spacing and relative depths of the footprints indicated that these migrating herds of mammals had been walking in a normal gait, as had the pair of hominids.

Just as experts today can identify animals by the tracks they leave, so, too, can paleontologists identify animals of long ago by their fossilized footprints. The impressions of antelope hooves, of elephant toe and footpads, and baboon hands and feet, have their characteristic imprints. Like the feet of other monkeys, as well as the feet of apes, the feet of the extinct species of Laetoli baboon had a big toe that projected away from the other toes of the foot. Consequently, the footprint of the Laetoli baboon looked very much like its handprint, in which the thumb also jutted away from the fingers.

The foot of *Homo sapiens* is distinctive among living primates in that the big toe lies parallel to the other toes. A human walking barefoot in mud or moist sand leaves a series of impressions of toe tips lying one next to the other. Set behind and apart from this series of indentations of toe tips is the imprint of the sole of the foot, which looks something like a question mark as it curves from the ball of the foot and down the outside to the heel. The gap between the tips of the toes and the ball of foot, which is so characteristic of human footprints, was also a prominent feature of the two foot tracks that paralleled each other in the fossilized Laetoli ash. Every paleoanthropologist who studied these humanlike footprints was forced to agree: The species that had left those tracks had to have been a habitually bipedal hominid.

Of the possible choices for the early hominid species that had left the footprints, paleoanthropologists were convinced that *Australopithecus afarensis* was the most likely candidate. The larger set of footprints must have been made by a male and the much smaller set by a female *afarensis*.

Why did it have to be *Australopithecus afarensis* who made the tracks? One compelling reason was that the very specimen upon which the species *A. afarensis* was based had come from Laetoli. In 1978, when they published their article, in the journal *Kirtlandia,* announcing this new species of early hominid, Don Johanson, Tim White, and Yves Coppens, France's preeminent paleoanthropologist and raconteur extraordinaire, who was initially involved in this work, could have chosen the Lucy partial skeleton, or any of a number of relatively complete specimens from Ethiopia, as the type specimen to represent their new species, *Australopithecus afarensis*. However, as Johanson, White, and Coppens stated in their article, they purposefully chose the mandible from Laetoli as the type specimen in order to make a specific point: *A. afarensis* had been a widespread species, ranging from what is now Ethiopia in the north to Tanzania, more than 1,500 kilometers to the south. Although most of the specimens had been found in the Afar region of Ethiopia, for which *A. afarensis* had been named, paleoanthropologists would never be allowed to forget Tanzania, because the type specimen of *A. afarensis* had come from Laetoli.

Another reason *Australopithecus afarensis* was believed to have been the source of the footprints at Laetoli was that in the 1980s, when the footprints were discovered, *A. afarensis* was the oldest known hominid species. One could have speculated that the footprints had been made by another species of bipedal hominid. After all, new species of dinosaur have been named on the basis of footprint tracks alone. And there was no lack of controversy over exactly how hominid *A. afarensis* really was.

The isolated foot bones of *Australopithecus afarensis* that had been discovered along with the jaws and teeth did not clarify the situation. There was much debate between paleoanthropologists at the State University of New York at Stony Brook and the Johanson team, and not only on the subject of how *A. afarensis* used its feet. More fundamentally, the two groups disagreed strongly about the precise shape of the foot bones, particularly the toes. If, as Johanson's team argued, the toe bones were relatively short and straight, as in the foot of a gorilla or *Homo sapiens,* then, like gorillas and *H. sapiens, A. afarensis* had probably been committed to a life of terrestrial locomotion. On the other hand, if the toe bones were very long and more curved than they are in *H. sapiens,* as the Stony Brook group contended, *A. afarensis* would have been at home primarily in the trees, as is the arboreal orangutan, which has long, curved toe bones. Long, curved toe bones would not be consistent with the maker of the Laetoli footprints.

The debate over just how bipedal *Australopithecus afarensis* really had been also came to engulf the other known parts of the fossil's skeleton. The pelvis of *A. afarensis* was shallow and bowl-shaped, as in other, previously discovered species of *Australopithecus* that had been found at the Sterkfontein site. Although it was not a dead ringer for a *Homo sapiens* pelvis, the *Australopithecus* pelvis was much shallower and more bowl-shaped than an ape's, which is long and narrow. The knee joint of *A. afarensis* was also like the knee joint of other known species of *Australopithecus.* The knees angled in under the body, rather than being straight up and down, as they are in a horse or an ape. The *Australopithecus* knee, however, was even more angled under the body than is the case in *H. sapiens,* which meant that these early hominids were extremely knock-kneed. The lower part of the vertebral column of *A. afarensis* curved inward, toward the stomach, as in other species of *Australopithecus* as well as in *H. sapiens.*

When these features were first discovered in the skeletal remains of *Australopithecus,* they caught the paleoanthropological community by surprise. A shallow, bowl-shaped pelvis, a tendency to be knock-kneed, and the development of lower spinal curvature, along with the inward position of the big toe, were among the salient features that early anatomists had come to associate with *Homo sapiens,* and, in particular, with our brand of bipedalism. It had been difficult enough for many scientists of the 1930s, '40s, and '50s to accept the reality of early hominid antiquity. It was even more difficult for many of them to accept that features that had for so long been held up as definitively human could be found in now-extinct species of possible hominid that were otherwise so un-*Homo*-like in their jaws, teeth, and skulls.

In the early phases of trying to make sense of how bipedal skeletal features could show up in an individual that also had massive jaws and teeth and a relatively small skull, paleoanthropologists tended to emphasize the differences between species of early hominid. This endeavor was made easier with the discovery of two types of early hominid: a larger, more massive-toothed and -jawed species, and a smaller, more lightly built species. In spite of these differences, both species seemed to be the same in skeletal anatomy below the neck. Although this similarity in postcranial skeletal anatomy could mean that the two species not only walked bipedally but also did so in the same manner, it took years for the paleoanthropological community to accept this reality. Consequently, the more robust early hominid species was thought to be only capable of waddling, not of walking upright. The more gracile species could walk upright but could not do so as well or in as energy-efficient a manner as we humans.

In the seventies and eighties, there was a shift of attitude. This shift coincided with an easing up of paleoanthropological reluctance to recognize species diversity in the human fossil record, at least in the early part of human evolution. Many paleoanthropologists now accepted the possibility that both the robust and the gracile species of *Australopithecus,* given the similarity of their skeletal anatomy, had walked bipedally in the same way and with the same degree of proficiency. In fact, for many in the profession, all species of *Australopithecus* were now seen as being able to walk as well, or at least as efficiently, as we do.

In its pelvis, knee joint, and vertebral column, *Australopithecus afarensis* was identical to other *Australopithecus.* For Johanson's group, this was further evidence that *A. afarensis* had been a fully upright, terrestrially committed biped. The Stony Brook group agreed that the shape of the pelvis, the inward angulation of the knee joint, and the curvature of the lower spine could reflect an upright posture. But, they argued, just because *A. afarensis,* or, by extension, any other hominid, had been able to hold its torso upright did not mean that the species in question had been a terrestrial biped. The Stony Brook group still maintained that *A. afarensis* had been primarily a tree dweller. In contrast to the ways in which monkeys and apes move about in trees—running, leaping, swinging—however, *A. afarensis* had shinnied up and down the trunks.

In shinnying up or down a pole, an individual holds its torso vertically and applies pressure to the support with the knees and the soles of the feet. *A. afarensis* would have held its torso vertically, with its bowl-shaped pelvis cupping its internal organs from below. Its strongly inwardly angled knee joints and long toes would have gripped the tree trunk. According to the Stony Brook group, all parts of *A. afarensis*'s skeleton were consistent with this species having been a shinnier of tree trunks. When, and if, this hominid did have occasion to come to the ground, it would have been obliged to walk bipedally because of the constraints of its skeletal anatomy.

In spite of the debates of the seventies and eighties, most paleoanthropologists came to embrace *Australopithecus afarensis* as the ultimate ancestor of all other hominids for the simple reason that this species was the earliest known hominid. As such, an equally simple answer to the question of who made the Laetoli footprints was that two individuals of *A. afarensis* had made them. This meant that, regardless of whether its toes were long or short, the big toe of *A. afarensis* was not divergent from the other toes; it was in line with them. *A. afarensis* must have been a biped in every way. This, in turn, meant that the humanlike features of the OH 8 fossil foot bones from Olduvai Gorge weren't an evolutionary novelty after all; OH 8 had merely retained these from a distant ancestor.

Upon being rearticulated, the foot bones of OH 8 gave the impression that they belonged to a hominid that was humanlike in every detail, including having an arch and a big toe that was aligned with the other toes. In spite of the fact that OH 8 had been discovered in the early 1960s, when the evolution of human bipedalism was still a hotly debated topic, the humanness of this fossil foot was readily accepted because it was supposed to be the foot of the earliest species of *Homo,* of which there were many cranial and dental specimens. Consequently, a major feature of human evolution had to have been the emergence at an early stage of humanlike bipedalism, which had been retained largely intact by *Homo* from *Australopithecus* with the sole addition of a little fine-tuning. With bipedalism already in place, evolution within the genus *Homo* consisted primarily of decreasing jaw and tooth size while at the same time expanding the size of the brain.

But now there's Little Foot. As Ron Clarke and Phillip Tobias pointed out in their *Science* article, this specimen does not conform to the expectation of what constitutes a primitive human foot. Little Foot's ankle bone was extremely humanlike in its proportions and its anatomical details. However, as you proceed down the series of articulated foot bones, each bone becomes more apelike in appearance, culminating in a joint that leaves no doubt that this purported hominid had once had a divergent big toe—almost as divergent as an ape's or a monkey's, in fact.

The unexpected happened when Clarke and Tobias compared the bones of Little Foot with their counterparts in OH 8. Little Foot's ankle bone was actually more humanlike than OH 8's. Even more surprising, the bones of the big toe of OH 8 were remarkably similar to those of Little Foot—which meant that OH 8 had had a divergent big toe, not, as had originally been thought, a big toe that was aligned with the other toes. When Clarke and Tobias compared the individual bones from Little Foot with the same bones of *Australopithecus afarensis,* this hominid also turned out to have had a divergent big toe.

What are we to make of all this? The implications are far-reaching, but they are also similar to the issues that were provoked by the discovery of *Ardipithecus ramidus.* Foremost among these issues is the question of how one identifies species. To which species of early hominid did Little Foot belong?

For decades, paleoanthropologists have maintained that only one species of early hominid, *Australopithecus africanus,* is represented in the deposits of the South African site of Sterkfontein. But, as Ron Clarke has argued on various occasions, the details of the cranial remains suggest the presence of more than one species at Sterkfontein. Since Little Foot is perhaps 1 million years older than most of the fossil hominids from Sterkfontein, there is really no compelling reason to assume that the bones are those of *A. africanus.* Clarke and Tobias did hint at the possibility that if Little Foot is not the earliest *Australopithecus africanus,* it could be another species of that genus. But why does it have to be a different species of that genus? In fact, why does Little Foot have to be a hominid at all?

The second question leads to another important implication: The discovery of Little Foot should impact the way in which paleoanthropologists define a hominid. In fact, the discovery of Little Foot, like that of *Ardipithecus ramidus,* should affect the way, or ways, in which paleoanthropologists think about what it actually means to be hominid. Is it by our teeth that we hominids are distinguished from apes and apelike fossils? If so, is it the shape of the teeth or the thickness of the enamel that matters? Or is it in the skull that the clues to being hominid are to be found? Is it the position of the vertebral column as it articulates with the skull base? The shape of the elbow joint? The orientation of the big toe? Is it all of these features? Or does the answer lie in another feature that is none of the above?

One would think that paleoanthropologists would have figured out more definitively which characteristics distinguish the larger taxonomic group to which humans and their potential fossil relatives have been allocated before arguing about which fossils represent species of *Australopithecus* and which *Homo,* and how many species of each genus there might have been. But this is not the case. Paleoanthropologists have tried to define what it means to be a "hominid." The problem, however, has been twofold: Human-related fossils were discovered well after comparative anatomists had settled on the features that distinguish living humans from other animals, and hominid fossils were treated as if they could be ordered in an evolutionary continuum from something apelike to something humanlike. Clearly, this has affected the way in which fossils thought to be hominid are interpreted, especially if the date of the fossil and its morphology do not conform to expectations. The discipline must free itself of the shackles of historical precedent, of the received wisdom of past generations of paleoanthropologists and anatomists. Otherwise, like the tail wagging the dog, the prevailing interpretation dictates the fate of the fossil, rather than a newly discovered fossil providing a test of a favorite evolutionary scenario. Since most early hominid fossils have been found in East African deposits, it is not surprising that a recent discovery from this region serves to illustrate this dilemma.

A Bipedal Ape of a Different Sort

Thirty-two years ago, in 1967, at the site of Kanapoi, in what is now the harsh, arid landscape to the west of Lake Turkana in northern Kenya, an American paleontologist from Harvard University, Bryan Patterson, was prospecting for fossils. Patterson, tall and thin, with a face deeply creviced from years of being exposed to such brutal conditions, was a superb field paleontologist. The deposits he was scrutinizing were believed to be more than 4 million years old. While combing the landscape for fossils, he came upon the lower end of a humerus, the upper arm bone. Even though Patterson was not a specialist in higher primates, it was obvious to him that this part of the elbow joint had come from a hominid. Shortly after returning to the United States, Patterson and William W. Howells, the senior physical anthropologist at Harvard University, coauthored a short article on this specimen that was published in the journal *Science.*

Howells is one of the pioneers in the use of complicated, multivariate statistical analyses in comparative metrical studies of living and fossil apes and humans. His own research relied on the use of cluster analysis of cranial measurements of modern humans. This approach leads to the grouping of individual skulls, and then groups of skulls, based on similarity of shape as determined by measurement. It is one of the ways of assessing patterns of population affinity and migration. Because Howells's work in this area has so greatly influenced the field, it is no surprise that one of his doctoral students, Henry McHenry, who specialized in the metrical analysis of early hominid fossils, decided to examine the Kanapoi humeral fragment statistically.

Eight years after Patterson and Howells's article on the Kanapoi humeral fragment, McHenry, who is now a professor at the University of California at Davis, in collaboration with his frequent collaborator, Rob Corruccini, a professor at Southern Illinois University, published the results of their statistical analysis of this specimen. They took measurements on the Kanapoi humerus and compared them with similar measurements on extant primates. The primates McHenry and Corruccini sampled were modern humans, gibbons, the common chimpanzee and the much rarer pygmy chimpanzee, gorillas, orangutans, and various Old World monkeys, such as macaques. For good measure, McHenry and Corruccini also included specimens of putative fossil apes, as well as the two other early hominid humeral fragments then known. One of these fossil specimens had been found at the South African site of Kromdraai, which is situated only a few kilometers from Sterkfontein. The other had come from Richard Leakey's site, Koobi Fora, which lies on the eastern shore of Lake Turkana, in Kenya. These two humeral fragments were at least 1 and possibly even 2 million years younger than the Kanapoi humerus. According to the paleoanthropological experts, the Kromdraai and Koobi Fora humeral fragments represented individuals of the same species of early hominid.

After crunching the numbers through their computer programs, McHenry and Corruccini came up with the following results. The three living great apes—the chimpanzee, gorilla, and orangutan—clustered together as a group. So did all the Old World monkey species. And so did all the humans. There was nothing surprising about these results. By the latter part of the nineteenth century, comparative primate anatomists had demonstrated that the three living great apes were very similar to one another anatomically, as were the other species of Old World monkey. In fact, Old World monkeys could even be grouped together solely on the basis of tooth morphology.

The fossil apes in McHenry and Corruccini's sample ended up being plotted in three-dimensional space between the living Old World monkeys and the living apes. This was okay, because the accepted scenario was that the limb bones of many fossil apes were more like those of Old World monkeys than living apes.

The early hominid specimen from Kromdraai emerged as being intermediate between humans and chimpanzees. This, too, was all right, because early fossil hominids were supposed to be evolving away from being apelike toward being more humanlike in their features. The Koobi Fora humerus was similar to that of the gorilla and the orangutan in only a few measurements. Overall, this specimen was totally unlike any living ape or any modern human. But, more surprisingly, the Koobi Fora humeral fragment differed significantly from the Kromdraai humeral fragment.

McHenry and Corruccini concluded that these two fossils could not represent the same species of early hominid. In subsequent years, in the light of more extensive fossil collecting in South and East Africa, paleoanthropologists concluded that each region of Africa had been occupied by different, but apparently closely related, species of robust *Australopithecus*. But this is not the implication of McHenry and Corruccini's analysis. If you think about it, their study leads to the conclusion that the Kromdraai and Koobi Fora humeral fragments not only came from individuals representing different species but that these species were not closely related. In fact, only one of these specimens—the one from Kromdraai—seems to have been a hominid. The shape and size relationships of the lower part of the humerus of the Koobi Fora specimen are not convincingly hominid.

These conclusions notwithstanding, the most unexpected result came from analysis of the Kanapoi specimen itself. As the oldest hominid specimen in the study, the Kanapoi humeral fragment should have been the most primitive, or apelike, in its shape and size relationships. At least that is what would have been predicted from the point of view that human evolution is a continuum of change that began with an apelike ancestor. But, being unaware of this expectation, the Kanapoi specimen clustered right in there with McHenry and Corruccini's modern human sample. Confronted with this fact, the two men had to admit that the "Kanapoi humerus is barely distinguishable from modern *Homo*."

In spite of McHenry and Corruccini's corroboration of Patterson and Howells's earlier conclusion about the similarity between the Kanapoi humeral fragment and modern humans, this rather amazing discovery went unacknowledged by virtually all other scientists in the discipline. However, Brigitte Senut, one of France's leading comparative primate anatomists and paleoanthropologists, restudied this specimen in 1980, and again in 1985. She, too, came to the conclusion that the Kanapoi humerus was geologically old and, in its anatomical features, definitely humanlike. Nevertheless, Brigitte's studies also failed to reach an appreciative audience.

As if it had a mind of its own, the story of human evolution proceeded along, building upon the idea that somewhat apelike–somewhat humanlike early hominids eventually became transformed into fully modern humans. The commonplace, but unsubstantiated, premise that the more ancient a fossil is, the more primitive it is—and, conversely, the more recent a specimen is, the more derived or specialized it is—continued to dominate the practice of fossil hominid taxonomy as well as the interpretation of the evolutionary relationships among fossil hominid species.

Then came the discovery of 1995. Meave Leakey, of the Kenya National Museum, who is continuing the paleontological work around Lake Turkana that she and her husband, Richard, had done together for years, and her collaborators announced in *Nature* that they had found eight more hominid specimens from Kanapoi, as well as twelve other specimens from the nearby site of Allia Bay. Meave Leakey, Craig Feibel of Rutgers University, Ian McDougall of the Australian National University, and Alan Walker of Penn State University introduced these new specimens as representing a new species of early hominid, which they named *Australopithecus anamensis.* The root of *anamensis* comes from the Turkana word *anam,* which means "lake." Although the sites of Kanapoi and Allia Bay are now in the middle of brutal, arid landscape, they had once been on the shores of an ancient lake.

Seasoned paleontologists and geologists that they are, Leakey, Feibel, McDougall, and Walker first made certain that they understood the geology and the dating of the different stratigraphic levels of these two sites. Based on its position in the stratigraphic section of Kanapoi, the humeral fragment that Patterson had found was not quite as old as he had thought. This specimen rightfully belonged to the Upper Kanapoi level, which is about 4 million years old, or slightly younger. The Upper Kanapoi level is more or less contemporaneous with the fossil section at Allia Bay, which yields a date of at least 3.9 million years before the present. The Upper Kanapoi sits stratigraphically above the Lower Kanapoi, which has been dated at about 4.2 million years before the present. Leakey and her colleagues chose a virtually complete lower jaw, with all its teeth preserved, from the Lower Kanapoi level as the type specimen of the new species, *Australopithecus anamensis.* But they also included the fossils from the Upper Kanapoi in this species.

The steeply sloping "chin" region of the type mandible of *Australopithecus anamensis* was more primitive, or apelike, than that of any previously known hominid. However, with minor differences, all teeth, from all three sites—the Upper and Lower Kanapoi, and Allia Bay—were quite similar to the somewhat geologically younger *A. afarensis.* The molars of *A. anamensis* also bore a thick covering of enamel. If *A. afarensis* was dentally a hominid, so was *A. anamensis.*

This is where the comparability ended, however. Below the neck, *Australopithecus anamensis* was unlike any other *Australopithecus.* But it was very much like a species of *Homo* in several very important ways. This observation did not rely solely on the humanlike humeral fragment that Bryan Patterson had discovered in the Upper Kanapoi. Leakey and her colleagues found other skeletal pieces, also from the Upper Kanapoi, that were more like *Homo* than any *Australopithecus.* The specific pieces of bone included the upper and lower ends, as well as parts of the shaft, of a tibia, which is the larger of the two lower leg bones. Although humans and *Australopithecus* have tibial features that distinguish them from apes, humans have a very particular tibial shape, especially in the upper part, that separates them from *Australopithecus.*

Modern humans, as well as other species of *Homo* and all species of *Australopithecus,* differ from apes in aspects of the upper and lower ends of the tibia, which contribute, respectively, to the knee and ankle joints. Basically, the shapes of the articular surfaces are very different. *Homo* and *Australopithecus* also differ from apes in that that they have straighter tibial shafts. *Homo* and *Australopithecus* differ from each other, however, in the shape of the upper part of the tibial shaft in the region just below the knee joint. In *Australopithecus* and apes, the broad articular surface for the knee joint sits atop a thin shaft that resembles a flat-capped mushroom. In *Homo,* however, the upper end of the tibial shaft fans out like a trumpet to meet the broad

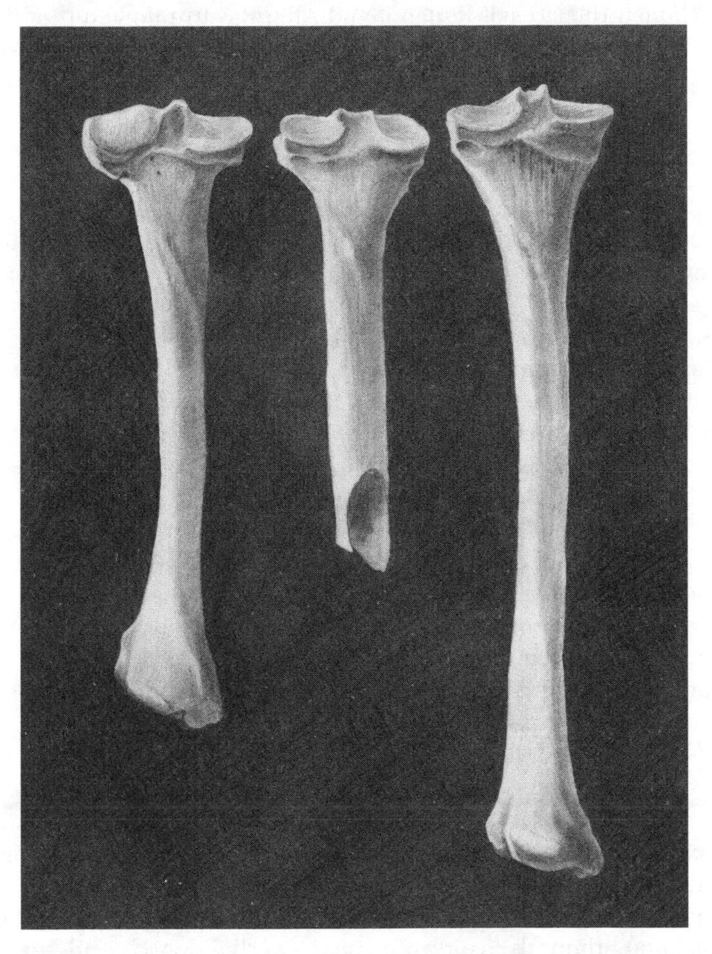

The tibia of an adult chimpanzee (left), an adult Australopithecus *(middle), and an adult modern human (right), all viewed from behind. The two surfaces on the tibia on which the two parts of the lower end of the femur sit in the knee joint are slightly concave in the hominids. In the ape, only one surface (the one on the right in the drawing) is concave; the other is convex. The shaft of the tibia is also straighter in the two hominids than in the ape. But the shaft of the human femur flares more markedly outward as it approaches the upper end than do the fossil and the ape.*

margins of the knee-joint surface. Given the geological age of *A. anamensis,* most paleoanthropologists would have expected this hominid to have the tibial configuration seen in apes and other species of *Australopithecus.* Contrary to expectation, however, the shaft of *A. anamensis*'s tibia presented the same shape as that of *Homo:* The upper end flared out in trumpetlike fashion to meet the margins of the knee-joint surface. If *A. anamensis* had been identified only on the basis of its humerus and tibia, there would have been no anatomical reason to exclude it from the genus *Homo.*

Peter Andrews, the fossil ape expert at the Natural History Museum in London (and the only person I know who can spend an afternoon scything down a wet field of grass without having to shower or change clothes afterward), wrote the commentary in *Nature* that accompanied Meave Leakey and her colleagues' *Australopithecus anamensis* article. To begin with, Andrews did not agree that the fossils from the Lower Kanapoi, the Upper Kanapoi, and Allia Bay belonged to the same species. He proposed that the fossils from the Upper Kanapoi and Allia Bay be kept together because they were contemporaneous. But he thought that the Lower Kanapoi specimens should be kept separate because they were older than the other specimens. Since, Andrews argued, the type specimen of *A. anamensis* came from the Lower Kanapoi, only specimens from that level could belong to that species.

When it came to the jaws and teeth, Andrews's comparisons between the Lower Kanapoi fossils and unquestionable fossil apes were quite compelling. The Lower Kanapoi fossils were similar to fossil apes in the severe slope of the front of the mandible, the shallowness of the hard palate, the size and orientation of the canine teeth, and the relative crown proportions of the upper molars. Since the molars of many fossil apes are covered with a thick layer of enamel, as are the molars of accepted hominids—including *A. anamensis*—this criterion does not help us to distinguish one species from the other.

In the absence of any associated skeletal bones, what we have from the Lower Kanapoi, Andrews pointed out, are fossil jaws and teeth that are more apelike than they are hominid-like. If, as he further suggested, we keep the pieces of humerus and tibia together with other bones, jaws, and teeth from the Upper Kanapoi and Allia Bay, we have a good case for the presence of a hominid at these two sites. Seen in this light, Andrews concluded, maybe the Lower Kanapoi specimens did represent an extinct species of ape, and not a hominid after all.

Is it possible that there were dentally hominid bipedal apes and dentally apelike bipedal hominids? Certainly, much of the identification of fossils as hominid, and the subsequent evolutionary interpretations of these fossils, have been seriously influenced by historical factors. Foremost among the difficulties is the fact that, well before humans were allowed to have a fossil record, an evolutionary past, and an evolutionary relationship to other organisms, they were compared with other organisms in ways that emphasized how different, and how great in that difference, they really are from the rest of the animal world. When fossils, beginning with Neanderthals, were admitted into human antiquity, they were viewed as being beneath humans, both in their morphology and their intellectual capabilities. These fossils were also interpreted within the context of a continuum that went from animals more primitive or "unevolved" than they to humans, who, by definition, represented the pinnacle of evolutionary achievement. When fossils chronologically older than Neanderthals were eventually discovered, they, too, were forced into this scenario. Consequently, the group "hominid," and the genera and species included in that group, have tended to be defined less as entities in nature than in the context of how closely they appear to approach the human, and to leave behind the ape, condition.

To further complicate matters, the preoccupation with the notion that there was a trend toward becoming increasingly humanlike has been wedded to speculations about the processes of natural selection and the sequence of adaptations that could have brought about this putative transformation from ape to human. While other disciplines in evolutionary science have expanded their appreciation of different possible mechanisms and processes of evolutionary change, the study of human evolution has been firmly planted in the traditional dogmas of Darwinism, including the belief that evolutionary change manifests itself only through an insensible series of infinitesimally small modifications. Clearly, the search for ancestors and intermediates that has dominated—and continues to dominate—the field of paleoanthropology attests to such beliefs. But what if this isn't the way it happened?

Genetics and Development of the Organism

Fossils provide one kind of information on the evolution of the features of organisms: They can potentially illustrate when, approximately, these features first appeared. Even if exact dates are not decipherable, the relative times at which features arose can be sorted out. For example, we know from the primitive fossil birdlike animal *Archaeopteryx* that the sequence of events in the evolutionary history of birds was the development of feathers first, then the loss of teeth. But *when* features first arose in groups of animals and *how* features actually develop are two entirely different things. As such, in the pursuit of a broad understanding of not only hominid but also animal evolution in general, it is important that we not put all of our eggs in the fossil basket. Fortunately, coincident with the recent unearthing of fossils that should force paleoanthropologists to reconsider the very basics of the story of human evolution, there has been an explosion of discoveries in the field of developmental genetics that are demonstrating for the first time, from the finest of molecular levels to the whole organism, precisely how anatomical features arise and develop.

Everyone has probably heard about chromosomes, genes, DNA, and the double helix, and maybe even RNA. But what about "eyeless," "tailless," "hunchback," "krüppel," "bicoid," or "nanos"? The first three words might appear on a list of deformities produced experimentally in animals by a mad scientist. The last three might as well as be in an extraterrestrial language. In reality, all six words come from a vocabulary that developmental geneticists are inventing on an almost daily basis. And the key to this new and ever-expanding lexicon of development genetics lies in distinguishing between genes that produce the proteins that ultimately make a structure or feature and genes that regulate the activity and interaction of these structural genes. The latter are the cornerstones of development. They are the regulators. They run the whole show.

"Eyeless" and "tailless" are representative of names that have been given to classes of regulatory molecules (some of which are genes and some the proteins these genes produce) that are being discovered faster than one can keep up with. These regulatory molecules and their products specify the fates of cells. Depending on where and when these molecules are silent or active, an organism's body can end up being organized around a longitudinal front-to-back axis or, like a wheel, around a central point. An organism can have a limb or a fin, an eye in a bony socket or neither, a lifetime of teeth or none at all. And because these regulatory molecules specify the fates of cells, they play a role, either directly or indirectly, in the determination of all aspects of the development of an organism's body, including the spatial organization of the body's numerous parts.

For example, as we know from studies on fruit flies, even before the cell that produces the egg cell does so, it is organized in such a way that it has a front, back, up, down, right, and left side or pole. Consequently, an egg has the orientations an organism's body would have even before fertilization creates the organism. Shortly after an egg is fertilized, when cells are replicating themselves like mad but before this clump of cells can be defined as an embryo, regulatory molecules are already orchestrating development along the earlier established axes of the organism: front to back, up and down, right to left. Later, other regulatory molecules will appear with more restricted roles in the development of specific structures, such as muscles, bones, and nerves, as well as sensory organs, such as teeth and eyes. What is most astonishing about these regulatory molecules is that either the same molecules, or their virtually identical molecular twins, are found in a diversity of animals, ranging from worms, to fruit flies and beetles, to frogs, zebra fish, chickens, and even to mice and humans.

Taxonomists might group worms and fruit flies together as invertebrates, which are animals that lack both a braincase and a backbone or some other internalized back-stiffening cartilaginous rod. But most people would not expect to learn that the same regulatory molecules can lead to the development of a worm, which lacks supportive skeletal structures altogether, as well as a fruit fly, which is encased in a hardened shell-like skeleton. Perhaps even more in the realm of science fiction is the fact that animals as visually dissimilar as invertebrates and vertebrates—the latter being animals with both a braincase and a backbone or an internalized back-stiffening cartilaginous rod—also share similar regulatory molecules. We know that worms, fruit flies, zebra fish, and humans are all, albeit distantly, related to one another as multicellular animals. But who would have thought that a squishy worm, a hardened cuticle-encased fruit fly, and an internally bone-supported human shared similar body-shaping regulatory molecules? Until just over a decade ago, nobody would have. Nowadays, virtually every discovery in developmental genetics further demonstrates how much of their regulatory molecules the diversity of animal life shares.

What features do invertebrates such as worms and insects, amphibians such as frogs, fish such as zebra fish, birds such as chickens, and mammals such as mice and humans have in common? Any schoolchild knows that these animals are all similar in the gross features of their cells. Plants, for instance, have rigid cell walls. Animals do not. But these multicellular animals share a lot of important features that are far more visible than those at the cellular level.

The bodies of many multicellular animals have a front, or anterior, end and a back, or posterior, end. Their bodies also have a right side and a left side that are close mirror images of each other. (Nature does not produce exact mirror images.) They have a concentration of nerve cells, called neurons, at their anterior end, which constitutes a primitive "brain," if you will. They have light-gathering sensory organs at the anterior end. And their mouths are at the anterior end. In short, they have a head. Less spectacular is the posterior end of these animals. But their anuses, being closer to the posterior end, are separated from their mouths. And most of these animals also have a tail. Not all multicellular animals—starfish or sea anemones, for example—have all of these features. However, the majority of multicellular animals do.

But there is another crucial aspect in which at least fruit flies, frogs, zebra fish, chickens, mice, and humans are similar. Their bodies are segmented. In fruit flies, as in other insects, as well as in crustaceans such as shrimp and lobsters, the musculature and the outer hardened shell also differentiate into series of segmented units that resemble the overlapping plates of a suit of armor. In vertebrates, the muscles of the trunk and even of the limbs arise as segmental units, as do the bones that lie internal to them.

The development of segmentation in such a diverse array of animals did not arise purely by chance in each animal or in the immediate ancestor of each animal. One of the biggest ongoing discoveries in developmental genetics is that all segmented animals studied to date share similar regulatory molecules, which orchestrate the development of segmentation. When these molecules were first discovered, they were called homeobox genes and referred to as *Hox* genes for short. The abbreviation *Hox* is still used to designate the regulatory genes that are involved in the development of segmentation and segmented structures, but the term *homeobox gene* now covers the entire gamut of regulatory genes that are fundamental to an organism's development.

Intuitively, we know that a mouse is a more complex organism than, say, a fruit fly. But how do increasingly more complex levels of organismic organization arise? The answer comes from the discovery that increasing complexity can result from the activation and interaction of already present homeobox genes as well as from the simple duplication or replication of homeobox genes to form larger and more numerous groups or clusters of genes.

In fruit flies and other insects, the *Antennapedia* homeobox gene cluster controls segmentation in the head region as well as in the anterior part of the chest region. This homeobox gene cluster also regulates the differentiation of the paired and segmented antennae of the head region and the paired and segmented legs of the anterior chest region. The name given to this class of *Hox* genes, *Antennapedia,* refers to antennae and feet. The fruit fly has only one *Hox* gene cluster of this particular class of homeobox genes.

A few years ago, a collaborative team of British and Italian developmental biologists discovered that mice and humans possess four *Hox* clusters of the *Antennapedia* class of homeobox genes. The four *Hox* gene clusters in mice and humans are involved in the differentiation of the segments of the head region. These *Hox* gene clusters control everything, from the development of the most rudimentary segmental structures of the head region through the development of those segments of the central nervous system that we identify as the midbrain, the hindbrain, and the spinal cord. Subsequent to these studies, a team of U.S. and Spanish neurobiologists demonstrated that the homeobox genes that control the pattern of head development in fruit flies are also represented in chicken and mice. In chickens and mice, and presumably all birds and mammals, these particular homeobox genes are associated with the differentiation and segmentation of the forebrain.

But as astonishing as these discoveries are, they illuminate only the tip of the developmental iceberg shared by multicellular animals. For example, the fruit-fly homeobox gene called *eyeless* has its vertebrate counterpart in the homeobox gene identified as *Small eye.* Both of these homeobox genes—*eyeless* and *Small eye*—control the development of an eye or light-gathering sensory organ in all animals, even though the eyes of insects and vertebrates are structurally so different: The insect eye is compound and the vertebrate eye is deformable and single-image-forming.

Insect wing development and vertebrate limb development are also controlled by similar homeobox genes. When the gene that controls wing development in insects was discovered, it was called *hedgehog.* When the homeobox gene for vertebrate limb development was discovered, it was referred to as the *Hoxd* complex. Subsequently, developmental geneticists found out that *hedgehog* and the *Hoxd* complex represent the same homeobox gene. Clearly, this means that the homeobox gene was retained in invertebrates and vertebrates from a common ancestor that lived aeons upon aeons ago. It also reflects the fact that this homeobox gene is involved in similar aspects of appendage development, even though the appendages—one a wing, the other a limb—are so different from each other in final form and structure. In both cases, the shape and basic structural patterning are laid down as the appendage elongates. In the insect wing, it is the detailed patterning of venation that pervades the thin membrane of the wing itself. In the vertebrate limb, it is the patterned sequence both of the bones as they become smaller yet more numerous and of the soft tissue that surrounds and supports them.

More recently, another amazing discovery in developmental genetics was made by a team of Harvard Medical School researchers. Although this collaboration focused on developmental deformities of the hand and foot in humans, the results have broader implications for understanding the evolution of the vertebrate forefoot and hind foot, which, in humans and other primates, are the hand and the foot.

At first glance, it may not seem that there is a great deal of external similarity between a fish and a mammal, but there is more here than meets the eye. Although a fish has fins and a mammal, like other tetrapods, has limbs, the fish's pectoral fins are in the same position as a tetrapod's forelimbs and its caudal fins are in the same position as the tetrapod's hind limbs. It turns out that in fish, such as the zebra fish, and in tetrapods, such as chickens, mice, and humans, the same homeobox gene clusters are involved in producing both sets of appendages, regardless of whether they are fins supported by rays or limbs supported by bones. The debate that still lingers among developmental geneticists is whether these homeobox genes were present in a distant ancestor and then turned on in specific fore and aft positions to produce two pairs of appendages; or whether the homeobox gene cluster for the front pair of appendages came first, then duplicated, giving rise to the posterior pair of appendages. Each explanation yields the same result, although the first one is the less complicated and perhaps more likely of the two scenarios.

But while fish, birds, and mammals share the same homeobox gene clusters that are responsible for the development of pairs of appendages fore and aft, the difference lies in two factors. First, in fish, these genes are active only along the back side of the elongating fin, while in tetrapods they are active along the back and across the front of the developing limb bud. Second, there is one tiny chemical difference that distinguishes one of the homeobox genes of tetrapods from the same gene in fish. In tetrapods, this homeobox gene—the *Hoxd-13* gene—has an additional molecular sequence inserted into it that specifies a repeated series of a particular amino acid, alanine. Tetrapods have fore*feet* and hind*feet.* Fish do not have this alanine-encoding insertion, and they do not have feet of any kind.

How do we know that this tiny molecular insertion is associated with the development of feet? A bird, which has a reduced number of wrist and forefoot bones in its wing compared with a mouse or a human, has about one-third more alanine repeats than a mammal does. Since we know, as much as we can know anything in the evolutionary past, that there was a reduction in the number of digits and wrist bones during bird evolution, it makes sense that a present-day bird would have a shorter sequence of alanine repeats than would a mammal that has more forefoot bones. The role of alanine in the formation of hands and feet is further indicated by the observation that humans born with developmental deficiencies in the number of hand and finger bones have reduced alanine repeat sequences.

If fins become limbs with feet at their ends merely through the turning on of homeobox genes in novel locations and the insertion of a short molecular sequence into one particular homeobox gene, then the evolution of primate hands and feet would be an even simpler evolutionary feat.

How This All Fits Together

Clearly, studies in developmental genetics should greatly impact the way we think about evolution and natural selection. The implications of developmental genetics for understanding evolution in general, as well as the details of the evolution of any group of organisms, including our own, cannot be underestimated. But the roots of Darwinism run deep, and the appeal of selection arguments is strong.

Let us consider the traditional, twentieth-century Darwinian approach to evolution. As is well known, genes produce features, whether this is a morphology, such as eye color, or a certain behavior, such as responding to a chemical stimulus. Typically, there is variation within a species from one individual to the next in the representation of that feature. Variation at the morphological or behavioral level is constantly added to and changed by small, basically insignificant mutations. Natural selection can pick and choose from the pool of variants. With the selection of certain morphological or behavioral variants over others, the representation of the genes responsible for those variants within the population will change over time. Eventually, the character of the species, and consequently the species itself, will change. This is how the whole process is supposed to work.

There have been many experiments that provided examples of how this selection process can work. The first of these was performed in the 1950s by the British biologist H. B. D. Kettlewell. Kettlewell demonstrated that factory smoke, which darkened the bark of the trees in the surrounding area, acted as a selective agent against the conspicuous light-colored moths while favoring the better-camouflaged dark-colored moths that also alighted in the trees. Under sooty conditions, a light-colored moth was an easily spotted prey for birds. But when factories were shut down or their output was cleaned up, the dark-winged moth became the more visible color variant. Eventually, bird predation shifted the frequency of the predominant coloration within the moth species to the lighter-winged variant. Clearly, natural selection can alter the representation of features within a population. From this kind of observation, the extrapolation is that, given enough time, a species can be modified to become a different species.

When Darwin was formulating his theory of evolution by means of natural selection—not, as is commonly misunderstood, *the* theory of evolution—he was unaware of genes and even of the simplest ideas of inheritance of characters. Darwin was not alone in this state of ignorance. None of his contemporary naturalists understood inheritance, either. It was only in the twentieth century, with events that led to the formulation of the so-called Synthetic Theory of Evolution, or the Synthesis, that certain ideas of inheritance, mutation, and Darwin's concept of evolution by natural selection were put together into the form that has dominated the evolutionary sciences. The field known as sociobiology would not exist if it were not for very particular ideas about the relation of genes to morphology, about mutation, and about the evolutionary processes that lead to change or novelty.

Darwin's idea was that natural selection could shift the character of a species by culling from among the variants within that species. Since Darwin did not know about genes and mutations, he thought that novelty was introduced into species through the recombination of existing variations. In addition, Darwin's belief that natural selection acts only on minuscule variations forced him to conclude that a totally new species will emerge only after a very long period of time: What we now see as a distinct kind of organism arose only after millions of generations of accumulating an unfathomable number of tiny changes. The perceived need for extraordinary lengths of time over which change could accrue was one of the factors that caused Darwin's ideas to fall rather quickly out of favor with many in the scientific community. In almost every new edition of his most famous book, *On the Origin of Species by Means of Natural Selection,* he called for greater and greater periods of time over which evolutionarily significant change could be manifested. The earth could not be made older to allow for this model of evolution.

Philosophers and historians of science often point to the irony that, while perhaps the greatest evolutionary theorist of all time was struggling during his later years with his ideas on natural selection and the origin of species, an Austrian monk, Gregor Mendel, was laboring to understand the essentials of inheritance. And, after years of experimental breeding of garden peas, Mendel did indeed unravel the basics of this mystery. He demonstrated that the heritable factors that were responsible for an organism's features are distinct, particulate entities, or units, that can be passed on intact to offspring. The eventual expression of these units of inheritance in future generations depends on how they reassociated in offspring and whether the expression of one unit is masked by another. Had Darwin been aware of Mendel's experiments he would not have undermined his case for evolution by invoking "blending inheritance," in which the characters of parents were not distinct but melded to produce the features of the offspring. But Darwin was not the only evolutionist—or biologist, for that matter—for whom Mendel's discoveries remained unknown. Thanks to a largely unappreciative scientific audience, and to its dismissal by Germany's leading botanist, Mendel's groundbreaking research remained submerged for decades.

With the rediscovery in the early 1900s of Mendel's principles of inheritance, the study of variation in populations took on new meaning. Population geneticists, as these early geneticists and those interested in this level of inquiry are called, could make sense of so much with the simple knowledge that inheritance was particulate—that there were units of inheritance, which eventually came to be identified as genes, that were transmitted intact from parent to offspring.

Since each parent passes on a variant of each gene, the offspring ends up with two versions of each gene. If the two inherited genes are for the same features—blue eyes, say—the offspring has blue eyes. If one gene is for brown eyes and the other is for blue eyes, the offspring has brown eyes, because the gene for brown eye color is dominant over the gene for blue eye color. The gene for blue eye color is not lost; it is just masked. Upon reaching sexual maturity, this offspring might pass the blue-eye gene on to its offspring, which could have blue eyes if it also inherited a blue-eye gene from the other parent. But this course of inheritance did nothing more than maintain the genes already present in a population or species. Geneticists suggested that a new variant would arise from a small mutation at the gene level.

Most of the experiments in population genetics that were done in the twenties, thirties, forties, and fifties used fruit flies. One of the most famous and influential of these fruit-fly population geneticists was Theodosius Dobzhansky. In fact, Dobzhansky was the chair of the section on genetics in the Committee on Common Problems of Genetics, Paleontology, and Systematics—that august group of scientists who assembled the Synthetic Theory of Evolution.

Among the features that fruit-fly geneticists studied were wing color, wing shape and size, the number of segmental plates covering the abdominal region, the number, distribution, and character of body bristles, and eye color. Since fruit-fly generations are short, and offspring develop and hatch within a matter of days, these geneticists could study hundreds of generations of fruit flies in a matter of months. By manipulating who (bearing certain features) bred with whom (themselves carrying similar or different features), population geneticists could shift the appearance of their experimental strains of fruit flies from, for example, red-eyed, straight-winged, and heavily bristled to purple-eyed, notch-winged, and sparsely bristled. These experiments demonstrated how population geneticists, acting as selective agents, could slowly shift the representation of features and their underlying gene frequencies from one generation to the next.

Although these experiments never produced a new species of fruit fly, they were taken as models for how, given enough time over which small changes could accumulate, natural selection could give rise to a new species. In the wild, the environment, however defined, would be the selective agent. Forests could open to grasslands, or tropical climates become temperate. In turn, species would have to change in order to remain adapted to their changing surroundings. At this level of speculation, Darwinian evolution appeared to be vindicated.

But while this particular evolutionary model—of organisms changing along with their ever-changing environments—could explain how a species is transformed over time, it did not offer an explanation for how the diversity of life is achieved. Darwin was, of course, aware of the need to deal with this problem. And, indeed, the only illustration in *On the Origin of Species* is a diagram of dotted lines depicting lineages changing through time and then branching into a number of species or lineages. But, in the end, in spite of having entitled his best-known book *On the Origin of Species,* Darwin was unable to address the issue of just how one species could split into two or more species.

In his notebooks, Darwin hinted at geographic separation of members of the same species as being of potential importance for the creation of more species. But he never pursued this thought fully. Part of the problem, it seems, was a general lack of understanding among naturalists where the principles of inheritance and mutation were concerned. What was needed was a way to combine Darwinian selection with Mendelian inheritance and imbue this union with a mechanism that would provide for the introduction of something that would eventually produce reproductive isolation among individuals that had descended from the same ancestral species. That something came in the form of mutation, which early population geneticists were discovering they could manipulate in their experiments on inheritance in fruit flies. Although there were numerous attempts during the early decades of the twentieth century to put this all together into a believable explanation of the origin of species diversity, the most popular model was formulated by Ernst Mayr, a bird taxonomist and evolutionary theorist who was also a member of the committee that had formulated the Synthesis.

Mayr based his ideas on species identification and the process of speciation on the observation that all individuals of a species are potentially able to contribute their genes to their species' collective gene pool. Accordingly, a species is a group of individuals that, given the chance, could breed and produce offspring, which, in turn, could produce offspring. By focusing on the ability of members of a species to breed successfully, Mayr was also promoting the idea, which had been touched on by various geneticists, that the splitting of a species into more than one species would involve the reproductive isolation of members of the parental species. If a species was defined by the potential sharing of genes among its members, then, in order to produce more species, gene flow between individuals had to be interrupted. A simple way to impede the exchange of genes between individuals was to put a barrier between subgroups of individuals of the same species. And the simplest way to do this was by separating these individuals from one another geographically.

If a barrier of some sort—a river, a mountain, a gorge, for instance—were to prevent some individuals from mating with other individuals of the same species, there would then be two groups of individuals on which selection could act. Since it is unlikely that the population on one side of this barrier would have exactly the same frequencies of genes distributed among its members as the population on the other side of the barrier, natural selection would be confronted with two different sets of variations from the very beginning. And since it is unlikely that the same small mutations that arose in one population would arise in the other isolated population, the pool of variation available in each group for natural selection would continue to be different. Eventually, over a long period of time, different combinations of morphological and genetic changes would accrue in each separate group. At some point in time, it would be genetically impossible for individuals of one group to breed successfully with members of the other group. In keeping with Mayr's initial premise that a species is a group of individuals that can breed successfully, there would now be two species, whose respective members would not be able to successfully interbreed. Since natural selection never sleeps, each new species would continue to be modified until it was transformed into something else. This is the essence of what became known as Mayr's biological species definition.

Mayr's concept of species formation—which always involved a component of linear transformation of something into something different, with splitting or branching evolution yielding the diversity of life—was generally accepted by evolutionists. But, while most evolutionary biologists were content with the theory that the history of life on earth is the result of a mixture of linear species transformation and spurts of diversification by way of branching, this evolutionary package was, and continues to be, not so easily embraced by those involved in human evolution.

Indeed, paleoanthropologists remain divided on the question of whether human evolution has been dominated by either branching or "straight line" evolution. Had there been lots of species of hominids? Or had there been only a few species, with human evolution essentially being a single continuum of change? In one of the most influential books ever to leave an impression on human evolutionary studies, *Mankind Evolving,* Theodosius Dobzhansky argued that most of human evolution had been straight-line, with one species evolving directly and gradually into another, and with natural selection leading the way. For hominids, Dobzhansky reduced what some taxonomists would identify as diversity—that is, species—to the mere level of variation between members of the same species. Although it was published in the early 1960s, the general thrust of *Mankind Evolving* still has a stranglehold on much of the discipline, especially when it comes to the evolution of the genus *Homo.*

In spite of the fact that Dobzhansky's very own experiments with fruit flies led him to believe that naturally occurring rearrangements of pieces of chromosomes—structures within the nucleus of a cell and on which genes reside—provide a mechanism by which new species can arise, he could not bring himself to accept the notion that humans are as susceptible to the whims of nature as other organisms. As Dobzhansky saw it, humans, and hominids in general, being culture-bearing creatures, fashion their own environment around them and continually modify it. By being able to do so, hominids were directing their own course of evolution. There was no impetus for speciation, for the multiplication of hominid species, when, after the initial appearance of the group, the capacity for culture came into its own.

Culture guided the slow and gradual transformation of one hominid into another until, eventually, *Homo sapiens* emerged on the scene. Since there is now only one culture-bearing hominid, it was inconceivable that there could ever have been more than one truly culture-bearing species of hominid. As for the fossils, it was primarily a matter of putting them into a chronological framework, so that one could actually see what the morphological and cultural transformation was and how much morphological and cultural diversity there had been at any given point in time. A fossil was assigned to a species in the continuum by virtue of its geological age. And once a fossil was pigeonholed into a certain species, whether or not it looked like other specimens already attributed to that species, its morphology was then interpreted as being part of that species' variability. The more specimens that were thrown into a species, the greater the variability of that species. The greater the variability within so few species, the less hominid evolution looked like the evolutionary pattern of other animals, and the less it seemed that humans were subject to the same rules of evolution that governed other organisms. It became a self-fulfilling prophecy: Humans are special, and so is their evolution.

Dobzhansky's legacy lives on. But his approach to human evolution—or the evolution of any group of organisms, for that matter—is the wrong way round. If you impose your view of the process of evolution on the fossils, then, of course, the picture you get will conform to your expectations. A paleontologist who believes that the predominant mode of evolution is for one species to be forever transformed will plug fossils into this framework and then explain how the collective morphology of the specimens at any point in time came to be the way it is. In order to explain how something came to look (or behave in) a certain way, one has to invoke arguments that rely on particular models of selection and adaptation. In paleoanthropology, it is popular to account for the extreme variability created by placing Neanderthals and a slew of other nonhuman-looking specimens into *Homo sapiens* as the result of adaptation to diverse climates or to the demand, or the lack of demand (depending on the specimen), of using the jaws and teeth as tools.

But if morphology is the result of an underlying genetic command, and changes in both reflect what we believe constitutes evolution, then it is morphology in fossils, and morphology and genes—not just genes—in living organisms that we should be looking at in trying to figure out how many species there are or might have been and what the potential relationship of species are to one another. Then, and only then, will we be able to stand back and reflect on the picture of evolution: Was the evolutionary history of a particular group slow and gradual, or was it rapid? Were there intense periods of speciation, or was there very little diversification?

Basic to any attempt to sort out the evolutionary relationships of fossils and living forms is the problem of defining species. What is a species, after all? How do we know that we are recognizing a species of living organism? And, more problematic, perhaps, how do we know that we can recognize a species in the fossil record? In terms of hominids, how is it paleoanthropologists can now admit that many species of *Australopithecus* once existed, when only a few decades ago many in the discipline would only grudgingly allow that, maybe, there had been two early species, but only one thereafter? How is it many paleoanthropologists can now admit that many species of *Australopithecus* once existed but still balk at the idea that there was ever only one continually evolving species of the genus *Homo?* Why should this be the case with hominids when taxonomic diversity is the rule in the evolutionary history of every other group of organisms that ever graced the face of the earth?

If the answer is that hominids are different, that they have taken over the role of natural selection in determining the course of their own evolution, then we might ask: How do we know that evolution works the way it is typically portrayed? Does natural selection actually drive evolutionary change? Are the anatomical features of organisms continually shaped and reshaped by natural selection in order to better adapt their bearers to changing conditions around them as well as demands placed upon them? If this is the way of evolution, how do we reconcile the fact that the difference between a fish's fin and a tetrapod's toed limb lies solely in when and where the same homeobox genes are activated, and in the insertion into one of these genes of a short sequence of molecules, with the traditional evolutionary expectation that a toe-bearing limb gradually evolved from a fin?

Since the morphologies that make up an organism ultimately derive from the turning on and off of homeobox genes, you would expect that evolutionary novelty would emerge abruptly, rather than through the accretion of minute building blocks to make a whole structure. As such, hominids may not have slowly come to walk on two legs by means of a process that transformed a creature like a typical quadrupedal monkey into a partially erect ape and then into a fully erect biped. Natural selection may have refined hominid bipedalism once it was realized, but it need not have been its creator. That role may lie only in the homeobox genes.

Perhaps our thinking on the processes of evolutionary change has been too unidimensional in its assumption that all we need to know is that mutation continually introduces variation into the pool of existing variation and that, by picking and choosing from this churning vat of variability, natural selection will somehow mold something new out of something old. There have been debates on just how fast natural selection can manufacture change. Darwinian gradualists maintain that mutations that matter to evolution are and always have been minor, and that morphological change is and always will be a cumulative process, even if it is speeded up at times. Another school of thought distinguishes between evolutionary change, which is seen as an abrupt process, and gradualism, whose focus on individual variation is relevant once a species has become established. For the latter evolutionists, who are often referred to as punctuationalists, the origin of species is a different phenomenon altogether. Historically, each side of this disagreement has viewed the other as having the wrong answer to what is thought to be the same question. What has not been widely considered, however, is the possibility that each side of the debate has something relevant to contribute to our understanding of evolution. The matter at hand, then, is to define the proper questions to which the available answers must be directed.

Although humans as subjects have been important in the historical development of evolutionary thought, research on human evolution will probably never give us any insights into the development of a worm or a zebra fish. The reverse, however, is true. Human evolutionary studies have for too long been pursued as if hominids were and still are exempt from the evolutionary processes that influence all other organisms. But developmental genetics is demonstrating, as was never before thought possible, that this is not the case. Even more profoundly, the nonstop discoveries in developmental genetics are making it eminently clear that we must expand our vision of evolution, and of evolutionary processes, beyond our present scope. We will have to think differently about—and perhaps totally reconceive—our basic notions of evolution.

In order to reach a point where we can integrate previously known facts into recent discoveries, and broaden our appreciation of the many dimensions of evolution, we must understand the historical background that often shackles us to the received wisdom of the past. Although humans and homeobox genes would seem to make strange bedfellows in this regard, we can learn much from dissecting the biases that lay in wait for such discoveries as Little Foot, *Ardipithecus ramidus, Australopithecus anamensis, eyeless, Antennapedia,* and *hedgehog.* As we proceed through this book, teasing fact from fantasy, we will be able to reassemble what we discover into a more expanded, and in many ways untraditional, evolutionary synthesis.

2

How Humans Distinguished Themselves from the Rest of the Animal World

At first, undifferentiated shapes or earth arose, having a share of both elements Water and Heat . . . but they did not yet exhibit a lovely body with limbs, nor the voice and organ such as is proper to men.

—Empedocles (500–430? B.C.E.)

When you look out upon the world, what do you see? Many of us, like me as I alternate between writing and staring out the window, see buildings, streets, a few trees, people of different sorts walking, dashing, or jogging, the occasional bird or squirrel or domesticated cat or dog. Without articulating how or why, we humans know that we are a very different animal from a squirrel or a bird, or a dog or a cat. Equally strongly we also know that, regardless of differences in height, head shape, skin, hair, or eye color, thinness or corpulence, and ability to move quickly, all those individuals out there are humans—the group of animals to which we belong.

For as long as humans have been around, we have been trying to figure out the world and how we fit into it: What truly makes a human different from a "beast," and how did so many varieties of humans come to be? Sometimes the inquiries into these questions have been spurred by genuine intellectual curiosity, and have been approached as scientifically as was possible at that point in history. At other times, such endeavors have not been motivated by a scientific drive at all. The Dark Ages would not have existed if the church hadn't squelched all activities except those that were biblically sanctioned and inspired by revelation. And at still other times, a pseudoscientifically based rationalization of human differences has led to imperialism and slavery, ethnic cleansing, and the promotion of eugenics, the selective breeding of the "best" of humankind. But even when the inquiry is a scientific and seemingly purely objective one, the answers obtained often pose a stumbling block. For it is one thing to strive to understand our place in nature but another to confront the reality of just how close to other animals we really are. The history of paleoanthropology illustrates this almost schizophrenic situation. Were this not the case, the very same comparative anatomists who claim to demonstrate the extreme morphological similarities between humans and other animals, particularly the apes, would not also elevate the uniqueness of their own species by allocating this species to the taxonomic category of family, which is a classificatory rank that is typically used to subsume larger groups of related species. That is tantamount to saying, "It's all right for humans to be one with nature, but only up to a certain point."

The dilemma of how we are to see and define ourselves relative to the rest of the world persists, largely because of the history that has preceded us.

In the Beginning

Because humans are found in virtually every corner of the world, one would expect that the intellectually inquisitive of every region would have had thoughts about the natural world surrounding them. But it is a curious twist of fate that so far the only writings and other historical records that preserve the earliest inquiries into the question of our place in nature are known from the Western world. Although it should be expected that early Western thought had been influenced by contact with other cultures, the extent to which this may have occurred is undecipherable. As Charles Darwin's grandfather Erasmus, a noted physician and one of the earliest evolutionists, speculated in his book *The Temple of Nature:*

> The mosaic history of Paradise and of Adam and Eve has been thought by some to be a sacred allegory, designed to teach obedience to divine commands, and to account for the origin of evil . . . and . . . that this account originated with the magi or philosophers of Egypt, with whom Moses was educated, and that this part of the history, where Eve is said to have been made from a rib of Adam might have been an hieroglyphic design of the Egyptian philosophers, showing their opinion that Mankind was originally of both sexes united, and was afterwards divided into males and females: an opinion in later times held by Plato, and I believe by Aristotle.

Although it is quite probable that many themes of early Western thought derived from Egyptian as well as Mesopotamian and Indian sources, only the inquiries and musings of the Greco-Roman philosophers and scientists have survived. Popular among Greeks of the fifth century before the Christian era (B.C.E.), such as Herodotus and Hippocrates, and also those of the fourth century B.C.E., such as Aristotle, was the belief that the environment played a major role in determining how an individual looked and behaved. Contributing to one's environment were influences of nature as well as those of one's society and culture.

The larger Greek world consisted of those portions of the European, African, and Asian continents that broadly circumscribed the Mediterranean region. Northern Europe, Asiatic Russia, the very Far East, and subequatorial Africa remained terrae incognitae. The Greeks were very aware of differences between the human groups known to them. But, instead of promoting these differences as signifying real biological differences between groups of humans, the Greeks explained variation in features such as skin pigmentation or configuration of nose, lip, and hair in terms of either environmental differences or racial mixture.

Herodotus, who is often acknowledged as the first historian, commented on differences between the skulls of Persian and Egyptian soldiers. Based on studies of the dead and wounded soldiers of their respective armies after battle, he described how easily, in comparison to an Egyptian's skull, a Persian's skull cracked. The supposedly more brittle Persian skull, he suggested, had become soft because of the Persians' custom of wearing felt hats. In contrast, from birth onward, the male Egyptian's head was shaved, which exposed the naked head to the sun. A life of full exposure to the sun was supposed to cause the bone of the skull to thicken.

Hippocrates, the father of medicine, also had his theories about human variation. For example, the Scythians, who lived in Asia Minor, were a "ruddy race," due to the cold and damp climate typical of the region. In general these people were supposed to be gross, moist, relaxed, and flabby because they did not swaddle their children or bind themselves tightly in clothing, as the Egyptians did, and they chose to ride in wagons rather than get about on foot. Moreover, the animals of this region were also supposed to be small because of the constantly cool, damp climate. The great diversity of physical characteristics and temperament among Europeans, on the other hand, was a direct consequence of the frequently changing and sharply contrasting seasons.

Aristotle, who gained wealth and status from tutoring Alexander the Great, was not only a great philosopher but also an early student of comparative animal as well as human anatomy. Like Hippocrates before him, Aristotle believed that the moist climate of Asia Minor had an effect on the inhabitants of the region. In one case, he pointed to the straight hair of the Scythians and the neighboring Thracians as resulting from the particularly damp environmental conditions of Asia Minor. In a similar vein of reasoning, Aristotle suggested that the "Aethiopians" of Africa had woolly hair because they lived in an arid region. Inasmuch as "Aethiopia" was supposed to be the land where the sun rose and set, the common explanation for the dark skin of the people was that it was sunburned.

It is in the writings of Aristotle that we find reference to similarities between humans and monkeys. In actuality, Aristotle made reference to the group of Old World monkeys we now identify as baboons. He also cited "the ape" as being similar to humans. However, since no African ape—neither the chimpanzee nor the gorilla—was known to the scientific world until the eighteenth century, it is unlikely that Aristotle was referring to a real ape. Rather, it is the Barbary ape, which happens to be a monkey with an unusually short tail, that probably constituted Aristotle's ape. As such, it is not surprising to find Aristotle describing what he considered to be a monkey as a tailed ape. (As comparative anatomists would later discover, true apes, like humans, really are distinguished from true monkeys by their lack of a tail.)

Aristotle was not only the first scholar to write about general anatomical similarities between humans and monkeys but also the first to distinguish humans from other mammals on the basis of bipedality. He defined a mammal as an animal that is a blooded and viviparous quadruped. This means that mammals are distinguished from bloodless animals such as insects because they give birth to live young and locomote via four limbs or points. He also defined a bird as an animal that is blooded and locomotes via four points; for Aristotle, two wings + two legs were functionally equivalent to having four legs. However, instead of bearing live offspring, a bird lays eggs, and it is because of this distinction that he did not group birds with mammals. We might wonder why Aristotle did not get sidetracked by the apparent similarity between birds and humans in being bipedal animals. He did not, because he believed that the legs of birds differed in the articular relations of their parts from the legs of humans. In describing birds, Aristotle wrote: "Two legs like man, being inward like quadrupeds and not outward like man." In reality, what looks like an inwardly bent knee joint in birds and nonhuman, quadrupedal mammals, such as cats, dogs, and horses, is the ankle joint. In all birds and mammals, including humans, the ankle joint is bent or flexed inward. In birds and quadrupedal mammals, the outwardly bent knee joint is much higher up along the leg than it is in humans, or most other primates, for that matter. But Aristotle's error in anatomy did keep birds and mammals apart, creating a separation that taxonomists would maintain for more than a millennium.

As for humans, Aristotle was able to sneak in a comment about braininess along with his definition based on bipedalism:

> Now, man, instead of forelegs and forefeet, has, as we call them, arms and hands. For he alone of the animals stands upright, on account of his nature and ousia [*ousia* can be translated from the Greek as "substantial being" or "defining character"] being divine, and the function of that which is most divine is to think and reason; and this would not be easy if there were a great deal of the body at the top weighing it down, for weight hampers the motion of the intellect and the common sense.

With this statement, written so long ago, in the fourth century B.C.E., Aristotle effectively delineated the two features—bipedalism directly, and the brain by inference—that humans would continue to exploit in distinguishing themselves from the rest of the animal world. Since Aristotle defined one particular group of animals, the mammals, in part on the basis of their locomoting via four points, he also ended up excluding humans, who walk on only two points, from this group, in spite of the fact that humans are not only blooded animals but also animals that bear live offspring. Taxonomists of the late eighteenth and nineteenth centuries would revive a version of Aristotle's groupings of animals by separating humans from other primates on the basis of two- versus four-handedness.

Scholars of the Renaissance and later periods also inherited from Aristotle the notion that inanimate and animate forms could be lined up in an ascending scale of increasing complexity. Aristotle's so-called Ladder of Life, or, as medieval scholars called it in Latin, Scala Naturae, was anchored on one end in the geologic, the mineral, the sediment on the bottom of a lake or swamp. From there, the Scala Naturae ascended as an unbroken graded series of forms of increasing complexity, from plants through animals. At times, Aristotle must have been hard put to differentiate between different kinds of similar organisms in his Scala Naturae. Perhaps this indecision reflected the difficulties taxonomists still have in trying to identify closely related species. Defining species and distinguishing between closely related species persist as some of the most perplexing conundra in modern taxonomy and systematics. Although these difficulties are complex enough when we are dealing with living organisms, they are multiplied many times over when the subject is fossils.

In spite of a history in which some scholars, including Charles Darwin, ascribed to Aristotle an appreciation of evolution and natural selection, it is quite unlikely that this is true. In Aristotle's worldview, some of life's forms arise by spontaneous generation—some out of thin air, and others, such as eels and insects, from mud and fecal matter. Humans, mammals, and birds, however, either start out as larvae or develop from eggs. From his comparative anatomical studies, which had to have involved some dissection, as well as from his comparative growth studies, Aristotle presented a masterful picture of the details of the diversity of life's forms—a picture that was in many ways so accurate that it remained central to discussions of comparative anatomy and taxonomy well into the latter part of the eighteenth century.

As Aristotle envisioned it, nature had design and purpose. In keeping with the teachings of Plato, the Greek philosopher and scholar with whom Aristotle had studied, the objects and organisms we see around us are imperfect. They are, however, in a constant state of flux, striving toward a final, perfect state. Since, as Aristotle was aware, some thing, or force, is required to produce motion, he hypothesized that there must be a Prime Mover and that it is responsible for all movement. Aristotle's Prime Mover was supposed to be a divine rotating sphere that surrounded the universe, with the earth at its center. Transmitting its motion ultimately to the moon, which, in turn, transmitted it to the earth, the Prime Mover conferred motion to all living as well as inanimate things on earth.

The Prime Mover's motion was responsible for pushing living, as well as nonliving, entities toward their destinies. For example, the destiny, or final cause, of a bird's egg is the adult bird. But Aristotle envisioned every living thing not only as a collection of bits and pieces but also as an individual with a psyche. It was the psyche that received motion from the Prime Mover. And it was the psyche that, in turn, drove each individual organism. Since Aristotle believed that everything on earth was imperfect, he imagined the psyche as driving the organism toward an ultimately perfect state. Because of the psyche's drive toward a perfect state, Aristotle conceived of life as a striving upward. It was not, as is sometimes misunderstood, that an organism was trying to become the next organism on the Scala Naturae but that each organism was struggling to become the ideal version of its own kind. Competition among organisms was, for Aristotle, a consequence of the imperfect state of nature.

In the Shadow of Aristotle

Among Greek and Roman philosophers there were two major themes about how their high standard of human society and civilization came to be. One view, which was fairly pessimistic, was that human history paralleled the life cycle of the individual. From a healthy, youthful beginning, the body ages and ultimately falls apart. Likewise, in the good old days, human civilization was in a healthy state, from which, unfortunately, it degenerated further and further into decay and corruption.

The alternative scenario was much more optimistic. Originally, human life was supposed to have been simpler, albeit more primitive. From these beginnings, technology developed gradually over an unknown period of time, leading, relatively recently, to an improvement of the human condition and to the levels of civilization the Greeks and Romans enjoyed. This theme incorporated a feeling of progress, the notion of which, centuries later, became entwined with various general theories of animal evolution, as well as with the more specific theories of human evolution. For example, in his only monograph on human origins and evolution, *The Descent of Man,* Charles Darwin was preoccupied with a scheme of progressive evolution that transformed the primitive races into the civilized ones.

The first-century-B.C.E. Roman poet and philosopher Titus Lucretius Carus (better known simply as Lucretius) also embraced the idea of progress during human history from a primitive to a civilized condition. In particular, he speculated that one of the first milestones in the development of human society was the attainment of articulate speech. The capacity for articulate speech, as with Aristotle's emphasis on intellect, eventually came to represent the outward expressions of the level of braininess that taxonomists would later use as a marker by which to set humans apart from the rest of the animal world. As we shall see, for taxonomists such as Linnaeus, articulate speech emerges as a critical criterion by which true humans were distinguished from subhumans—subhumans being those species that warranted being placed in our genus, *Homo,* but were otherwise not quite human enough to be placed in our own sapient species, *sapiens.* Philosophers of the eighteenth century, as well, such as Rousseau, would also focus on the capacity for language in their deliberations on the subject of just what, precisely, distinguishes humans from other animals.

Lucretius's picture of early, uncivilized humans is reminiscent not only of cartoon images of "cave men" but also of late-eighteenth- and early-nineteenth-century reconstructions of "early man." Lucretius's early humans were supposed to have had larger, harder, and stronger bodies and muscles than the typical Roman. These early humans were essentially unperturbed by extremes in temperature. They could tolerate unknown foods, and they rarely became ill. Because they were only slightly different from other animals, Lucretius's early humans traveled in bands and lived in the woods and in caves, sleeping on the ground and taking refuge from inclement weather in thickets. They lacked fire and clothing. But they did use tools, bringing down wild animals with clubs and stones.

Like other Roman philosophers, Lucretius subscribed to the Greek belief that differences between human groups in such features as skin color, susceptibility to disease, and general physical attributes were correlated with the climates and regions in which these different populations lived. Unfortunately, the Greco-Roman idea of a naturalistic correlation between environmental and racial differences was abandoned with the spread of institutionalized Christianity and its replacement of the scientific method with divine inspiration.

Enter the Dark Ages

One of the truly intellectually admirable qualities of the Greeks and Romans was their approach to acquiring knowledge about the world and its inhabitants. If, like Aristotle, you wanted to learn about the growth and development of a chick or a human, or the anatomy of a whale or an insect, you went out and tried to make direct observations. You might even engage in a kind of experimentation. For early Christian scholars, however, knowledge about the world and the heavens about them was obtained not through personal experience but through the vehicles of prayer and revelation. This religious rather than scientific route to knowledge persisted with little interruption throughout the medieval period and on into the Renaissance. In the midst of this scientifically stultifying atmosphere, it is of some interest that, during the late sixth through the early seventh centuries, Isadore, the bishop of Seville, could in parallel with Christian dogma embrace the Greco-Roman idea that populational differences in personality and physique are correlated with location and environment. As for human origins, Isadore of Seville noted that the Latin word *homo* could be derived from another Latin word, *humus.* Inasmuch as *humus* refers to the soil or earth, or, more specifically, to the organic component of soil, Isadore of Seville took this linguistic similarity as support for the story of the creation of Adam by God as told in the Book of Genesis of the Old Testament.

Perhaps the most important part of this period in history for gaining perspective on the development of human evolutionary stories is the melding of Aristotle's idea of a Ladder of Life with the story of creation in the Book of Genesis. If the story of the Book of Genesis was correct, there had been a divine creation of each kind of organism on the face of the earth. Arranged in a Scala Naturae, the different kinds of organisms represented an ascending scale of complexity, leading to humans. Given that humans were supposed to have been created in God's image and were therefore the only animals perfect in both mind and body, the different kinds of organisms in the Scala Naturae were believed to represent a progression of increasing perfection that culminated in humans. But because each life form was supposed to be immutable, each was inextricably fixed as the kind of organism it was, being forever stuck in a ladder of increasing complexity with nowhere to go.

This biblically grounded interpretation of the world and its inhabitants became known as the Great Chain of Being. Because, according to this doctrine, life in its myriad guises had been created by a Divine Being in a certain way and with a certain intention, it became the goal of the taxonomist to sort out the correct sequence of creation, leading from the lowliest organisms to humans. Since the steps or stages in this progression of increasing perfection of living forms were assumed to have been small, taxonomists were also preoccupied with filling in the gaps—finding the missing links—between forms that had already been assigned their position in the Great Chain of Being. Unfortunately, engaging in this pursuit resulted in many taxonomies in which taxonomists not only arranged their perceived groups of humans along an imagined scale from the most primitive to the most perfected but also chose to relegate their hierarchically ranked races of humans to the level of distinct species.

The motivation behind these different taxonomies—on the one hand allowing even supposedly primitive humans to be classified with humans of presumed perfection, and, on the other, demoting all but the most perfected of humans to the ranks of species closer to the "brutes"—came from different interpretations of the biblical story of creation: Either there had been one creation for all humans, regardless of the diversity among people living in different places and environments, or there had been a separate origin for each perceived race of human. A strictly biblically based interpretation of creation yields a single-origin model for humans that places racial differences in a monogenic rather than a polygenic context. Polygenesis, or the separate-origin model of human diversity would, however, become embodied in the classifications of some of the most prominent of eighteenth- and nineteenth-century taxonomists. The fear of being accused of being a polygenecist has probably been behind the reluctance in paleoanthropology to recognize species diversity in the fossil record, especially when we are dealing with potential close relatives of *Homo sapiens*.

The Age of Questioning

Beginning with the Holy Crusades, which lasted from 1095 to 1291, the stranglehold that the Catholic Church had on scientific research began to crumble. By way of the Crusades, medieval Europeans were exposed to different cultures as well as to the Islamic emphasis on science and learning. With this exposure came an eventual questioning of the Christian dogma that forbade inquiry of any sort beyond the pages of the Bible and a repudiation of the leaders that represented such inquiry.

When the split in the Catholic Church that led to two popes—one in Rome and the other in Avignon, France—yielded to the Reformation, it was only a matter of time before some of the energy that went into questioning the authority of the church on religious grounds spilled over into a scientific questioning of religious dogma. For example, Nicolaus Copernicus boldly published his rejection of the earth's being the center of the universe, showing, instead, that the sun actually held this central position. Fortunately for Copernicus, his natural death in 1543 preempted his being the victim of religious persecution. Within a hundred years, Galileo Galilei laid to rest the notion that unchanging planets orbited the earth in perfect circles. He did so by pointing his telescope toward the heavens and demonstrating, for example, that the sun has spots of intense activity, Jupiter has its own orbiting moons, and Venus passes through different phases.

Although other doctrines also eventually crumbled, one that has not is Aurelius Augustinus's calculation of a six-thousand-year-old earth. Augustinus lived from A.D. 354 to 430 and was later canonized as St. Augustine. His six-thousand-year-old earth would be cited with authority in the latter part of the nineteenth century in the anti-Darwinian tirades of Bishop Samuel Wilberforce in his debate with Thomas Huxley and, in the first quarter of the twentieth century, in the anti-evolutionist oratory of William Jennings Bryan during the Scopes trial. A biblically based calculation of the earth's age is still embraced by creationists. At least Augustinus's notions of a flat earth—on which all inhabitants had already been discovered—would be dashed by the exploratory and venture-capitalistic voyages of the fourteenth and fifteenth centuries. With the Dutch, British, and Spanish sending out missions to all parts of the world in order to secure trade routes, resources, goods, and commodities, as well as far-flung lands and their inhabitants, the age of discovery was well under way. It turned out, as Herodotus had predicted, that the earth was indeed round. And, the clout of Augustinus notwithstanding, there also happened to be previously unknown people living on the other—the non–Old World—side of it.

But exploration and the discovery of new and different animals, as well as of different humans, became a double-edged sword. On the one hand, finding first different kinds of humans living in the Americas, then an ape inhabiting two of the major islands of Southeast Asia, and eventually two other apes ensconced in the depths of Africa, certainly added to questions and theories about human origins. However, since even as late as the nineteenth century the task of the taxonomist remained wedded to an elucidation of a Great Chain of Being, the discovery of potential missing links of the chain also led to some rather unsavory scientific conclusions.

Through the work of such notable human anatomists as Albertus Magnus, who lived in Cologne, Germany, during the thirteenth century, nonhuman primates came to be seen as providing the links between humans at the top of the Great Chain and the beasts and brutes at the bottom of it. However, because the Great Chain was supposed to be a progressive sequence from the imperfect to the perfect—perfection in body and mind separating humans from the rabble of the animal kingdom—there had to be a progression from one link to the next. According to medieval comparative anatomists, the unenviable link between "perfect" humans and "imperfect" nonhuman primates was filled by the "pygmies" of Africa. The discovery of more and different kinds of humans, as well as of more and different kinds of nonhuman primates, only gave the illuminators of the Great Chain more to work with in terms of refining the links in their envisioned chain of perfection. By the middle of the nineteenth century, pygmies (properly called the San), Hottentots, Australian Aborigines, and even the Tierra del Fuegians had been offered up by various taxonomists (as well as later evolutionists such as Charles Darwin and Thomas Huxley) as the most primitive of humans. And as the most primitive of humans, they made perfect links to the apes.

A Matter of Degree

The discovery of unexpected groups of humans in the New World did more harm than good to the biblical basis for believing in monogenesis. If, as the Bible said, all humans had descended from Adam and Eve (and even Eve was descended from Adam), then, from a taxonomic standpoint, all humans, regardless of their apparent differences, belonged to the same species. Unfortunately, there was nothing to prevent taxonomists, or out-and-out racists, from ranking humans in hierarchical fashion within their own species. However, as the polygenecists saw it, the differences between Europeans and other people were so great that it was not possible that all humans had descended from the Adam of the Bible.

One of the earliest proponents of polygenesis, Theophrastus Bombastus von Hohenheim, a mid-sixteenth-century German philosopher, made several arguments against New World groups having descended from the original Adam. These human groups could have descended from another Adam, but not from the Adam from which he claimed descent. Among other things, von Hohenheim suggested that God would never have allowed these newly found humans, who supposedly lacked a soul and were equivalent to parrots in speech, to have descended from the same ancestors he had.

Among taxonomists who favored polygenesis and the belief that each group of humans had descended from a separate Adam, perhaps the most outrageous was the Frenchman Jean Baptiste Genevieve Marcellin Bory St. Vincent. In a work published as recently as 1825, Bory St. Vincent took the polygenic position to the extreme. He proposed that there were no fewer than fifteen distinct species of human beings. In addition, each one of Bory St. Vincent's human species subsumed a number of varieties. Not surprisingly, only a few western European groups topped Bory St. Vincent's list of separate species, while the Hottentots were relegated to the species that was supposed to be closest to the apes.

Whether the taxonomist favored monogenesis or polygenesis, it is clear from the taxonomic literature that by the seventeenth century the upper end of the Great Chain of Being had settled into a hierarchical arrangement of monkeys, apes, and humans. Of course, whether humans were derived from one or more Adams, they were placed in a group apart from other animals. But while taxonomists endeavored to diminish the size of the gaps between links, the taxonomies they generated also emphasized the differences between the groups they formally named. The very nature of taxonomy—pigeonholing—inevitably leads to such an outcome.

Among taxonomists of the sixteenth and seventeenth centuries, the simplest way to emphasize the differences between humans and other animals was not to classify humans at all. Such was the approach taken by Konrad Gesner and Francis Willughby. Although he has been referred to as the German Pliny, Gesner, a Swiss by birth, was actually a compiler of biological information rather than a practicing field naturalist and observer. His major work, *The History of Animals* (*Historia Animalium*), published in the mid-sixteenth century, was one of the first attempts to distinguish between biological fact and myth. Although he did describe various so-called lower primates, not once did he mention humans. Among Gesner's longer-lived legacies, however, was putting groups of organisms, which he did not refer to specifically as species, into a higher category that he called the genus. Thanks to Linnaeus, who coopted this taxonomic category, we still use the rank of genus in our classifications.

Just over a century after Gesner, the British taxonomist Francis Willughby produced a popular classification of animals that also failed to mention humans. It is likely that Willughby had been influenced by the Cambridge University professor John Ray, who would be remembered not only as one of the greatest natural philosophers but also as the father of modern systematic zoology. In his taxonomy, Willughby attempted a natural classification of animals. To his credit, he did not follow the lead of many other taxonomists in grouping organisms on the basis of where they lived and how they locomoted: whether they flew through the air, swam in the seas, lakes, and rivers, or walked or crawled on land. Instead, he used real anatomical features to define groups. Although the word *primate* would not be introduced into the taxonomic literature for another ninety years, by Linnaeus, it is obvious that Willughby was describing primates when, in 1668, he wrote: "Man-like; having *faces* and *ears* somewhat resembling those of *Men,* with only four broad *incisors,* or cutting teeth, and two short eye-teeth, not longer then the other, their *fore-feet* being generally like *hands,* with *thumbs,* going upon their heels; whether the *Bigger kind;* either that which hath a *short tail:* or that which hath *no tail.* 1. BABOON, *Drill.* APE-*Jackanapes. Lesser kind;* having a *long tail* . . . 2. MONKEY, *Marmosit.* SLOTH, *Haut,* Ay." Clearly, there is nothing either in the general description ("Man-like," etc.) or the specific features of lacking a tail that would exclude humans from Willughby's classification. Yet humans were never discussed.

How strangely schizophrenic it must have been to live during a time when the details of anatomical similarity, not just between humans and apes and monkeys but between humans and other animals, were becoming increasingly well known and undeniable. Yet these similarities had to be denied or diminished in importance. Even intellectuals who were not taxonomists, and would have been free of this dilemma, were inevitably influenced by the Great Chain. For example, although the great fifteenth-century artist and anatomist Leonardo da Vinci could write clearly about the obvious similarities between humans and other animals, in the end he could not in his mind bring humans any closer to the rest of the organic world.

"It is an easy matter," da Vinci wrote in his *Notebooks,* "for whoever knows how to represent man to afterwards acquire . . . universality, for all the animals which live upon the earth resemble each other in their limbs, that is in the muscles, sinews and bones, and they do not vary at all, except in length or thickness as will be shown in the Anatomy." In his sweeping prose about the anatomical similarities among animals, da Vinci even corrected Aristotle's misconception of how a human's knee joint differed from that of other animals. "Show a man on tiptoe," he proclaimed, "so that you may compare a man better with other animals . . . [and] . . . [r]epresent the knee of a man bent like that of the horse."

But if humans are anatomically so comparable to other animals, in what does the difference lie? According to da Vinci, "man does not vary from the animals except in what is accidental, and it is in this that he shows himself to be a divine thing . . . for where nature finishes producing its species there man begins with natural things to make with the aid of this nature an infinite number of species." Like so many others before and after him, da Vinci was drawn away from the anatomical similarities between humans and other animals to the mental, the manipulative, the control of nature.

One of the first works in which humans were discussed in conjunction with other primates, as well as with other animals in general, was published in 1632 by Joannes Jonstonus. But it would be another one hundred years, with the publication of the first edition of Linnaeus's taxonomic treatise *Systema Naturae,* before humans were classified in a group with other animals. Although today we refer to our taxonomic group as Primates, and correctly attribute the origin of this group name to Linnaeus, the first name that he gave to us was Anthropomorpha, which John Ray had coined for the nonhuman primates alone. Although *Anthropomorpha* means "man-shaped," Ray still kept humans out of that group. It was in the tenth edition of the *Systema,* which was published in 1758, that Linnaeus changed the name of our order from Anthropomorpha to Primates.

Linnaeus, the Swedish botanist who had been christened Carl von Linné, published the first edition of the *Systema* in 1735, when he was only twenty-eight years old. Heavily influenced by the work of such pioneer taxonomists as Konrad Gesner and John Ray, Linnaeus went further than they and, in the very first edition of the *Systema,* developed the basic form of classification that we still use today. In the mid-sixteenth century, Gesner had grouped species into genera. But it was not until the end of the seventeenth century that John Ray tried to instill concrete meaning into the definition of a species. Until then, the concept of a "species," which had not been formally defined in taxonomy, could be anything the classifier imagined.

Linnaeus put the two taxonomic terms together. For him, following Ray, species were real entities—indeed, species were the basic units of life in nature. As such, a species should be identifiable and recognizable by something real, such as a particular morphology or behavior. Because of this quality, Linnaeus made the species the fundamental unit of his classification scheme. If there were two species, each distinguishable from the other by a specific feature, but the two had some other characteristic or characteristics in common, they would be placed in the same genus. The genus would then be defined on the basis of the feature shared by the species. The genus would be analogous to a surname, which can be held in common by a number of individuals. The species would be analogous to a first name, which is typically unique to an individual among relatives of the same surname. Since Linnaeus demanded that every species be placed in a genus, if there was only one species, the characteristic of the species would also define the genus, at least until another species was discovered that might share that genus. As species would be grouped into genera, so, too, would genera be grouped into higher categories—such as families—which, in turn, would be grouped into larger groups, and so on. The farther up the hierarchy of a Linnaean classification one goes, the more general becomes the information pertaining to each broader taxonomic group.

If Linnaeus had done nothing else except challenge authority and classify humans in a group with other animals, that would have been sufficient to secure his place in history. But he did much more in terms of organizing the organic world in ways that are still accepted by taxonomists. One milestone was his recognizing that mammals constitute a real group in nature.

In the first through the ninth editions of the *Systema,* Linnaeus grouped his orders of mammals, of which Primates was one, into the next larger group in the hierarchy, the class Quadrupedia. In the tenth edition of the *Systema,* Linnaeus changed the name from the class Quadrupedia to the class Mammalia. As originally conceived, the class Quadrupedia was defined on the basis of its members being four-footed and possessing body hair. In addition, quadruped females gave birth to live young and lactated via mammary glands. By delineating these and other features as being characteristic of the class Quadrupedia, Linnaeus became the first taxonomist to realize that although whales are aquatic, they should be grouped with the hairy, terrestrial quadrupeds. And by classifying humans with other animals in the order Anthropomorpha and placing Anthropomorpha in the class Quadrupedia, Linnaeus also acknowledged the fundamental anatomical similarities between humans and the rest of the animal world. Linnaeus's Quadrupedia is equivalent to the group we know as Mammalia.

Linnaeus developed his classification using all the information that was available to him. At times, he relied on his own observations. At other times he based his work on the letters and publications of colleagues and even the reports of mariners and travelers. This meant that not everything that he classified was real—such as monstrous humans or humanlike apes, which often figured in the stories travelers told and became incorporated into the *Systema.* Bearing in mind that for most of his life Linnaeus worked under the constraints of the Great Chain of Being, it is not hard to understand why he might embrace these mythological creatures. After all, they did provide convenient links that would bridge gaps in the taxonomic hierarchy. Nevertheless, it took no small measure of courage for Linnaeus to declare not only that humans are sufficiently like other primates to be grouped with them but also that, more generally, humans are basically quadrupeds anatomically, even though they stand on two legs.

Linnaeus stated his argument for humans being quadrupeds as follows:

> No one has any right to be angry with me, if I think fit to enumerate man amongst the quadrupeds. Man is neither a stone nor a plant, but an animal, for such is his way of living and moving; nor is he a worm, for then he would have only one foot; nor an insect, for then he would have antennae; nor a fish, for he has no fins; nor a bird, for he has no wings. Therefore, he is a quadruped, has a mouth made like that of other quadrupeds, and finally four feet, on two of which he goes, and uses the other two for prehensive purposes.

Sounding very much like the expectations that twentieth-century paleoanthropologists would later have of their fossil "missing links," Linnaeus's argument for grouping humans with apes drew on comparisons with both real and mythical apes:

> [A]nd indeed, to speak the truth, as a natural historian according to the principles of science, up to the present time I have not been able to discover any character by which man can be distinguished from the ape; for there are somewhere apes which are less hairy than man, erect in position, going just like him on two feet, and recalling the human species by the use they make of their hands and feet, to such an extent, that the less educated travellers have given them out as a kind of man.

As far as Linnaeus was concerned, the only features that distinguished humans from apes and other animals were the possession of speech and the ability to reason. But even these qualities would not have been sufficient cause for Linnaeus to place humans in one genus, *Homo,* and apes in another, *Simia*—as, indeed, he did. It was only his reluctance to formalize the anatomically obvious—humans and apes are *very* similar—into taxonomy that prevented him from putting humans and apes in the same genus. As Linnaeus wrote in 1747 to another noted taxonomist, J. G. Gmelin: "I demand of you, and of the whole world, that you show me a generic character—one that is according to generally accepted principles of classification, by which to distinguish between Man and Ape. I myself most assuredly know of none. . . . But, if I had called man an ape, or vice versa, I should have fallen under the ban of all the ecclesiastics. It may be that as a naturalist I ought to have done so."

Linnaeus's students, most notably Ole Söderberg and Christian Emmanuel Hoppe, elaborated further upon the anatomical as well as various behavioral and social similarities between humans and apes. They, too, concluded that it was difficult to justify putting these primates into different genera. Following in Linnaeus's footsteps, theirs was definitely a minority position among contemporary and even later naturalists and taxonomists. One of the most eminent of those contemporaries of Linnaeus who could not bring themselves to close the gulf between humans and apes was Georges-Louis Leclerc, Comte de Buffon. Buffon, the great French natural philosopher, headed the museum of natural history in Paris and established on the neighboring grounds the famous Jardin des Plantes. He was France's premier natural scientist. He produced treatises that remain exquisite examples of anatomical description and illustration, and expounded at length and with great authority upon everything geological, zoological, and botanical that he could get his hands on. He also sponsored and encouraged intellectual curiosity in others, even if doing so resulted in his hiring scientists of diametrically opposite scholarly and religious persuasions: Baron Georges Cuvier and Chevalier de Lamarck.

Cuvier came from the higher echelons of French aristocracy. He was the champion of a particular biblically grounded interpretation of earth history. Instead of a single flood, Cuvier and his followers believed that repeated floods had been responsible for the discontinuities not only between the geologic strata but also between the fossil forms contained in each stratum. For Cuvier and other catastrophists, as they were called, each species, represented in fossil form in a particular stratum, had been placed on earth by a Divine Creator. With each flood or catastrophe, virtually all life was wiped out. After each catastrophe, the Divine Creator repopulated the earth with life forms. This succession of catastrophes and creations produced the disjunctions between strata and the fossils preserved in each stratum. The occasional fossil form that persisted from one stratigraphic layer to the next was supposed to have escaped the ravages of the last flood by taking fortuitous refuge in an eden of sorts.

Although Cuvier was committed to a biblical interpretation of earth history, he was, nevertheless, one of the greatest comparative anatomists of the later eighteenth and early nineteenth centuries. He was the first scholar to interpret the biology of extinct species using comparable living forms as models, a practice that is still employed in functional paleomorphological studies. Cuvier also has the dubious distinction of being the first paleontologist to identify a primate in the fossil record, which he called *Adapis*. Although the word *"Adapis"* literally means "toward the bull" (*Apis* actually referring to the Egyptian bull god), Cuvier concluded in his description of the fossil that the species represented a type of extinct pachyderm. On both counts—the name and the interpretation—he erred. Adapis was a fossil primate whose evolutionary relationships lay with the lemurs of Madagascar.

Cuvier's nemesis at the natural-history museum in Paris was Chevalier de Lamarck. Lamarck, a member of the lower nobility, had been saved from a scientific life of struggle by Buffon, who gave the younger scholar the task at the natural-history museum of studying worms and other invertebrates. Although he brought distinction to the investigation of invertebrate life, and was the first scientist to use the term *zoology* in reference to the study of animal life, Lamarck is perhaps best known for having formulated and popularized the notion that an animal's desires can cause anatomical change. Or, to put it another way, the use and disuse of organs can engender their change.

This concept, which is more commonly phrased as "the inheritance of acquired characteristics," was, however, generally popular among early evolutionists, of which Lamarck was one of the first. Inasmuch as the basics of genetic inheritance were not known until the turn of the twentieth century, evolutionists had to invoke a variety of now seemingly outrageous ideas about how organismal change might occur. Even Darwin invoked notions of use and disuse to explain evolutionary change. How else, if not through lack of use, could he explain why humans have a minuscule appendix while other mammals have a large one? According to the doctrine of use and disuse, humans did not need a large appendix, so it gradually shrank in size.

Lamarck's evolutionary idea was that change occurs within lineages that had originally been established by a Divine Creator. Since the number of lineages was constant, and change occurred only within each, there was always the same number of species on earth, from the beginning until the present. As such, while Cuvier and his colleagues envisioned a Divine Being creating a constant parade of life forms, Lamarck and his followers invoked transformational organismal change to explain the succession of organisms. Although he did invoke a Divine Creator at the beginning, Lamarck's otherwise not strictly biblical view of life was anathema to Cuvier. As such, it is a bit of historical irony that Cuvier ended up presenting the eulogy at Lamarck's funeral. At this solemn occasion, Cuvier took the opportunity to excoriate Lamarck soundly for conjuring up a notion as heretical as the inheritance of acquired characteristics. When the social Darwinists of the twentieth century needed to expunge these notions from social science, who better to point the finger at than Lamarck. Pitting Lamarckianism against Darwinism also made it easy to sidestep a rigorous scrutiny of the premises upon which neo-Darwinism had become based.

Although Buffon was a scholar of enormous capacity for knowledge and inquiry, in the end he remained a believer not only in a Divine Creator but also and firmly in the fixity of species. True, he was not as restrictive as many of his predecessors in just how much variation he believed was permissible within a species. And, compared with many of his colleagues, he did allow for a considerable amount of variability within a species, especially the human species. He ardently subscribed to the theory of monogenesis and invoked environment as a primary cause of human variation. But, in spite of all his scientific accomplishments and intellectual insights, Buffon could not embrace the idea that species changed over time. Nor could he embrace the idea that the Great Chain of Being was composed of links—known or still to be discovered—that connected humans intimately with the rest of the animal world, even though many of his own studies actually reinforced the similarities between humans and other primates.

At one level of inquiry, Buffon was able to describe a continuum of morphological similarity among a series of animals and even plants that proceeded down from humans to "quadrupeds, cetaceous animals [whales], birds, reptiles, insects, trees, and herbs." In this sequence, Buffon accorded the four-handed primates (the Quadrumana) a position between bipedal humans (Bimana) and the lower animals. He was even able to go to a finer level of discrimination in his Great Chain, making a distinction between a true monkey, such as the baboon, and a true ape, such as the orangutan, and suggesting that the Barbary ape—that short-tailed monkey known to Aristotle—was the intermediate species between the former two primates.

Although the African apes would later play a central role in paleoanthropologists' speculations on what the earliest human ancestors might have been like, it was the Asian ape, the orangutan, that struck Buffon as being most humanlike. He did not elevate this ape beyond the level of a brute animal, but he did allow that the orangutan was "a brute of a kind so singular, that man cannot behold it without contemplating himself." Nevertheless, in spite of the marked degree of anatomical similarity between humans and orangutans, Buffon was more impressed by the differences. As he pointed out, these two species were dissimilar in the relative proportions of the brain and, more important, in the capacity for speech and the ability to think. "Can there be more evident proof than is exhibited in the orang-outang," Buffon concluded, "that matter alone, though perfectly organized, can produce neither language nor thought, unless it be animated by superior principle?" It was a Divine Creator who was responsible for this extra animation of matter. And it was a Divine Creator who conferred a "soul . . . on man alone, by which he is enabled to think and reflect."

In Buffon's opinion, the gulf between humans and, through the orangutan, the apes and then monkeys would never be bridged by a missing link.

There would always be a gap in the Great Chain of Being between human and brute: "The passage is sudden and from a thinking being to a material one, from intellectual faculties to mechanical powers, from order and design to blind impulse, from reflection and choice to ungovernable appetite."

Buffon was not alone in placing emphasis on intellect and language as the attributes that separated humans from other primates. Linnaeus had invoked these properties in his definition of *Homo sapiens.* However, Linnaeus also delineated other species within the genus *Homo,* even though these species did not have language. The difference between the two naturalists—Buffon and Linnaeus—lies in the level of importance each placed on intellect and language and how this emphasis was reflected in a classification. The same kind of subjective approach to classification goes on today. On the one hand, more than fifty species of living Old World monkey and all of their potential fossil relatives are crammed into a single taxonomic family. But, on the other hand, our species, *Homo sapiens,* and a handful of potential fossil relatives luxuriate in their own taxonomic family. Humans have always emphasized their singular importance and difference from other animals, even while proclaiming their unity with them.

Toward a Better Definition of *Homo sapiens*

More than a hundred years after the British taxonomist Francis Willughby produced a classification of animals—which did include primates but in which there was no mention of humans—another British taxonomist, Thomas Pennant, produced a classification that carried on this outdated tradition. Although Pennant's treatise, which was published in 1781, dealt with all quadrupeds, the quote most frequently cited reflects his outrage at Linnaeus's classifying humans along with other animals: "I reject his [Linnaeus's] first division, which he calls *Primates,* or Chiefs of Creation, because my vanity will not suffer me to rank mankind with *Apes, Monkies, Maucaucos* [lemurs], and *Bats,* the companions Linnaeus has allotted us even in his last System" [the one published in 1758].

Fortunately, in stark contrast to the likes of Thomas Pennant and others of his persuasion, there was Johann Friedrich Blumenbach, whom many refer to as the father of anthropology. Like Buffon and Cuvier, in particular, Blumenbach was a superb, hands-on comparative anatomist. He was also one of the first anatomists to express an interest in physiology. In his published works, beginning in 1775 with his doctoral dissertation, *On the Natural Varieties of Mankind,* Blumenbach followed closely Linnaeus's classification of animals. But he differed from Linnaeus in his approach to studying the anatomy that formed the backbone of the classification.

At times, Linnaeus had focused solely on dental features to unite species within their respective orders. Blumenbach, on the other hand, was astutely aware of the fact that apparently closely allied species could differ markedly in the kinds and morphologies of the teeth they possessed. As such, through comparative anatomy and physiology, he strove to broaden the bases upon which species were defined and subsequently grouped into larger assemblages.

His comparative approach spilled over into the study of human varieties, or races. Again beginning by following Linnaeus, Blumenbach argued that all humankind was united as a single species—not, as polygenist such taxonomists as Bory St. Vincent thought, that different races were actually different species. Although Pieter Camper, an eighteenth-century Dutch comparative human anatomist and a proponent of studies of physiognomy, was himself not a believer in polygenesis, his studies on the so-called facial angle had been coopted by polygenists as support for their contention that human races should be regarded as separate species. Blumenbach soundly criticized the use of facial angles—which were only a measurement in profile of the degree of protrusion or verticality of the snout or jaws of an animal. For Blumenbach, it was not significant that some native Africans, when viewed in profile, had a protrusive jaw that could be likened by polygenists to the superficially similar protrusion of an ape's jaw. Rather, as developed first in his dissertation and in subsequent publications, what was important for Blumenbach was the abundance of anatomies common to all humans to the exclusion of other animals, including other primates.

It was Blumenbach's appreciation of shared unique qualities, regardless of perceived individual differences, that guided him toward delineating a suite of soft- and hard-tissue anatomical features that stand as a testament to the unity of all *Homo sapiens.* True, some of these features, such as having two hands or standing erect, had been noted by naturalists and comparative anatomists before him. But it was Blumenbach who, for the first time, set out not only to list but also to document the many distinguishing features of his own species. And as such, it is to Blumenbach that we may trace the beginnings of not just anthropology, as is most frequently acknowledged, but also of paleoanthropology, even though it would be another century before fossil hominids were recognized as real entities.

Just as paleoanthropologists today often cite one of Darwin's suggestions to give their own particular evolutionary idea a boost, so, too, did Blumenbach repeatedly refer to a particular Linnaean idea to set the stage for one of his arguments. In the introductory paragraphs of the first section of his 1795 treatise, *Of the Difference of Man from Other Animals,* Blumenbach quoted a few passages from the *Systema* that reflected Linnaeus's belief: There is essentially no "character, by which man can be distinguished from the ape." But then, recalling the fable of the emperor's clothes, Blumenbach pointed out that Linnaeus—the acknowledged father of modern classification—had actually not presented a single character upon which to base either the genus, *Homo,* or the species, *sapiens.* In truth, although Linnaeus provided albeit scanty morphological traits in support of other genera included within Anthropomorpha and then Primates, he had defined the genus *Homo* not on the basis of any tangible anatomical feature but in terms of a motto: "*Nosce te ipsum,*" which means "Know thyself." As for the species *sapiens,* Linnaeus provided a list of races and described each by a mixture of physical, cultural, and behavioral properties.

Blumenbach's goal was to fill in the information and gaps that remained in Linnaeus's work. In orderly fashion, he discussed "the external conformation of the human body," "the internal conformation," "the functions of the animal economy," "the endowments of the mind," "the disorders peculiar to man," and, finally, "those points, in which man is commonly, but *wrongly,* thought to differ from the brutes."

In terms of the first subject of scrutiny—"the external conformation of the human body"—Blumenbach began by enumerating the four anatomical features he considered to be paramount in attempting a definition of *Homo sapiens:* posture, erect; pelvis, broad and flat; hands, two; and teeth, close-set and serially related. While elaborating upon each of these four areas, Blumenbach also discussed other details of anatomy by which he thought humans could be distinguished from other animals.

In approaching the topic of erect posture, Blumenbach recognized that he had to demonstrate that this feature was both natural for and specific to *Homo sapiens.* As for its being natural, Blumenbach cited accepted cases of children being raised in the wild. Among such popular stories were those about Peter the wild boy, the girl of Champagne, and the Pyrenean wild man. According to popular accounts, although these individuals were raised by animals and were therefore wild in nature, they walked upright, and did so naturally and spontaneously. As for erect posture being unique to *Homo sapiens,* Blumenbach was quite specific in his anatomical comparisons, detailing features that continue to be central in current paleoanthropological debates on what, precisely, makes a hominid distinct from apes and apelike fossils.

Human infants crawl on their hands and knees, not on all four feet as quadrupeds do. Developmentally, the ankle bones of human infants ossify earlier and more rapidly than their wrist bones. Relative proportions of parts of the human body—most notably long, strong legs compared with a shorter trunk and shorter and weaker arms—reflect the anatomy of an animal that is meant to walk on two legs. The human chest is different from that of most other animals; it is compressed from front to back. The chest of a quadruped is compressed from side to side. Humans have widely separated shoulder joints, with the shoulder blades intervening from behind rather than being on the side of the trunk. Humans have short breastbones, and the viscera are not encased in an extensive rib cage. Quadrupeds have long breastbones and a greater number of ribs enclosing their viscera.

Blumenbach stuck his neck out and declared that of all the animals only humans possess a true pelvis. What he meant by this assertion was that compared with other animals, including apes, humans alone have a pelvic structure in which the large bones of the right and left sides (the ossa coxae, or, as Blumenbach called them, the ossa innominata), together with the fused por-

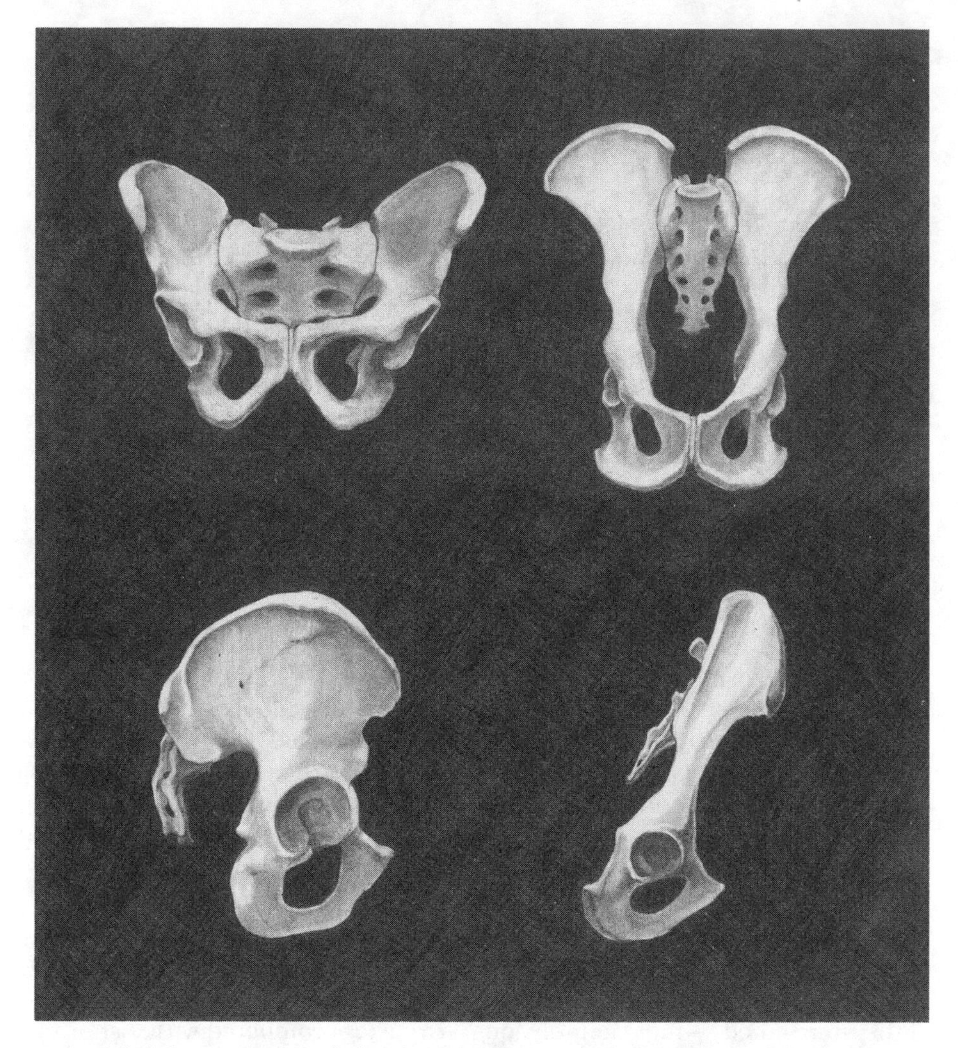

An adult human pelvis (left column) and an adult chimpanzee pelvis (right column). The human pelvis is clearly bowl-shaped, being broader from side to side, deeper from front to back, and, overall, more squat. The bladelike parts of the ape pelvis, the ilia, are taller, narrower, and face back. In humans, the ilia curve around, from the back toward the front. In side view, the human sacrum—the fused set of vertebrae between the right and left ilia—is clearly curved out and then down and in. In the ape, the sacrum is essentially straight.

tion of the vertebral column (the sacrum) and the vestigial tailbones (forming the coccyx) that articulate between the two ossa coxae, form a basin in which the viscera are cupped. In further detail, the upper bladelike part of each ossa coxa, the ilium, is broad and expanded, and the coccyx at the end of the sacrum continues the inward sacral curvature of that platelike bone.

Adding soft tissue to the bony pelvis, Blumenbach pointed out that the distinctive forward angulation of the human vagina paralleled the curvature of the sacrum and the coccyx. In quadrupeds, including apes, the straight vagina lies parallel with the long axis of the pelvis. Although Blumenbach admitted that the forward orientation of the human vagina may make giving birth more difficult than a straight trajectory, he also suggested that this particularly human arrangement mitigates against possible problems in the carriage of the fetus that could arise in a bipedal individual during pregnancy. In attempting to address with discretion Lucretius's speculations on the optimum copulatory position for humans ("How best to prolongate the soft delight?"), Blumenbach alluded to the frontal position, citing an etching by Leonardo da Vinci as evidence of the natural orientations of male and female genitals in this copulatory position. As for the details of soft-tissue anatomy, Blumenbach noted, as Aristotle and Buffon had done earlier, that humans are easily distinguished from the apes and other animals by their possession of very large gluteal muscles at the back side of the pelvic region. Since these massive and protrusive muscles are swaddled in a thick layer of fat, only humans possess a true buttock, which, as Blumenbach described it, is of "fleshy, useful, and semicircular amplitude."

Blumenbach's emphasis on the human hand as something that is anatomically distinct in the animal world can be traced to Aristotle. Aristotle called the human hand the organ of organs. Blumenbach extolled the human hand as being the most perfect hand because of its long thumb. According to Blumenbach, the thumb of an ape is short and almost nailless. This last piece of information indicates that Blumenbach was not referring to all apes but only to the Asian orangutan. The thumb of a chimpanzee, though it is not as long compared with its fingers as ours is, is not as reduced in size as the orangutan's, and it is not distinguished by a reduced nail. The thumb of a gorilla is closer to ours in relative length, and it, too, bears a well-developed nail. An orangutan's thumb is very short and often lacks a nail.

As for the foot, here Blumenbach reiterated the well-known difference between humans and other primates. Humans have a big toe, and this feature is supposed to be specific to bipedalism. Apes and other primates have a thumblike big toe that is adapted for grasping. Because of the differences between humans and apes, Blumenbach suggested that it would be more accurate and appropriate to refer, as Buffon had done, to nonhuman primates as four-handed or quadrumanous, rather than as four-footed or quadrupedal, animals. In his classification, he placed nonhuman primates in Quadrumana. He classified humans with their two hands as Bimana. For years afterward, Bimana and Quadrumana were the taxonomic names most frequently used by primate taxonomists.

Blumenbach also set out to distinguish humans from apes in features of the jaws and the teeth. For example, humans have vertically implanted lower incisor teeth. Human canine teeth, which are not long, are aligned with, rather than set apart from, the incisors. And human molars are distinctive in that they are rounded and lack pointed cusps. The human

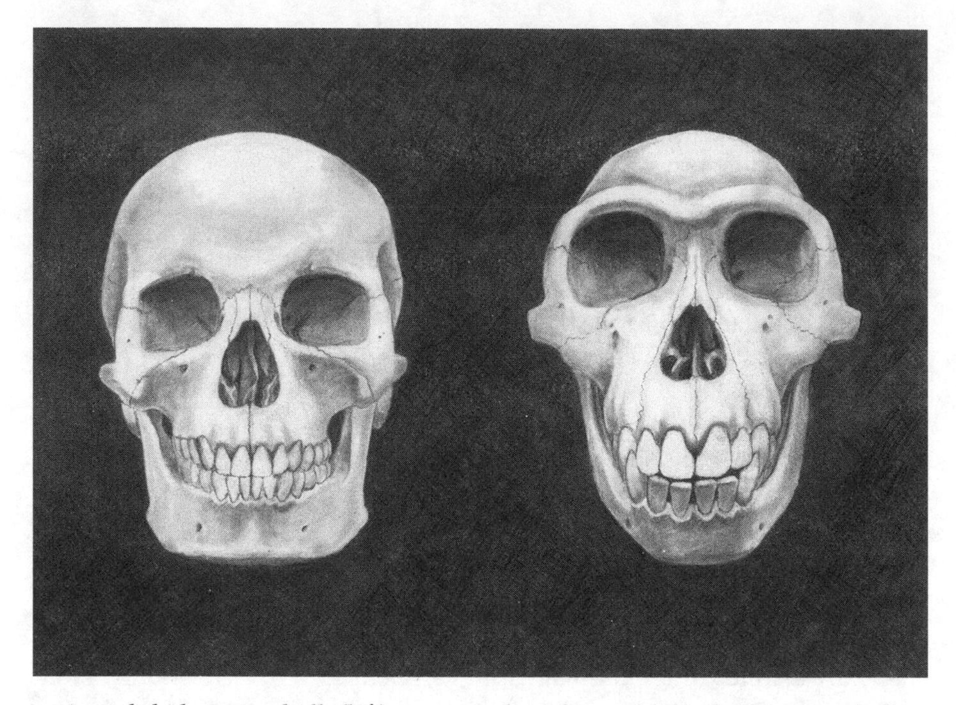

An adult human skull (left) compared with an adult chimpanzee skull (right). The human jaw bears a chin, whereas this region is sloped back in the ape. The human's teeth are smaller, shaped differently, and rooted more vertically at the front of the jaw than an ape's. The ape clearly displays a pair of sutures coursing down from the side of the opening for the nose along the front of the snout to the tooth region. Such a suture is not found in adult humans.

mandible, or lower jaw, is distinguished from an ape's in that it is very short and bears a prominent chin. In addition to these features, Blumenbach cited the articulation of the mandible with the skull, noting that its configuration not only differs from other animals but is also correlated with the fact that humans are truly omnivorous animals.

Other features of the head region that Blumenbach thought were diagnostic of *Homo sapiens* include soft-tissue features such as earlobes (which, it turns out, are sometimes present in chimpanzees) and swollen lips. In their upper jaws, humans possess only a single, relatively small opening through the palate, just behind the upper incisors. Other animals possess twinned, or double, apertures perforating the palate, and these are often much larger. Perhaps it was then known to Blumenbach, who actually performed numerous dissections, that the nerves and arteries from the right and left sides, which are kept separate in animals with the double palatal apertures, intermingle in the single palatal aperture of *H. sapiens*.

The final feature Blumenbach addressed in his comparisons was whether or not humans have one bone or two bones contributing to each half of the upper jaw. Other mammals have two bones on each side, with the larger maxillary bone being clearly separated by a suture from the smaller premaxillary bone in front of it. In the older literature, the premaxillary bone had been identified as the incisive bone or, in Latin, the *os incisivum,* because the upper incisors are rooted in it. However, ever particular about what things are called, Blumenbach suggested that this bone, residing as it does between the maxillary bones, is best referred to as the intermaxillary bone, or *os intermaxillare.* Because sutures clearly delineating an os incisivum or os intermaxillare are not identifiable in *Homo sapiens,* various scholars had thought that the human upper jaw was composed entirely of the right and left maxillary bones, which met at the midline and in which the upper incisors were anchored. Pieter Camper (of facial angle fame) and others before him, including the great anatomist Andreas Vesalius, maintained that humans have only a single maxillary bone.

At first, it appears that Blumenbach was mounting a demonstration of why humans are the same as other animals in having two bones that make up their upper jaw. He began by pointing out that vestiges of sutures between what could be interpreted as an intermaxillary bone and the surrounding maxillary bones can be found in the human fetus. However, Blumenbach then launched into an argument in support of his contention that other primates are similar to humans in *lacking* an intermaxillary bone. In his various anatomical studies and dissections on specimens of a *Cercopithecus* monkey, as well as on orangutans, Blumenbach could not find a discrete intermaxillary bone—which actually exists in these primates. Because of this failure in discovery, he was forced to conclude that the *absence* of an intermaxillary bone is not unique to *Homo sapiens* and cannot be used to distinguish humans from other animals. The only major difference Blumenbach noted was that nonhuman primates have more protrusive jaws.

Blumenbach's contemporary Johann Wolfgang von Goethe disagreed. Goethe, the author of *Faust* and other major literary works, had become obsessed with demonstrating that humans were similar to other animals not only anatomically but in every way except one: Humans can reason. As for the intermaxillary bone, Goethe was dead set on proving Camper wrong, probably because the latter had invoked absence of the intermaxillary bone as evidence that humans were not a part of nature. Goethe's anatomical investigations eventually led him to identify an intermaxillary bone in *Homo sapiens.* As he wrote in 1784, in letter to a friend, Johann von Herder: "I have found—neither gold nor silver, but what gives me inexpressible delight—the *os intermaxillare* in man." In this way, Goethe, and others who later reconfirmed the existence of an intermaxillary bone in *H. sapiens,* sought to eliminate the gap between humans and other animals.

What for Blumenbach was of paramount importance in separating humans from the rest of the animal world? When it came down to it, Blumenbach was in agreement with Goethe: The singular human feature is the use of reason. As Blumenbach declared, reason is "that prerogative of man which makes him lord and master of the rest of the animals." Next in line in importance for Blumenbach in distinguishing humans from other animals was the ability of humans to invent. For Blumenbach, the meaning of *invention* encompassed both tool-making and language, both of which were attributes of *Homo sapiens* that were also admired by Benjamin Franklin. As for human language, Blumenbach portrayed it as differing from the conventional sounds produced by animals in its "arbitrary variety." In their gift of reason, humans were also capable of arbitrariness.

The Legacy

It is said that there is nothing new, just reformulations of the same old thing. But, somewhere, sometime, someone has to be the first. As concerns the foreshadowing of paleoanthropology, we can single out Johann Friedrich Blumenbach. Although his predecessors—Aristotle, Vesalius, Linnaeus, Buffon—did have their say about certain features or attributes being distinctly in the human domain, it was Blumenbach who laid out the entire suite of characters, every one of which played, and continues to play, a major role in one debate or another on human evolution. As should be, most of these features are anatomical in nature: erect posture, bipedalism, and the skeletal correlates, including a bowl-shaped pelvis, with specific features of the ilium and sacrum as well as of the attendant musculature; a hand with a long thumb; teeth of certain sizes, shapes, and spatial relationships to one another; number of apertures perforating the hard palate; degree of protrusion of the jaws; presence of an intermaxillary bone (which we now refer to as the premaxilla); development of a chin; and articulation of the mandible with the skull, and the implications of this articular relation for specific diets. Among nonanatomical attributes, reason, language, and the propensity for tool-making come to the fore.

When one thinks about it, even though details can be added to each category, there really isn't a great deal missing from the gamut of features that are relevant to questions concerning human origins. But before scholars could begin to consider how and why humans parted company with apes evolutionarily, the reality and antiquity of fossil hominids had to be accepted.

3

Coming to Grips with the Past

Where, then, was the human species during the periods in question? Where was this most perfect work of the Creator, this self-styled image of the divinity?

—Edward Pidgeon (1830)

Discoveries of fossil hominids in Africa began early in this century, the first being reported in 1925. But public awareness of the importance of this continent in human evolutionary events did not become heightened until the 1950s and especially the 1960s, when Louis and Mary Leakey made their important finds at Olduvai Gorge, in Tanzania. Thanks to the efforts, during the sixties and eighties, of the paleoanthropologists Richard Leakey and Donald Johanson, and their outstanding collaborators—most notably Meave Leakey, Alan Walker, Bill Kimbel, Tim White, and Yoel Rak—the role of Africa in human evolutionary events has been cemented firmly in the public's consciousness.

Before Africa assumed the role it now enjoys in the history of human origins, Asia had been the chief contender during the latter part of the nineteenth and well into the twentieth century for being the seat of human antiquity. By the mid-1800s, Neanderthals had been accepted as having some relationship to us, *Homo sapiens*. But Neanderthals had inhabited Europe, and there seemed to be nothing hominid in Europe that was either older or more primitive than Neanderthal. Where, then, to seek the potential ancestors of Neanderthals? For most paleoanthropologists the natural place to search for yet more ancient hominids was the Far East. After all, Asia had witnessed the longest continuous record of the most developed aspects of civilization: art, science, architecture, religion, war. The great French naturalist Georges-Louis Leclerc, Comte de Buffon, had postulated that humans had originated in Asia and from there had invaded the rest of the world. Indeed, Buffon had speculated that the different races had come into existence through a process of degeneration from a hypothesized early Asian predecessor.

It took some time, however, for the scientific community to be able to admit that not only did humankind have a past of some time depth but that there had also been a variety of different kinds of now-extinct human relatives. From a biblical standpoint, such as that adopted by the French paleontologist and catastrophist Cuvier, it was possible to allow for the existence of extinct plants and animals. But it was a much more difficult matter to allow for the existence of "antediluvian man."

Accepting the Reality of Fossils

Although it took centuries for the Western world to accept that humans were like other animals in having a prehistory, their fossilized physical and cultural remains had long been known. It was their significance that remained undeciphered.

Megalithic structures, of which perhaps the most famous and well known is the five-thousand-year-old monument at Stonehenge, England, were undeniably unnatural, man-made additions to the earth's landscape. Such structures were known as far north as the Orkney Islands, which are well separated from the northern coast of Scotland, as well as outside the British Isles, in Scandinavia, for instance. Huge slabs of upright stone, sometimes capped with horizontal hewed rock beams, that had been arranged to form perfect circles were a source of amazement and speculation to the traveler who came upon them. Sometimes, as at Stonehenge, only one circular structure commanded the surrounding landscape. Elsewhere, as in the Orkneys, a triad of these circles of stones had to have been strategically placed: One structure stood on top of a hillock, another lay in a meadow, and the third straddled a thin strip of land that separated the salty water of the North Sea from the fresh water of a lake.

But it was not only the size of these structures, and the size of the boulders from which they had been constructed, that was astonishing. Equally noteworthy was the fact that nowhere in the immediate vicinity of any monument was there a potential source for the mammoth stones that were used. They stood alone in rock-free space. How could these circles of stones, constructed of boulders twice as tall, if not hugely taller, and as thick, if not many times thicker, than a mere human being, have been put in place? One of the earliest archaeologists on record, the Dane Saxo Grammaticus, mused on their origin in his writings of the latter part of the twelfth century. In these transcripts, he speculated that the world had once been inhabited by giants with superhuman strength. How else, he asked, could these enormous stones have been transported not only across level land but to the tops of hills? Surely puny, present-day humans could not have constructed these monuments.

For centuries, people had also been puzzled by what we recognize today as various kinds of stone artifacts representing different kinds of lithic industries. The stone tools from western Europe of the Early Paleolithic period most frequently took the form of large hand axes, which were sometimes much larger than a person's hand in length and breadth. A hand ax can be distinguished from other kinds of stone tools because one end is blunt or semirounded and the tool tapers to a point at the other end. Typically, it was hewed from a larger piece or block of stone by a process of reduction in which thin flakes were knocked off from opposite sides of the original hunk of stone. By taking off flakes from alternative sides, it is possible to reduce the thickness of the implement while, at the same time, creating a cutting edge. Because material was flaked from two sides, archaeologists refer to this kind of implement as a biface.

The materials of choice for Early as well as Middle and Upper Paleolithic stone toolmakers were obsidian, flint, and chert, which are of volcanic origin. Obsidian is the glassiest of these minerals in consistency and the freest of impurities; chert is on the low end, and flint is in the middle in terms of these properties. All three rock types, however, can be flaked to produce a tool of uniform shape, often with an aesthetically pleasing surface. Another source for toolmaking is quartz, whose properties do not permit thin flakes to be percussed off. But a quartz hand ax undoubtedly had to have been functional. Otherwise, so many of them would not have been produced throughout western Europe. However, the overall appearance of a quartz hand ax is crude and irregular compared with a hand ax that is hewed from obsidian, flint, or chert.

In contrast with the large Early Paleolithic hand axes, the typical Middle Paleolithic stone tool was manufactured from the flake itself, rather than from the block from which the flake had been struck off. Archaeologists define the Middle Paleolithic primarily on the basis of this technique, which is referred to as the prepared core-flake technology. In order to make a flake-based tool, the original block of flint, for example, is prepared by first taking off a series of flakes from around the entire surface. This removes the skin, or cortex, of unusable material. Flakes are then removed from around the core. This fashions the block in such a way that, eventually, a flake can be knocked off that is pretty much the shape of the intended tool, with the particular cutting or scraping edge, or edges, already defined. After the prepared flake is removed from the core, all that might be needed is a little fine-tuning along the edge. Flake tools can be large in size. However, many were quite small and bore elaborately prepared surfaces. Hand axes made using the core-flake technique were typically smaller and more finely manufactured than those of the Early Paleolithic.

Prehistoric archaeologists characterize the Upper Paleolithic in western Europe as being dominated by a blade technology. A blade is just a flake taken off a prepared core. But it is long, narrow from side to side, and thin from front to back. Flakes of the Early and especially the Middle Paleolithic were usually much thicker and broader. The prepared Upper Paleolithic blade core was often columnar. Reminiscent of the body of a Greek or Roman fluted column, its surface would come to bear the long, shallow

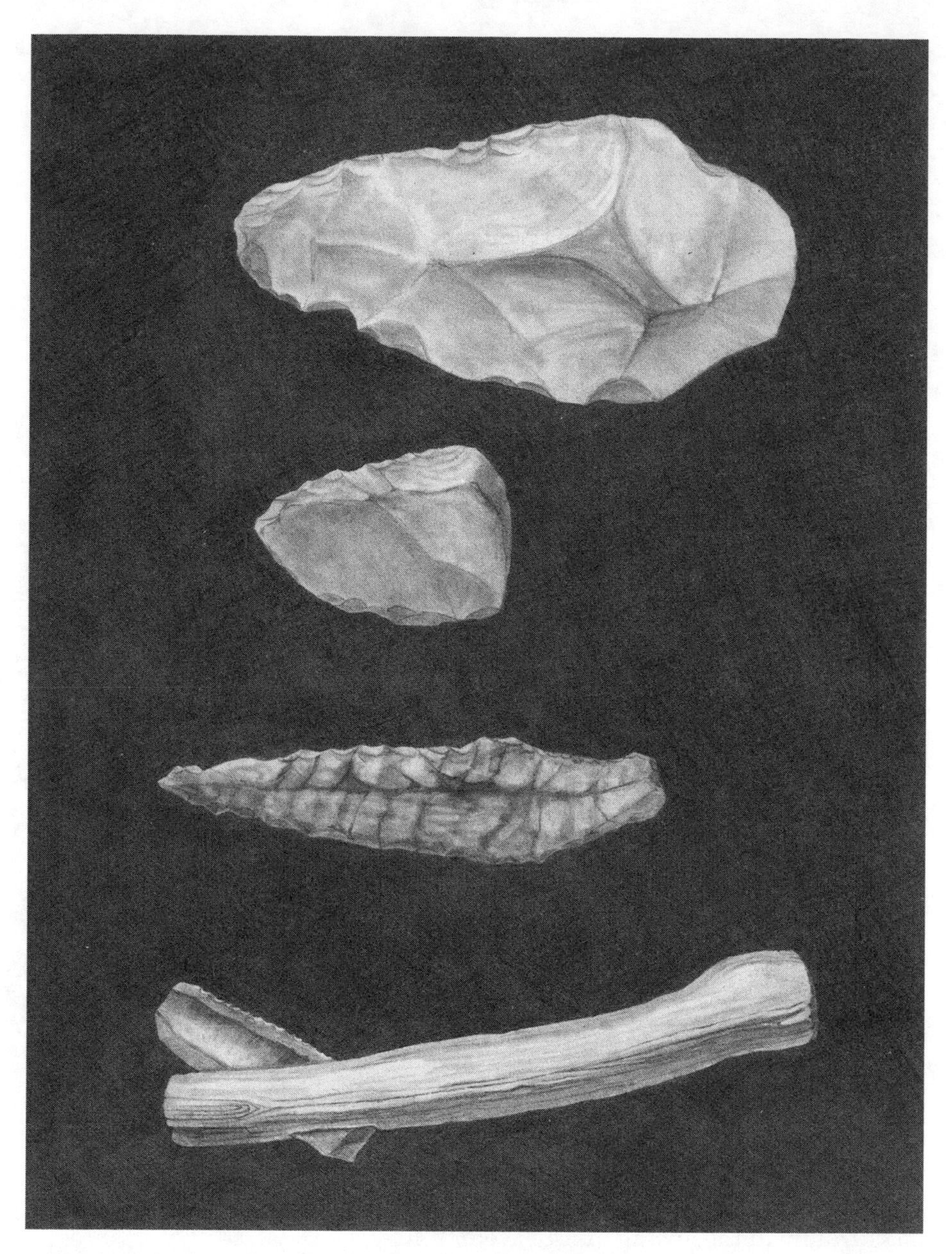

Early Paleolithic hand ax (top), Middle Paleolithic flake (top middle), Upper Paleolithic blade (bottom middle), and Neolithic sickle blade hafted to a wooden handle (bottom). The Early Paleolithic tool is made by reducing a block of stone to the desired size and shape. The other tools are removed from a block that has been prepared in such a way that the shape and size of the piece taken off are fairly well predetermined. The Early Paleolithic technique of preparing the edges of the tool is crudest.

depressions left behind by removed flakes. A blade from such a prepared core will bear a few of these depressed scars, or flutes, on one surface, while its other surface will be essentially smooth and slightly convex. Once a blade is removed from its core, it can be kept as it is, or its edges can be retouched in order to produce a slightly serrated margin. A blade can also be broken to a smaller length and finely flaked by retouching to produce an edge that goes horizontally across or obliquely down, depending upon the task for which the implement is intended.

With the emergence after the Upper Paleolithic period of the Neolithic period, small stone tools became the dominant elements in the tool kit. Blades were broken or snapped to form blanks for the production of other kinds of tools. Particularly common among Neolithic stone tools was the sickle blade, which usually had both ends snapped off to produce straight, short sides that would abut firmly against one another when a number of these blades were lined up end to end. One of the long edges of a sickle blade would be chipped to produce a series of points along a scalloped margin. The opposite edge, which was sunk into or otherwise attached to the wooden handle, was either kept in its original state of sharpness or blunted accordingly. After prolonged use in harvesting, the denticulate cutting edge of a sickle blade would take on a sheen from being polished, stroke after stroke, by the cellulose in the cells of the plants that were being harvested.

Imagine if you were living during the European Dark Ages (or any time, for that matter, when the idea of human antiquity was inconceivable) and, as you were plowing a field, digging into a hillside, or walking along a riverbank, you found one of these stone objects: Paleolithic or Neolithic stone tools, as well as the cores from which they had been struck. What would you make of them? Lay as well as supposedly scientific folk dealt in different ways with the existence of the things you and I take for granted as being the artifactual remains of now-extinct hominids.

Konrad Gesner, the inventor of the taxonomic rank the genus, was among the scientists who proposed that these mystifying stone tools belonged to a class of objects called ceraunia. The word *ceraunia,* which comes from related words in both Greek and Latin having to do with thunder, reflected a belief among some scholars that these unnatural objects were thunderbolts or thunderstones, which, like meteorites, had come crashing to earth. The unearthed flint projectile points and arrowheads were "elf arrows." But not all scholars, the sixteenth-century Italian geologist Ulisse Aldrovandi among them, accepted an extraterrestrial or mythical explanation for the creation of so-called thunderstones. As far as Aldrovandi and others were concerned, these curiosities had been formed by the same processes that created all other geologic objects. Because natural stones, rocks, and minerals were believed to grow in the earth, the same must be true of these geologic oddities. Whatever their true origin, ceraunia came to be seen as a natural category of mineral—as natural as agate, turquoise, and even potter's clay, with which ceraunia were often collectively discussed.

In addition to treating stone tools as objects that were products of nature, Aldrovandi, like many others, believed that the fossilized remains of extinct organisms were also inorganic objects to which the earth had given birth. Collectively, stone tools and fossils were commonly referred to as things "dug up," and for centuries they were treated as objects of similar origin and meaning. Although to our eyes there is no similarity at all between a stone tool and a fossilized snail shell, belief in the commonality of their origins led Aldrovandi, in his publication of 1648, to illustrate stone tools and fossils side by side, as objects of comparable and interchangeable noteworthiness. Although it was not long after the mid-1600s that stone tools and fossils were accepted as being of dissimilar origin and recognized for the implements and remains of the once living that each, respectively, represents, this possibility had actually been proposed decades earlier.

Because sixteenth-century western Europeans knew of the existence of people elsewhere in the newly explored world who used flint and other nonmetallic sources to make tools for everyday use as well as for war, there was a ready-made precedent for recognizing the dug-up stone tools for what they really were. Among the scholars upon whom the implications of present-day people using stone tools were not lost was the Italian Renaissance scholar Michele Mercati. Although Mercati's manuscript remained unpublished until the beginning of the eighteenth century, it was widely read by his sixteenth-century contemporaries as well as by later scholars of the seventeenth century. Gaining support from Lucretius's centuries-older scenario concerning the lifeways of early humans ("men chased the woodland beasts with stone and heavy clubs"), Mercati had been rather matter-of-fact in his speculations on how pre-metal-using humans of the past had modeled flint into everyday implements, such as knives.

One would think that Mercati's statement of the obvious would have had an impact on how his contemporaries, especially the supposed scholars among them, interpreted the curious and unnatural objects divulged from the earth: Of course, people, recent and past, make and made stone tools. This explains the similarities between recently manufactured stone tools and those found in the ground. Since a recently manufactured stone tool is obviously not the same thing as the bone or shell of a living animal, the stone-tool-looking ceraunia and the bone- or shell-looking objects should no longer be considered the products of the same phenomenon. The logic of this argument seems reasonable. But for some reason Mercati's articulation of the obvious did not have the desired effect. Abetted by the work of Aldrovandi and others, stone tools and fossils remained objects of similar advent in people's minds.

Twenty-one years after Aldrovandi's opus, *Musaeum Metallicum,* was published (with its illustrative plates showing stone tools and fossils intermingling as one and the same kind of object), the multifaceted Danish geologist and anatomist Nicolaus Steno (née Niels Steensen, or Stensen) managed to create one of the first serious cracks in the mythology that fossils and ceraunia were similar things. By combining chemistry with the as yet undefined disciplines of geology and micropaleontology, Steno, who was familiar with Mercati's manuscript, was able to demonstrate with sheer simplicity and elegance that the structure of geologically formed mineral crystals was totally unlike that of the biologically mineralized shells of both fossilized and extant clams and mussels. But Steno did not stop there. Perhaps even more impressive, and more important, he demonstrated that the detailed structure of the shells of fossilized invertebrates was identical to the shells of living invertebrates.

Realizing that fossils and ceraunia were not the same thing was the first big step in the public's coming to grips with the fact that life, even human life, had a past of some antiquity. The newly appreciated knowledge that fossils represented the remains of once living but now extinct organisms also paved the way for a recognition of stone tools as the objects they really were: objects made by people. One of the earliest scholars to discuss at some length the subject of stone tools as implements that had actually been manufactured by earlier humans was the English natural historian Robert Plot. Although he had written many natural histories on different regions of England, it was in his treatise of 1686, on Staffordshire, that Plot waxed eloquent on the manufacture of stone implements by earlier humans. In marked contrast to Aldrovandi's (by then) almost forty-year-old publication, not only did Plot discuss stone tools as actually being stone tools but he also illustrated these stone tools in the same pictorial plates as the more contemporary man-made objects of the region of Staffordshire. And Plot really did go all the way in defining man-made objects in the broadest sense. For, on the same plates, he illustrated stone tools together with seventeenth-century implements, toys, sculptures, and monuments.

As so often happens when the obvious becomes obvious, no one could understand how in the world "the obvious" had not been recognized earlier for what it was. Such was the way with stone tools. Once the blinders were removed, no one from the turn of the eighteenth century onward could imagine how a product that was so clearly of human activity could have been mistaken either for an object that had fallen to earth from the sky or for an odd-looking concretion that had been formed in the bowels of the earth. During the 1700s, it became quite commonplace to see publications on how stone-tool manufacture could be fit into a sequence of technological sophistication in which the more primitive technology was on the bottom and the obviously more sophisticated processes of metallurgy were on the top.

Among those who took every opportunity to denigrate their predecessors for accepting the reality of ceraunia was the English natural historian John Woodward. In his last publication, which appeared in 1728, the very year in which he died, Woodward made clear the implication of a sequence of technological sophistication that proceeded from stone tools to metallurgy. For Woodward, as well as for many of his intellectual peers, stone tools were equated not just with being technologically unsophisticated but also with their manufacturers being a barbaric and savage sort of human. Consequently, Woodward and his intellectual kin wove this hierarchy of technological lowliness and barbarism versus technological sophistication and developed civilization into a reading of people of the past as well as the present. Early humans, who, by definition, must have been barbaric and uncivilized, had to use stone tools, because, through no fault of their own, they had lacked knowledge of metallurgy. By analogy, those groups of living humans who had only recently been discovered by European travelers, who used stone rather than metal implements, must also be barbaric and uncivilized.

At this point in the history of coming to grips with the reality of human antiquity, it is worthwhile to leap forward a century and a quarter, and a few decades thereafter, in order to get a glimpse of how this dichotomy between barbaric, uncivilized humans versus civilized humans was played out in the writings of one of the most important figures in the development of evolutionary thought, Charles Darwin.

Charles Darwin and Human Origins

Although Charles Darwin's family did not belong to English aristocracy, they were sufficiently well-off that he had the freedom to engage in intellectual pursuits. As can be said of all of us, Darwin was informed by the history of science that preceded him and by the intellectual and social circles in which he traveled. Although he did not succumb to the implications for British imperialism that the philosopher Herbert Spencer promoted in his formulation of what became known as social Darwinism, Darwin did embrace the commonplace Victorian notion of a racial hierarchy. For Spencer, the differences between the uncivilized, the slightly civilized, and the most civilized humans were due to a kind of natural selection that favored the more superior societies over the more inferior groups. Social Darwinism fed into imperialism by fostering the idea that the more superior races had an obligation to "protect" the less civilized, less privileged, and, of course, the more naive races. So, the argument went, it was for the good of the less civilized that the more civilized took over the former's lands and natural resources and orchestrated their fates for them.

Darwin's primary argument for human evolution, which involved the transformation of uncivilized humans into civilized humans, was grounded in a racial hierarchy. In fact, the case for the transformation from uncivilized to civilized humans was the thrust of most of part 1 of his only publication to be dedicated to human evolution. This book was not his most famous, *On the Origin of Species by Means of Natural Selection,* which was published in 1859. That distinction belongs to *The Descent of Man,* which appeared in 1871. Present-day evolutionists, especially sociobiologists, whose theoretical models are predicated on Darwin's selection arguments, often refer to *The Descent of Man.* Their focus, however, has been on the second, not the first, part. For it is in part 2 that Darwin outlined in detail his hypothesis for a special category of natural selection, which he called sexual selection.

Although Darwin had introduced the concept of sexual selection in *On the Origin of Species,* he did not fully develop this model until he wrote *The Descent of Man.* Simply stated, the model of sexual selection postulates that features in male and female members of a species can be, and are, selected for by one or the other sex. A frequently used example of sexual selection is the amazing tail plumage of the male peacock, which is the most obvious physical feature of the male associated with courting and mating behavior. The more impressive, as a human would see it, the male peacock's fan of tail feathers, the more successful the male will be in gaining access to and ultimately impregnating a female. The female, in turn, is seen by selectionists as choosing the fortunate male from among the available competitive males. Darwin felt compelled to develop the theory of sexual selection to account for the presence of features—such as the male peacock's outlandish tail—that are of no obvious adaptive significance, and are irrelevant to his arguments of evolution based on utility of structure, competition, and selective advantage.

Whatever notoriety *The Descent of Man* currently has, it is due to Darwin's presentation of minutiae to bolster his model of sexual selection, and not to his discussion of the descent of man in a deep evolutionary sense. It is in the less popularly well-known part 1 of *The Descent* that Darwin sets out his scenario of human evolution, which is grounded in his contention that humans evolved in Africa, not in Asia.

In the eighteenth century, Buffon, along with other scholars, argued that since it is in Asia that one finds the longest record of sophisticated human civilization, it was in Asia that all important events in human history began. Inasmuch as Buffon did not subscribe to Lamarck's or to any other early evolutionist's version of transformation, it was not human evolution that he was trying to describe. Rather, Buffon and his colleagues envisioned Asia as the source of humans who later migrated west—in Buffon's scenario, to the region of the Caucasus Mountains—and subsequently became the progenitors of all modern races. For Buffon, racial differentiation was the result of degeneration from an Asian prototypical human to all other perceived variants of *Homo sapiens.* Buffon's theory of degeneration also dealt with the origin of the "white race," which became known as "Caucasian" because these humans were supposed to have remained in the area of the Caucasus. Although Buffon recognized different races, his perspective was similar to Greek and Roman notions that the form of one's body and spirit was correlated with one's environment.

Darwin rejected the long-standing theory that humans had originated in Asia. Since he thought of human origins in an evolutionary context, and natural selection was his proposed force behind evolutionary change, Darwin could not accept Asia as the center of human origins, primarily because this part of the world was not "dangerous" enough to provide the selection pressures necessary to produce something as unique as a human being. Not that Asia could not have been the locale in which human evolution began. Another of Darwin's long-held beliefs was that we should seek the fossil ancestors of closely related living species in the region where these closely related living species are found today. Orangutans are found in Asia, and chimpanzees and gorillas in Africa. Since humans are globally so widespread, a couple of options were available to Darwin, the choice among them being dependent upon the specific theory of human-ape relatedness he chose to embrace.

Linnaeus may have been the first taxonomist to place humans in an order of mammals with other animals. But it was Darwin's longtime supporter and England's leading anatomist, the irascible Thomas Henry Huxley, who was the first to argue the taxonomic association of humans and a specific group of mammals on the basis of comparative anatomy and embryology. Huxley presented his case in 1863, in an essay entitled "On the Relation of Man to the Lower Animals."

In this landmark essay, Huxley went much further taxonomically than Linnaeus had by arguing that not only did humans belong with a particular group of mammals, Primates, but also that, among primates, humans were most similar overall to the great apes. Because, in the 1800s, the gorilla was the best known publicly of the three great apes—in large part because it was a popular zoo animal—Huxley made most of his anatomical comparisons between humans and this large African primate. In addition, as Huxley admitted from the outset, he felt that it was the gorilla alone among the three great apes that most closely approached humans in its anatomical organization. On occasion, when he lacked anatomical or developmental details on the gorilla, Huxley sought comparisons between humans and one of the other two great apes, the chimpanzee and the orangutan. Although Huxley documented a suite of features (most notably in various body proportions) in which humans and gorillas were most similar to one another, he ended his essay "On the Relation of Man to the Lower Animals" by putting all three great apes together in their own family group and humans alone, and apart, in another.

What was important to Huxley was that he had provided the information that was lacking when Linnaeus classified humans with other primates: that he had demonstrated morphological and developmental similarity first between humans and other animals, then between humans and primates, and then between humans and a small group of primates. But, having done so, he backed away from what could have been an argument for evolutionary relationship. In addition, as Huxley stated clearly in 1896, in the preface to the reprinting of "On the Relation of Man to the Lower Animals," the basic motivation for his tackling this problem in the first place was to provide the details of comparative anatomy and embryology that he felt his dear friend and colleague Charles Darwin was not adequately trained to obtain for himself. By doing so, Huxley believed that he was providing data that Darwin would need when he finally came to argue his case for human evolution.

Had Huxley been interested in figuring out the evolutionary relationships of humans and apes, and had he used the criterion of overall similarity as the arbiter in such matters, he would surely have grouped humans and gorillas together. Nevertheless, it is an inextricably entrenched myth in human evolutionary studies that Huxley had actually argued that humans are most closely related to both African apes, the chimpanzee and the gorilla. This misunderstanding derives primarily from paleoanthropologists having incorrectly translated Huxley's statements of similarity between humans and the African apes into a theory of relatedness, which Huxley never intended to argue.

The passage from Huxley's essay that twentieth-century paleoanthropologists cite in supposed historical support of a close evolutionary relationship between humans and the African apes is the following: "It is quite certain that the Ape which most nearly approaches man, in its totality of organization, is either the Chimpanzee or the Gorilla." But it is obvious that this passage is devoid of any evolutionary meaning. This statement was the brash Huxley's way of setting the stage in his favor before entering into his detailed argument: He chose a nonhuman primate for comparison with humans that would immediately be appreciated by his readership as looking more humanlike than either a monkey or a lemur.

Overlooked by twentieth-century paleoanthropologists who seek in Huxley an evolutionary argument is his conclusion: "The structural differences between Man and the Manlike Apes certainly justify our regarding him as constituting a *family apart* from them; though, inasmuch as he differs less from them than they do from other *families* of the same *order,* there can be no justification for *placing* him in a distinct *order.*" By italicizing key words, we can see clearly what Huxley was up to. He was arguing, as Linnaeus had done more than a century earlier, that even though there may be differences between humans and even those primates of greatest anatomical similarity, humans should be placed with them in the same order of mammals. The differences between humans and the great apes, however, were sufficient to warrant placing the former in a separate family within the same order. Period.

Although the popular notion, at least in the popular culture of Darwinism, is that Thomas Huxley—that brazen and self-confident orator and debater—had embraced and defended every one of Darwin's evolutionary musings, this is hardly the case. In the last pages of "On the Relation of Man to the Lower Animals," Huxley raised concerns about the "tenability, or untenability, of Mr. Darwin's views." Not about the reality of evolution, for Huxley was himself an evolutionist, but about Darwin's theory of evolution by means of natural selection. For if natural selection as envisioned by Darwin was the major force in accomplishing evolutionary change—by first creating and then picking and choosing from among myriad tiny differences among individuals—such a process would require huge amounts of geologic time for change of any evolutionary significance (such as would lead to the emergence of a new species) to accrue. Better suited to Huxley's at times volcanic temperament than to Darwin's gradualistic view of evolutionary change was the model of saltation. In contrast to Darwin's view of evolutionary change as being slow and gradual, saltationists hypothesized that major evolutionary differences, such as those that exist between species, evolve suddenly and abruptly. Huxley was a staunch saltationist.

When Darwin wrote *The Descent of Man,* he paralleled Huxley in his approach to setting the stage for human origins. First, he reviewed the available real and anecdotal evidence for the unity of animals with backbones as a group (Vertebrata), for the unity of mammals as a group of vertebrates, and then for the unity of primates as a group of mammals. Like Huxley, Darwin strove to point out the ways in which humans were anatomically similar to other vertebrates in general and to mammals more specifically. Again like Huxley, Darwin argued that although humans differed from other primates in many obvious ways, they were also quite similar to these particular mammals. Darwin did not, however, delve into the embryological and developmental evidence as Huxley had done. But he did make every effort to cover the published as well as anecdotal comparative anatomy, even to the extent of referring to one anomalous feature or another in the occasional human or great-ape specimen that seemed to point to these primates having a common evolutionary heritage. For example, Darwin cited the case of a fetal orangutan that had developed a slight peak to the top of its ear. Because various monkeys, most notably Old World monkeys, such as baboons and macaques, have ears that are very prominently peaked, Darwin viewed this case as an evolutionary atavism—a throwback to a more primitive phase in the evolutionary past of apes, and, by implication, of humans.

This is where Darwin's use of comparative anatomy—real, imaginary, or spectacular—differed from Huxley's. As Darwin had set out his "view of life" in *On the Origin of the Species by Means of Natural Selection,* so, too, was he trying to lay the groundwork for his view of human origins in the early chapters of part 1 of *The Descent of Man.* Whereas Huxley had argued on the basis of overall similarity for the taxonomic recognition of smaller and smaller subgroups of mammals, leading to a subgroup of humans and great apes within which humans constituted their own taxonomic group, Darwin focused on humans in an evolutionary context. And because he firmly believed that one would find the ancestors of closely related living species in the region where these species currently live, Darwin had to broach the subject of to whom, among the living great apes, humans were most closely related before he could unfold his scenario of human evolutionary history. The catch here is that there are three great apes—one living on the islands of Borneo and Sumatra, in Southeast Asia, and the other two in the heart of Africa—and no one, not even the great comparative anatomist Thomas Huxley, had provided Darwin with the fodder he needed in order to anchor human origins in one particular region of the world. What to do?

Darwin fell back on arguments predicated on the power of natural selection. In which region of the world—Africa or Southeast Asia—would there be the kinds of selection and, consequently, survival pressures that, to Darwin's way of thinking, would be able to produce a human, an animal so distinct both in intellect and in mode of locomotion? Clearly, Darwin concluded, the evolution of such a unique creature could not have taken place in lushly forested, tropical Southeast Asia. What would be the selection pressure to become human?

On the other hand, there was Africa—particularly southern Africa, to which Darwin specifically pointed—with its hot, arid, grassy expanses and its array of speedy predatory animals. Now, this was the sort of setting, bereft of havens of safety and lacking in an abundance of readily available food resources, that could provide the kinds of pressures natural selection needed to work with. And where else in the world but in Africa, Darwin prodded his reader along, do you find both apes and humans living in conditions that would favor the evolution of those most fundamental of human attributes: making tools with one's hands, being fleet on two feet, translating thoughts into speech?

At this point in his argument, Darwin found himself in a bind. In order for him to proceed with his scenario of human evolution, he had to convince his readership that humans really are more closely related to the two African apes than they are to the orangutan alone, or to a group consisting of all three great apes. The latter grouping is where Thomas Huxley had left things in 1863. And it was not until later, in the 1870s, that a world-famous and academically multifaceted German, Ernst Heinrich Haeckel, would suggest that orangutans were more similar to humans than the other great apes in external brain configuration. Haeckel was an expert in evolution, comparative anatomy and embryology, and paleontology, and his volumes were translated into many languages and, for the last decades of the nineteenth century, vied with the King James version of the Bible for being the best-selling reading of the time. Haeckel was also the first to publish a diagram of an evolutionary tree—literally, an illustration of a tree with species names at the ends of branches.

To be sure, Darwin was an expert in many different arenas of scientific inquiry. His treatise on the taxonomy of barnacles, for example, remains a classic reference work. But because he was not a practicing, well-versed, comparative vertebrate anatomist himself, Darwin had to rely on the information others provided in journals, books, articles, and even personal letters. The problem, of course, was that nobody, not even Thomas Huxley, had made the leap from discussing anatomical and embryological similarities between humans and the great apes to interpreting these similarities in an evolutionary context. As the self-assured Huxley explained in the preface of the reprinting of his essay "On the Relation of Man to the Lower Animals" in 1896: "[I]nasmuch as Development and Vertebrate Anatomy were not among Mr. Darwin's many specialities, it appeared to me that I should not be intruding on the ground he had made his own, if I discussed this part of the general question." (In *On the Origin of Species,* Darwin's only comment on human evolution was restricted to one sentence: "Light will be thrown on the origin of man and his history.") "In fact," Huxley continued, "I thought that I might probably serve the course of evolution by doing so."

Given that Huxley had avoided altogether any evolutionary speculation in his essay, we can only surmise from this last quote that he was leaving this task to Darwin, who would put Huxley's comparative studies into an evolutionary context. But Huxley's main conclusion—that although among primates humans are most similar to the great apes, they should be placed in one taxonomic family and the great apes in another—would not work for Darwin's scenario of an African locale and an African apelike ancestor for human origins (although it was obviously easier to overlook the fact that the African apes are found in the evergreen forests of central Africa, whereas the harsh savanna environment of southern Africa, so necessary in order for Darwin's natural selection to evolve a human, lies several thousand kilometers to the south).

The other possible interpretation of Huxley's comparative studies, which relied on humans being more similar overall to the gorilla than to either of the other two great apes, would not work for Darwin, either. A translation of overall similarity into evolutionary relatedness would lead to the conclusion that humans are more closely related to the gorilla than to either the chimpanzee or the orangutan. Darwin's lack of interest in this evolutionary hypothesis probably reflects the fact that he could not convince himself of the possibility that the chimpanzee and the gorilla are *not* more closely related to each other than either is to any other primate, including *Homo sapiens.* A close evolutionary relationship between the chimpanzee and the gorilla is the only one that makes sense in terms of comparative anatomy. Beginning in the latter part of the nineteenth century, with the detailed work of German comparative anatomists, the list of similarities between the two African apes had been brought from the general to the specific. These two apes, alone among primates, have developed very unique configurations of their soft- and hard-tissue anatomy, especially in the arms, wrists, and hands.

If Darwin had imposed an evolutionary meaning on Huxley's comment about similarity between humans and chimpanzees and gorillas ("It is quite certain that the Ape which most nearly approaches man, in its totality of organization, is either the Chimpanzee or the Gorilla"), he did not mention it at all in his writing. Not citing another voice of authority at a pivotal point in a critical argument—when it could so easily have been done—was very unlike Darwin. Consequently, Darwin's failure to cite Huxley's work here—or elsewhere in *The Descent,* for that matter—strongly indicates that he had not taken this liberty of interpretation.

Usually when Darwin built a case in support of a particular notion, he provided detailed, lengthy listings of minutiae. He did no such thing when it came to the question of to whom among the apes humans are supposed to be most closely related. Having apparently satisfied himself that natural selection could best produce a humanlike progenitor in the dangerous setting of southern Africa, rather than in luxuriant Southeast Asia, and that the forested locales of living chimpanzees and gorillas so far away from southern Africa did not constitute an obstacle to his argument, Darwin made the following declaration:

> In each great region of the world the living mammals are closely related to the extinct species of the same region. It is therefore probable that Africa was formerly inhabited by extinct apes closely allied to the gorilla and chimpanzee; and as these two species are now man's nearest allies, it is somewhat more probable that our early progenitors lived on the African continent than elsewhere.

This was it. Humans now had their closest living relatives in the African apes, the chimpanzee and the gorilla.

Elsewhere in *The Descent,* the only time that Darwin specifically used the chimpanzee and the gorilla as potential examples of human evolution was when he pondered whether "man is descended from some small species, like the chimpanzee, or from one as powerful as the gorilla." Darwin opted for a small human ancestor because he could not imagine that an animal as social as *Homo sapiens* had evolved from something as large, strong, and ferocious as a gorilla, which can defend itself against enemies. Darwin speculated that "it might have been an immense advantage to man to have sprung from some comparatively weak creature," because being relatively defenseless would lead, via natural selection, to sociality. In turn, being social would lead to "higher mental qualities"—and, of course, sociality and intelligence belonged solely in the human domain.

As Blumenbach had done almost a century earlier, Darwin also argued that all human races belong to the same species. However, either because he adhered to a notion of evolutionary change marked by a gradual transformation from one evolutionary grade to another or because he had embraced first a hierarchical ranking of species and of races within species, Darwin viewed human origins and evolution in terms of a progressive transformation from the ape to the human, and from the most primitive to the most civilized of humans. His argument for the evolution of human bipedalism from the common form of locomotion among primates, quadrupedalism, clearly reflect this: "[W]ith some savages . . . the foot has not altogether lost its prehensile power." In the context of a perceived gradual evolutionary transformation through a series of intermediate stages, this presumed fact about the grasping foot of some "savages" accorded well with Darwin's belief that "as the hands became perfected for prehension, the feet should have become perfected for support and locomotion." The implication, of course, according to Darwin, is that the architecture of the human foot would not have been perfected during the transition from fossil human relatives to primitive

Homo sapiens, but only during the transition from extant primitive savages to extant civilized humans.

Darwin's adherence to a progressive sequence from nonhuman primate to "savage" to "civilized" human is seen throughout *The Descent.* For instance, having in one passage summarized the presumed reproductive tendencies in the "lower" primates, Darwin later declared that "the reproductive power is actually less in barbarous, than in civilized races." Whereas imitation is not a strong behavioral feature of most mammals, it is typical of "higher" primates, beginning with monkeys, "which are well known to be ridiculous mockers." As others, including Darwin himself, had recorded as truth, imitation was supposed to be a strong behavioral feature in humans, but it was supposed to be especially strong in savages. At one point, Darwin was very explicit about the significance of imitation: "The strong tendency in our nearest allies, the monkeys, in microcephalous idiots, and in the barbarous races of mankind, to imitate whatever they hear deserves notice."

As for the process of reasoning, Darwin gave an imaginary example of what a savage and a dog would do versus what a civilized human would do when, in want of liquid refreshment, each finds a potential source of water at an inconveniently low level. In his answer to this imaginary dilemma, Darwin assured his readership that "a civilized man would perhaps make some general proposition on the subject; but from all that we know of savages it is extremely doubtful whether they would do so, and a dog certainly would not."

Darwin was so convinced of "the greater differences [in mental faculties] between the men of distinct races" that at one point he stated that further discussion of the matter was hardly necessary. Nevertheless, a few pages after making this comment, he was moved to affirm the commonly held belief that there is a "close relation between the size of the brain and the development of the intellectual faculties" and that this relation "is supported by comparison of the skulls of savage and civilized races, of ancient and modern people, and by the analogy of the whole vertebrate series."

In this last example, Darwin was taking the often exploited generalization about increasing brain complexity in vertebrate groups (that brain complexity is greater, on average and across a diversity of species, in amphibians than in fish, in reptiles than in amphibians, in birds than in reptiles, and in mammals than in birds) and laying it atop his perception of a progressive transformation from ape to primitive human to civilized human. He believed that this gradient of increasing complexity applied to the brain and, by extension, to one's mental faculties, because colleagues of his had proved that Europeans ("civilized" races) have the largest average brain size, whereas "savage" races (Australians, for example) have the smallest average brain size of living humans.

Darwin's reference to ancient people derived from another study (one by the famous French neuroanatomist Professor Paul Broca) for which earlier nineteenth-century skulls, exhumed from their graves, had been measured and compared to later nineteenth-century skulls. According to Broca, not only had there been an increase in overall brain size during this time (from 1,426 to 1,484 cubic centimeters) but this increase had been in the frontal region of the brain, "the seat of intellectual faculties." As for truly ancient human specimens, the only potential fossil human relative then widely known to the general public was represented by the Neanderthal skullcap, which had been discovered in the 1850s in a limestone cave in the Feldhofer Grotto, near Düsseldorf, Germany. In direct contradiction of Darwin's envisioned transformation series of increasing brain size and mental faculty, the Neanderthal skullcap had been calculated to have a brain size that was much larger than what was typical of Europeans, who supposedly have the largest brains of living humans.

But Darwin dismissed the enormous cranial capacity of that Neanderthal specimen. Modern humans, especially civilized humans, he argued, were actually more like domesticated breeds of animals than like animal populations in the wild. Citing the case of domesticated rabbits, he pointed out that rabbits kept in captivity are less bulky than their wild kin. This overall reduction in size (from wild to domesticated rabbits) he attributed to the practice of keeping domesticated rabbits in close quarters, which, in turn, restricted their use of "intellect, instincts, senses and voluntary movements." Something similar would apply to humans.

This particular argument of Darwin's would seem to go against the grain of his general theme: a progressive evolution of just about any and every feature from ape to savage to civilized human. But Darwin was not afraid to invoke a contradiction if it supported a specific case within a larger argument. In this particular instance, he fell back on his belief that examples of selective breeding of domesticates can serve as analogs for how evolution would work under the direction of natural selection. In detail, he was convinced that civilized humans could be likened to domesticated animals. Since natural selection engenders evolutionary change by picking the most beneficial and advantageous features, Darwin's human evolution enacted this in the transformation of the wild savage to the civilized domesticate.

Clearly, Darwin invoked various kinds of information in order to distinguish civilized from savage humans. But it was not until nearly the end of part 1 of *The Descent* that he presented the stone-tool analogy as a basis for identifying savage, barbarous humans. As John Woodward and his contemporaries had concluded almost 150 years prior to the publication of *The Descent,* so, too, did Darwin presume that, like earlier humans, only barbaric, uncivilized living humans would still be using stone tools. Earlier humans had no choice; they did not have knowledge of metallurgy. Since modern humans do have access to metallurgy, those who continued to use stone tools must still be barbaric.

Although various scholars had suggested that originally humans had been civilized, and that from there some groups retrogressed into barbarism, Darwin felt secure in coming to the opposite conclusion: "[C]ivilized nations are the descendants of barbarians." In support of this contention, he cited the retention in civilized peoples of traces of barbaric traits (such as counting on one's finger), and the ability of savages to "raise themselves a few steps in the scale of civilization." As Darwin concluded in chapter 5 of part 1 of *The Descent:* "It is apparently a truer and more cheerful view that progress has been much more general than retrogression; that man has risen, though by slow and interrupted steps, from a lowly condition to the highest standard as yet attained by him in knowledge, morals and religion."

In Darwin's Wake

It seems that whenever you pick up a popular article or read a piece in a newspaper about evolutionary theory, the topic is described as "Darwin's theory of evolution." This, of course, is an incorrect attribution on various counts. Evolution is not a theory. It is a phenomenon. What evolutionists, whether Chevalier de Lamarck, Erasmus Darwin, Charles Darwin, Niles Eldredge, or Stephen Jay Gould, strive to understand are the processes that make evolution tick. This is not an easy task, because evolutionary events occur over greater periods of time than any scientist, or generations of scientists, could observe—assuming they would know that such an event was taking place. Even Darwin was aware of the distinction between observing the results of the phenomenon of evolution and trying to understand the way or ways in which the results of evolution came about. Darwin's realization of this dichotomy—accepting the reality of a phenomenon and trying to understand the workings of the phenomenon—is obvious in the title of his book *On the Origin of Species by Means of Natural Selection.*

In light of the amount of publicity given to Darwin, and (although this is incorrect) to his being *the* supposed discoverer of evolution, one would think that Darwin had remained at the forefront of evolutionary studies from the very first edition of *On the Origin of Species.* This, however, is hardly the way it was. The events that conspired against him were manifold.

One of Darwin's worst enemies was himself. In the revised editions that followed the 1859 publication of *On the Origin of Species,* he weakened his original arguments in two major ways. First, his increasing reliance on conjectured scenarios of the "use and disuse of organs" (otherwise known as the principle of inheritance of acquired characteristics) to explain evolutionary change became increasingly difficult for even his most ardent supporters to swallow. Second, his insistence on evolutionary change resulting from the accumulation of minute differences from one generation to the next—revised in each edition of *On the Origin of Species* to require natural selection to pick and choose among increasingly minute differences—forced him to plead for increasingly longer periods of time over which any change of evolutionary significance could occur and, as a consequence, to invoke increasingly more ancient, and less believable, estimates of the age of the earth. Eventually, Darwin lost virtually all the support that still remained among the ranks of fellow natural historians and geologists.

But Darwin alone did not undermine the success of his theory of evolution by means of natural selection. Paleontology, too, could not sustain his predictions of what should be expected of the fossil record. Before its failure, however, the field of paleontology seemed the field of salvation for Darwin's theory.

Given that evolution, according to Darwin, was in a continual state of motion, with ongoing but slow and gradual change accruing over long periods of time, it followed logically that the fossil record should be rife with examples of transitional forms leading from the less to the more evolved. Not only had Darwin put these thoughts into words but he had also illustrated them in a diagram that consisted of hypothesized ancestors giving rise over time to hypothesized lineages of descendant organisms. In various places in this diagram, Darwin indicated the extinctions of hypothetical lineages as well as the origins of a multiplicity of species from the same ancestor. In words and in illustration—the only illustration in *On the Origin of Species*—Darwin breathed new life into the discipline of paleontology, which was the only field of study that could provide the scientific world with an actual picture of his view of evolution.

Fueled in no small way by the role that paleontology could assume—reconstructing and also demonstrating the course of evolution—the world's leading museums of natural history focused on fossil collecting. When these institutions were first founded, they were envisioned as the forums for displaying, in often overfilled and poorly labeled cabinets, unorganized geologic and biological collections of plain old specimens. Now, armed with the possibility of being able to exhibit not just an array of fossils but the drama of evolution itself, museums vied with one another to secure the best fossil localities and discover increasingly older representatives of the lineages of now-extinct animals. In the American West, fossil hunting took on the stereotype of the ruthless Old West. Fossil localities were kept secret and guarded by men with rifles. Armed guards also accompanied the trainloads of plaster-protected and crated fossils that often traveled by night in order to avoid detection. Sometimes, however, these attempts at secrecy did not work, and gangs from rival museums would successfully raid and loot the paleontological spoils of the competition.

But when the dust settled, and the fossils were assessed in terms of whether they validated Darwin's evolutionary predictions, a clear picture of slow, gradual evolution, with smooth transitions and transformations from fossils of one period to another, was not forthcoming. Instead of filling in the gaps in the fossil record with so-called missing links, most paleontologists found themselves facing a situation in which there were only gaps in the fossil record, with no evidence of transformational evolutionary intermediates between documented fossil species. Without fossil intermediates to back up Darwinian predictions of how evolution works, the turn of the century saw both paleontology (an evolutionary discipline) and gradual change via natural selection (an evolutionary model) fall on hard times. Even the paleontologists' special plea—that the gaps in the fossil record were the consequences of poor preservation, the loss of fossils through erosion or other destructive processes—did not work.

If one could not see evolution revealed before one's eyes through the lens of paleontology, where else could one look? At the time that Darwin and his contemporaries produced their works, the word *evolution* was not charged with the present-day notions of evolutionary change. Rather, the word *evolution* was part of the vocabulary of embryologists, scientists who studied the development, or ontogeny, of organisms. For them, *evolution* meant an "unfolding," as in the unfolding of development from embryo to fetus to newborn to juvenile to adult. Because of this difference in meaning between the nineteenth century and now, it is not surprising that Darwin used the word *evolution* only once in the first edition of *On the Origin of Species.*

The scholar who attempted to put together into a meaningful package the domains of evolution (as we would now use the word), fossils, and embryology was Ernst Haeckel, the most prominent professor at the University of Jena, in Germany, and one of the most visible scientists of the Western world. The result of his mental labors was a theory formally called the Biogenetic Law, or the Law of Biogeny, which is otherwise known by the phrase "ontogeny recapitulates phylogeny." Simply stated, the Biogenetic Law posits that it is possible to see the evolutionary history of an organism unfolding through the developmental stages of that individual: The ontogeny of an individual repeats, or recapitulates, the individual's evolutionary history, or phylogeny. By studying the embryology of a human, for example, one would be able to document the various stages in human evolution. As most embryologists envisioned it, humans, being mammals, would develop through a fish stage, then a reptile stage, before reaching the mammal stage.

Haeckel differed from other embryologists of the nineteenth century in how he interpreted the various developmental stages through which an organism passed. As most embryologists saw it, the ontogenetic stages that were shared by organisms represented only a basic level of development. Consequently, when a human embryo was at the "gill stage," it reflected a phase common to the embryos of fish, amphibians, reptiles, birds, and other mammals. The human embryo was not at that stage a miniature adult fish. Even a fish embryo in the gill stage does not resemble an adult fish. The features that characterize the adult organism emerge after the embryo veers away from the commonly shared phases of ontogeny to develop into the features that are specific to its kind. Haeckel, however, thought that at each stage of development the embryo was a miniature version of an adult organism. At the gill stage, for example, the embryo was basically a tiny but full-grown fish. When Haeckel placed this interpretation in an evolutionary context, he proposed that it was possible to see the adult ancestors of an organism in the various stages of its embryonic development. Fish came before mammals. So the gill stage obviously represented that phase in mammal evolution. Clearly, Haeckel concluded, it was possible to see the recapitulation of an organism's evolutionary past by studying its ontogeny.

Armed with this general outline of the relationship between ontogeny and phylogeny, Haeckel sought to explain how evolutionary change might be brought about. He proposed that alterations to the sequence of development would be sufficient to modify an organism, and that the most straightforward mechanism would be to add a new stage to the end of the developmental continuum. With each terminal addition, the preceding developmental stages would be pushed to earlier and earlier positions in the individual's ontogeny. Eventually, the compression of the earliest phases of development might not even be observable to the embryologist. Of course, each new terminal addition would be the miniaturized adult of that ontogenetic stage.

The twentieth century had little use for Haeckel's Biogenetic Law and his untenable proposition that the developmental stages through which an organism passed were the adult phases of its ancestors. Unfortunately, since Haeckel was associated with the most well-known attempt to understand evolution by studying development, it took many decades (until the 1970s, in fact) for evolutionary biologists to once again accept the possibility that this was a viable avenue of investigation. As will become clear, the interplay between development and evolution does not have to be burdened with Haeckelian overtones. In fact, an appreciation of developmental biology and genetics is critical to unraveling the mysteries of evolution.

In the virtual absence of fossil-hominid specimens during the latter part of the nineteenth century, Haeckel postulated the existence of now-extinct forms that would have spanned the evolutionary void between something apelike and something humanlike. One of these forms, which Haeckel represented at the level of the genus, was *Pithecanthropus,* which means "ape-man" (*pith* refers to "ape," and *anthropus* to "man"). The other hypothesized human ancestor, also represented at the genus level, Haeckel called *Alalus,* referring to "speechless primitive man." Although he did suggest these two different names, Haeckel tended to use *Alalus* as the name of the hypothetical transitional form between the tailless apes (the most "manlike" of apes) and true *Homo.* But as is obvious from Haeckel's writing, *Pithecanthropus* and *Alalus* were really one and the same in his mind.

It is with Ernst Haeckel, then, that the nascent paleoanthropological world would be presented with an image of something concrete in the fossil record of "man." Something that could be more clearly appreciated as a true fossil intermediate between the manlike apes and man. Something to which, even in its conjectural state, Haeckel had given a name to reflect its transitional status—"ape-man"—between apes and humans:

> I have applied this name *[Pithecanthropus]* to the speechless Primitive Men (*Alali* [the plural of *Alalus*]), who made their appearance in what is usually

> called the human form, that is, having the general structure of Men (especially in the differentiation of the limbs)—but yet destitute of one of the most important qualities of Man, namely, articulate speech, as well as of the higher mental development connected with speech. The higher differentiation of the larynx and of the brain occasioned by the latter, first gave rise to the true "Man."

The themes here are familiar: intelligence (via a highly developed brain) and speech (via a highly developed brain and a transformed larynx). Even Linnaeus had made up subcategories of *Homo* to accommodate those missing links in the Great Chain of Being that were transitional to becoming truly human. One of Linnaeus's "speechless" forms of *Homo* was the orangutan, which he classified in the species *H. sylvestris.* But Haeckel's transitional ape-human form had a particular twist. Although it was less evolved than living humans from the neck up (at least with regard to mental capabilities and language), it was essentially the same as living humans from the neck down (in terms of the features of the arms and legs). As far as the history of paleoanthropology is concerned, this dichotomy—of what's going on in the head region, or cranium, versus what's going on in the rest of the skeleton, the postcranial skeleton—played no small role in how fossils came to be interpreted taxonomically and how the supposed picture of human evolution became embedded in people's minds.

Coming to Terms with Human Antiquity

Although various scholars of the latter part of the nineteenth century produced quite a few treatises on human evolution, there were actually very few humanlike remains of potential antiquity then known, and none was given fossil status. The humanlike bones that were known were Neanderthals, and they were all from European sites. As for the ways in which Neanderthals were received by the scientific community, the seeds of disparate interpretations were sown right from the beginning.

The first public report on the anatomy and potential significance of the humanlike remains from the Feldhofer Grotto in the Neander Valley was presented by the eminent professor of anatomy at the University of Bonn, Germany, Hermann Schaaffhausen, and by Dr. Carl Fuhlrott, the local schoolteacher to whom the fossils had been given by the miners who found them. The announcement of this find was made at the meeting of the Lower Rhine Medical and Natural History Society, in Bonn, in 1857 (apparently the same year as the discovery, but the discovery may actually have been the year before). Fuhlrott spoke first, reviewing the geologic setting in which the bones had been found. His presentation was bold, especially given the intellectual atmosphere at the time. For he argued that these bones were not the usual old bones, meaning old but still of reasonably recent origin. They were, he declared, real fossils. His conclusions were based on their being in a level of the grotto that had been sealed off from the present by a geologic deposit of some depth, as well as on their probable association with the bones of ancient animals.

The import of Fuhlrott's interpretation of the antiquity of the Neanderthal bones was not trivial. The recognition of fossils as the remains of once-living organisms of some antiquity may have become somewhat commonplace by the mid-nineteenth century, but the perceived sequence of the origin of life-forms was still interpreted within a biblical context, even if the succession of organisms was now allowed to have some time depth. As such, humans, the last product of a Divine Creator, could only be of recent origin. When the story of the biblical Flood was incorporated into the scenario, extinct animals were interpreted as being antediluvial, whereas humans and other living organisms were of the diluvium. There could be clamlike fossils, horselike fossils, fossil anything, as long as the fossil remains were not humanlike.

In order to prove human antiquity, one had to demonstrate, unequivocally and to the satisfaction of an audience that was skeptical to begin with, that the bones of humanlike animals were truly contemporaneous with the bones of other mammals, such as mammoths and cave bears, which had already been accepted by scientists as being extinct species. Since most human remains that had been either excavated or unintentionally dug up were usually clearly associated with modern species of animal, the case was biased against a demonstration of human antiquity. A good response to claims of contemporaneity between human bones and the fossilized remains of extinct animals was that the obviously more recent human bones had become intermingled, either artificially or accidentally, with the older bones.

Incontrovertible demonstration of the reality of humanlike fossils that the disbelievers could not dismiss on scientific grounds came from Neanderthals, and rather late in the nineteenth century, for that matter. But it was not forthcoming from further investigations either of Fuhlrott's Neander Valley specimens or of the few other Neanderthals known by the mid-1800s. The indisputable proof of the reality of extinct humanlike forms was made public quite late in the nineteenth century—in 1886, to be precise—with a publication by the anatomist Julien Fraipont and his colleague Max Lohest on the remains of two adult Neanderthal individuals from the cave site of Spy, in Belgium. There, for the first time, was the proof of extinct humans, or, at least, of extinct humanlike forms. What Fraipont and Lohest demonstrated clearly was the contemporaneity of the Spy 1 and Spy 2 individuals both with the acknowledged fossilized bones of extinct animals and with Paleolithic stone tools.

Until 1886, then, any claims of great human antiquity were met with unabashed opposition. And this is exactly the fate that befell Fuhlrott's arguments for the Feldhofer Grotto Neanderthal remains. Opposition to his claims came not from an outspoken antagonist among the delegates at the meeting in Bonn, however, nor from an outraged colleague who vented his spleen at a meeting elsewhere or through a published letter. Rather, it was

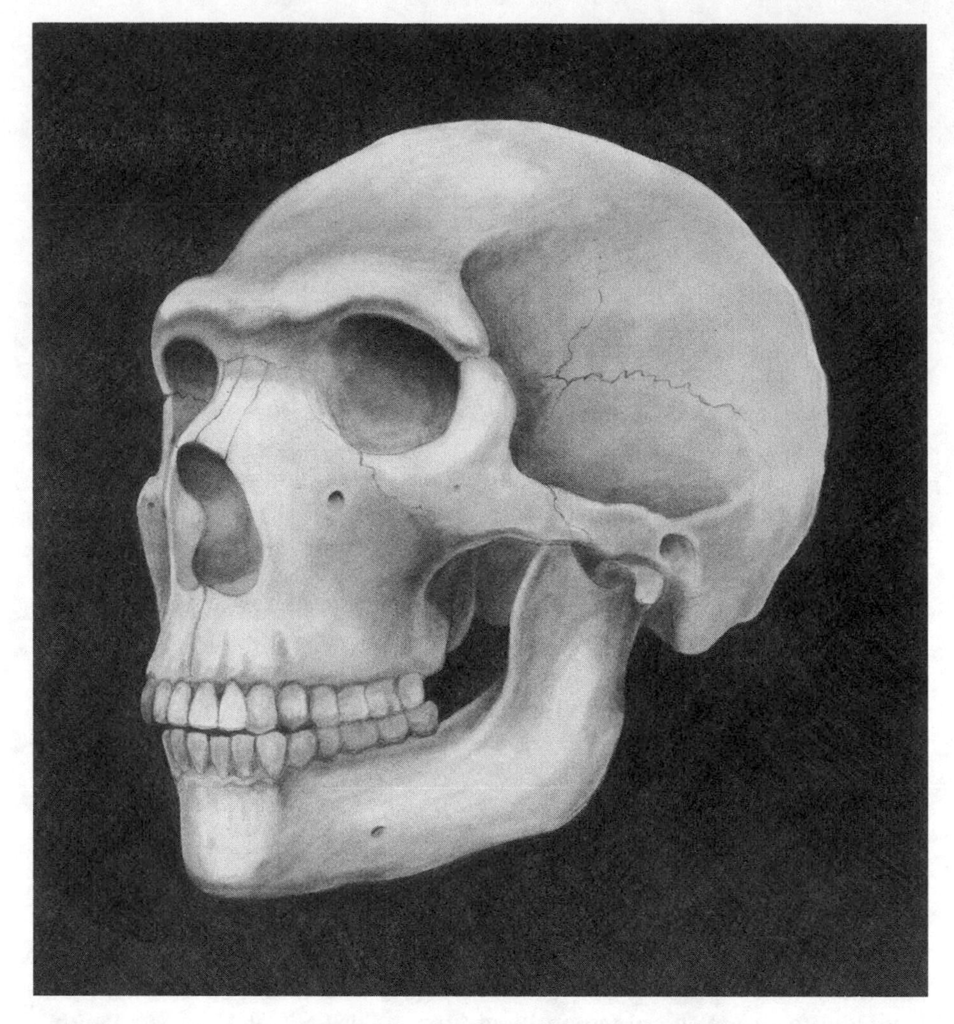

Reconstruction of a typical Neanderthal skull and lower jaw. The original Neanderthal skullcap had a slightly thicker brow ridge, while the Spy Neanderthals looked more like this illustration.

Fuhlrott's own collaborator, the esteemed Professor Hermann Schaaffhausen, who rejected the fossil status of—and, as a consequence, a great antiquity for—the Feldhofer Grotto specimens.

Schaaffhausen brought different arguments to bear on his denial of the Feldhofer Grotto Neanderthal bones being both fossilized and ancient. For one thing, he believed that these bones retained a large amount of their organic material. If they had become fossilized, they would have been turned completely to stone. Even though they had been found under four or even five feet of muddy deposit, their apparently still largely organic state was clear evidence of recent origin.

Another of Schaaffhausen's arguments relied on the writings of a colleague, the honorable Herr Professor Doctor Hermann von Mayer. Von Mayer had studied the degrees to which animal and human bones from different historical periods, as well as animal bones from antediluvian epochs, had become mineralized. He had focused on the presence or absence of a particular mineral deposit that forms a thin black branching pattern, which sometimes looks like miniature seaweed adhering to a surface. Because of the typical branching pattern, geologists called these formations dendrites. Geologists also believed that the black coloration of dendrites came from the mineral manganese, which will indeed stain fresh bone a dark brown or black. Von Mayer did discover well-formed dendrites on the bones of acceptable antediluvian fossil mammals, such as extinct cave bears, mammoths, and species of horse. But he also found dendritic deposits on the skull of a Roman period dog that were as prominent as those on real fossil bones, as well as faint indications of dendrites on the skull of a Roman, although other Roman skulls bore no such crystalline deposits. Von Mayer's conclusion was obvious: Dendrites alone do not a fossil make. In the case of suspected intermingling of antediluvian bones and bones from the diluvium and later historical periods, one could not, according to von Mayer, use the presence of dendrites to settle the dispute. As for the Feldhofer Grotto Neanderthal remains, Schaaffhausen drew on von Mayer's criteria when he declared that there was no basis for thinking that the bones had become fossilized. As such, they were not ancient. Last, and in great brevity, Schaaffhausen dismissed out of hand Fuhlrott's claims that the Feldhofer Grotto Neanderthal bones had been found with the bones of extinct animals. He was not even convinced of the association of these remains with the bones of more recent, diluvian animals.

Even as he rejected Fuhlrott's claims for the Neanderthal skeletal remains, Schaaffhausen also had to address the question of who, exactly, this primitive-looking humanlike individual could have been. Perhaps, he speculated, this individual had been one of the wild races to which Roman historians had made mention. If this was so, the invading modern Germans would have encountered these indigenous but more primitive people. Whatever or whoever the Feldhofer Grotto Neanderthal really was, it definitely represented a savage and barbarous race—but, Schaaffhausen was quick to add, not necessarily more savage and barbarous than the most savage and barbarous of known living races, such as Australian Aborigines. Among European comparative anatomists, the African Hottentot and the Australian Aborigine were frequently offered up as the most primitive of living races of *Homo sapiens*. Schaaffhausen preferred the Hottentot, while Huxley made his comparisons with the Aborigine. Darwin, on the other hand, who encountered the Tierra del Fuegians while serving as captain's companion and amateur naturalist on the HMS *Beagle*, thought that these peaceful and handsome people, of whom he developed an immediate fear, best fit the preconception of the world's most primitive, savage, and barbarous race.

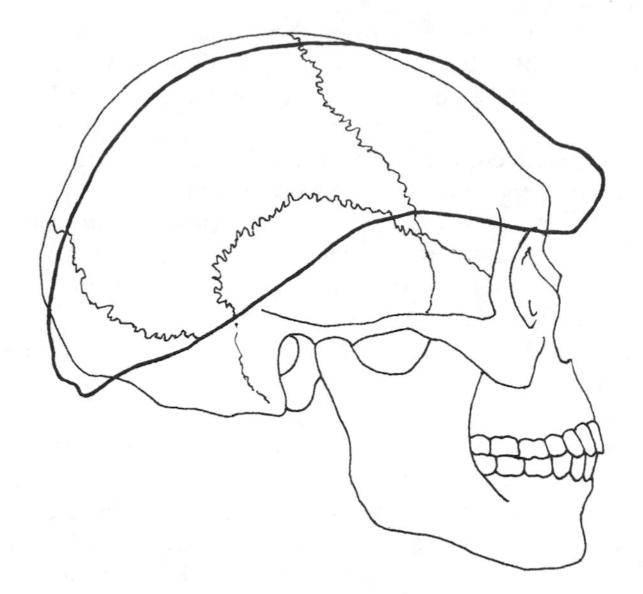

In his essays of 1863, Thomas Huxley argued that the skull of an Australian Aborigine (the skull with lower jaw in the drawing) could be easily transformed into a skull that conformed with the Neanderthal skullcap (superimposed on the complete skull). (Adapted from Huxley)

Thomas Huxley believed that the Feldhofer Grotto Neanderthal, with its long and low skull and prominent brow ridges, represented nothing more than an addition to the low end of a perceived human racial hierarchy—just a mere anatomical hop, skip, and a jump down the line from the Australian Aborigines. In his essay "On Some Fossil Remains of Man," in which he dwelled at some length on the position of Neanderthals in a sequence of human racial ascendancy, Huxley presented an etched illustration of the side view of an Australian Aborigine's skull with the outline of the longer and less rounded Feldhofer Neanderthal partial cranium drawn over it. Anyone looking at this illustration by itself would wonder why two such anatomically dissimilar specimens were illustrated together in the first place. If, however, you believed in a racial hierarchy within *Homo sapiens,* as Huxley and virtually every other western European comparative anatomist did, then you would be easily convinced by his explanation: "A small additional amount of flattening and lengthening, with a corresponding increase of the supraciliary ridge [the brow ridge], would convert the Australian brain case into a form identical with that of the aberrant fossil."

In spite of all the objectively real anatomical differences by which Huxley and Schaaffhausen easily distinguished the Feldhofer Neanderthal bones from those of living *Homo sapiens*—for instance, the Neanderthal had thicker cranial and postcranial bone, a longer, wider, and lower skull, thickened and continuous brow ridges, more circular ribs, a much longer collarbone—in the end, these scholars could not be as dispassionate in their taxonomic assessment of the hominids as they could with any other organism they had studied. Rather than let the anatomy lead them to the obvious conclusion—that the Feldhofer Grotto Neanderthal represented a species apart from *H. sapiens*—Huxley and Schaaffhausen took the position, as the former scholar put it, that "the Neanderthal cranium is by no means so isolated as it appears to be at first, but forms, in reality, the extreme term of a series leading gradually from it to the highest and best developed of human crania."

But the clear-cut differences in cranial and postcranial skeletal anatomy between the Feldhofer Grotto Neanderthal and *Homo sapiens* were not lost on everyone who studied the material. William King, a professor of geology at Queen's College, in Galway, Ireland, was not swayed by Schaaffhausen and Huxley's arguments that the Feldhofer Grotto Neanderthal was merely a variant of *Homo sapiens*. He presented his opinions orally at the 1864 meeting of the British Association for the Advancement of Science, with the published version appearing later that year.

King, who was initially in synch with Huxley and Schaaffhausen, remarked on how some Neanderthal features, especially of the partial cranium, could be viewed as ape- or simianlike. Together with a few comments of his own, he repeated Schaaffhausen's repudiations of suggestions that the unusual Feldhofer Grotto Neanderthal skullcap could have come to look the way it did by being artificially deformed either before or after death. For his rejection of another claim—that the odd configuration of the Neanderthal skullcap was due to its bearer's being the victim of "idiotcy"—King relied almost completely on Huxley's discussion of how the shape of the skull would have been a normal configuration for its bearer. But then King broke company with both Huxley and Schaaffhausen.

Although King admitted that there are indeed "general features of resemblance between the Australian, Neanderthal, and ancient Danish crania," which had formed the basis of Huxley's comparisons, he boldly stated that, in truth, "a closer resemblance is assumed than really exists." Drawing on his own studies of modern human cranial shape, King pointed out that "it has yet to be shown that any [present-day] skulls hitherto found are more than *approximately* similar to the one under consideration," the Feldhofer Grotto Neanderthal. (Even today, disagreements among paleoanthropologists in the recognition and interpretation of anatomical similarity remain central to the ongoing debate on the taxonomic status of Neanderthals, and of other fossil hominids, too, for that matter.)

King then proceeded to take on Darwin. Specifically, he addressed Darwin's invocation of the artificial selection practiced by animal breeders, who choose which varieties of cow or dog to favor over others, as a metaphor for natural selection. As Darwin had often noted, everyone can see that individuals of a wild species, as well as individuals of breeds of domesticated species, differ from one another. Consequently, the kind of manipulation of variation that an animal breeder can engender when dealing with domesticated animals provides a clear example of how natural selection works on species in the wild. Pursuing the topic of domestication even further, Darwin proposed that humans conform less to the picture of variation in wild species than they do to cases of domestication, such as that of the dog, where individuals can differ as greatly from one another in size and various physical attributes as Chihuahuas and Great Danes. Indeed, Darwin argued in *The Descent, Homo sapiens* should be regarded as a domesticated animal. A consequence of this point of view was that different groups of humans could be regarded as similar to different breeds of other animals. And, in turn, this conclusion could have tremendous implications for interpreting the human fossil record, especially with regard to how the newly discovered Neanderthals were dealt with.

One could imagine Darwin figuring out a way in which to incorporate the Neanderthal into his scenario of the transformation of a savage into a civilized human. But even before Darwin could articulate the argument in print, King had rejected it: "It is . . . to be apprehended that, however clearly the Neanderthal fossil may be shown to be inadmissible into the human species, an attempt will be made to set aside the consequent conclusion by an appeal to the fact alluded to"—an attempt that would introduce examples of artificial selection and the potential of artificial selection to produce wildly differing individuals of the same species in support of the argument that if Chihuahuas and Great Danes, for example, can be members of the same species, so, too, can Neanderthals and humans. But, King reminded his audience, "[t]hese [domesticated] breeds, so remarkably differentiated by cranial peculiarities, are *artificial,* whereas the varieties of mankind are *natural* . . . [and as such] [t]he dissimilar skulls met with in the former are merely striking illustrations of organic or structural modifiability, produced by what Darwin calls Natural Selection, nothing more."

In the phrase "what Darwin *calls* Natural Selection, nothing more," King recognized that the notion of natural selection was the construct of an individual, in this case Charles Darwin. Although Darwin was deeply committed to elucidating the processes by which evolution might work, a theory of evolution by means of natural selection was still a theory based on a particular model. Other evolutionists could agree with Darwin that evolution was real, but they did not have to agree with Darwin on the matter of how evolution worked. Thomas Huxley, for instance, was one scientist who did not accept Darwin's case for natural selection outright, as he made very clear in *Man's Place in Nature.* Darwin did understand that his theory of how evolution worked was based on a particular model of process, in this case the machinations of what he called natural selection. But he did build his case on this model, which, in the very last paragraph of *On the Origin of Species,* he referred to as his "view of life."

King was more like Huxley in that he did not tie an acceptance of the evidence of evolution to a particular model of what produces evolution. He was also similar to Huxley in that he did not confuse an attempt to figure out evolutionary relationships with the practice of taxonomy. Consequently, for King, even if Huxley had been correct and Neanderthals had at some point in time been transformed into modern humans, a debate over the reality of this evolutionary possibility was an activity to be kept separate from the nitty-gritty of sorting out species. Consequently, a straightforward comparison of the Feldhofer Grotto Neanderthal and extant *Homo sapiens* convinced King that the number of anatomical details by which the two hominids differed was sufficient for them to be recognized as distinct species. And, indeed, he boldly stated as much in his oral presentation of 1864, when he proposed that the Neanderthal should be allocated to its own species, *H. neanderthalensis.* In the published version of that paper, King suggested that similarities between the Feldhofer Grotto Neanderthal and the chimpanzee—for example, in the protrusion of the brow ridge and in the disposition of the low-lying braincase—which further emphasize the differences between the Neanderthal and extant *H. sapiens,* might even warrant recognition of a separate genus for the former hominid. But for King this unnamed genus, like the living apes, would have been destitute of the qualities of *H. sapiens:* language and moral attitudes.

In one fell swoop, King performed two historically important services. First, he introduced the species name that still remains valid for Neanderthals, *Homo neanderthalensis.* Second, he put his newly named hominid species between modern humans and apes (the latter being represented by the chimpanzee). The first act—distinguishing Neanderthals as a valid species—still has its detractors among paleoanthropologists. The second act set the stage for others to seek the fossil missing links that might fill in the gaps between humans and apes.

4

Filling in the Gaps of Human Evolution

The transition from the ancestral animal, still squirming within us, to Man is too recent for us to be able to understand the ensuing conflicts which often seem disconcerting and incomprehensible.

—Lecomte de Noüy (1947)

Which Came First, the Body or the Brain?

Although Neanderthals seemed to represent a primitive version of *Homo sapiens,* in the minds of many scholars they were not as primitive or brutish in appearance or as ancient geologically as one would expect of a true human ancestor. There had to have been something older and more apelike in our evolutionary past than Neanderthals. It was just a matter of time, and of a paleontologist being in the right place at the right time, before this anticipated missing link was found.

The first person who actively set out to discover a more pithecoid, or apelike, fossil relative of *Homo sapiens* was the Dutch physician Marie Eugène François Thomas Dubois. He did so in 1887, when he embarked by seagoing vessel on the arduous voyage to the Southeast Asian island of Sumatra, which lies more or less parallel to, and just to the west of, the Malay Peninsula. Why seek such fossil treasures in Asia rather than in Africa? Charles Darwin's suggestion of the importance of Africa notwithstanding, the reigning popular theory of human origins still focused on Asia. But by the end of the nineteenth century, the potential role of Asia in human origins had expanded from the discourses of Buffon and others on the origin of modern races to a consideration of human evolution. In the years immediately before and after the turn of the century, the most influential figure in reaffirming, at least vocally, the central role of Asia in human origins was Henry Fairfield Osborn.

In 1891 Osborn, who was then a young paleontologist at Princeton University, accepted offers to move to New York City and join both the faculty of Columbia University and the curatorial staff of the American Museum of Natural History. He established both the department of zoology at Columbia, of which he was the professor, and the department of paleontology at the American Museum, of which he was the curator. Osborn eventually became director of the American Museum and catapulted that institution into the forefront of studies in natural history. His dual positions at Columbia University and the American Museum became the norm for many of the paleontologists and natural historians who followed in his footsteps. Although Osborn worked personally and professionally to promote scientific inquiry in his own areas of expertise, including sponsoring student research as well as field expeditions, he was dogmatic and always correct. Osborn, who was particularly interested in the early evolution and diversification of mammalian groups, as well as in human evolution, became convinced that most, if not all, of the important events in mammalian evolutionary history had occurred in central Asia. In fact, he was adamant about it. Of course, since central Asia was still a scientific no-man's-land, no one could (or probably would dare) refute Osborn's proposition. The Gobi Desert, which would later emerge as a gold mine of early fossil mammals and dinosaurs, became the sought-after land of paleontological plenty. This, Osborn was certain, was the place in which to search for early human fossils.

Being a prominent scholar and administrative figure, however, does not in and of itself guarantee the success of one's predictions. As of 1892, only one fossil mammal tooth—from a relative of the living rhinoceros—had been discovered in the vastness that was central Asia. This lone specimen was discovered not by the American geologist Raphael Pumpelly or the German naturalist Ferdinand Freiherr von Richthofen, who had traveled extensively throughout central Asia, including the Gobi Desert. It was discovered by the Russian geologist Vladimir Obruchev. Obruchev made this singular fossil discovery along the ancient caravan route that stretched from the border between China and Mongolia to the capital of Mongolia, Ulan Bator.

In spite of what was obviously a significant dearth of fossil material, central Asia and the Gobi Desert in particular remained at the heart of wistful paleontological endeavors. In keeping, it seems, with the popular mythology of there being a Shangri-la waiting to be discovered in the farthest reaches of the Tibetan Himalayas, paleontologists continued to view the Gobi Desert as a place that concealed in its sands the record of a now-extinct Garden of Eden. All one had to do was walk the terrain long enough looking for fossils as they weathered out of the sands. And, indeed, dinosaurs aplenty as well as a diversity of unknown tiny ancient mammals were forthcoming in repeated paleontological expeditions to Mongolia. But this lucky strike was never borne out for fossil hominids.

The hype notwithstanding, central Asia did not seduce Dubois with its promises of paleontological wealth. But Southeast Asia, with its areas of lush tropical greenery and a long record of continuous civilization, did catch Dubois's attention, as it had Buffon's. Sumatra and the islands of Indonesia in general would also provide a welcoming and safe haven for a traveling Dutchman. Spurred on by the riches and power that trade with Asia brought, the Dutch had held political control of the islands of Indonesia for almost two centuries. The Dutch East Indies Trading Company, which is perhaps best known to students of American history for its involvement in the fur trade prior to the Revolutionary War, had long before been a prominent player in the development of East–West trade relations. The home base of the Dutch East Indies Trading Company, and the center of Dutch control in Southeast Asia, was the capital of Java, which was then called Batavia but is now known as Jakarta.

After reaching Sumatra, Dubois established himself professionally on the staff of a hospital. In his spare time, out he would go in search of human ancestors. His first two years in Southeast Asia came and went, however, without reward. But then, in 1889, he heard tell of the discovery of a human-looking fossil from the nearby Indonesian island of Java. His spirits elevated, Dubois immediately took off for Java. Upon seeing the fossil, however, his spirits were just as quickly dashed. Although it was of some antiquity, this skull from the site of Wadjak was not pithecoid at all. In fact, the features of the Wadjak skull were clearly similar to those of present-day inhabitants of the Pacific and, in particular, to the skulls of some Australian Aborigines.

Not one to be easily derailed from his quest, however, Dubois set up shop in Java and began excavating at a site called Trinil, which was in the Ngawa district of the Madium Residency. Dubois's perseverance was rewarded in late 1890 with the discovery of a fossil tooth, a third molar, or wisdom tooth. It was larger in its dimensions and more pithecoid in its general shape and anatomical details than that of a recent human. Inspired by this discovery, Dubois persisted in his excavations at Trinil and in 1891, only a few meters from where he had found the molar, he uncovered a calvaria (the technical term for a skullcap) that was the answer to his dreams. In Dubois's eyes, this fossil calvaria was demonstrably more pithecoid than any known human fossil. On he pressed with his excavations at Trinil, following the path of the stratum that had yielded the fossil molar and calvaria. In August of 1892, Dubois hit pay dirt yet again, fifteen meters away from where he had found the first two fossils. This time, however, the part of the skeleton that he discovered came from below the neck. It was a femur, the upper part of the leg.

Within the same year that Dubois found these fossils, he wrote and submitted to a Batavian journal an announcement and a brief description of his discovery. There he referred his Javanese fossils to the genus *Anthropopithecus,* which was the genus name most frequently used by nineteenth-century taxonomists for the living chimpanzee. As he would illustrate photographically in subsequent publications, Dubois was struck by what he perceived to be the great degree of similarity between the Javanese fossil calvaria and the skull of a modern chimp. Whereas *Homo sapiens* charac-

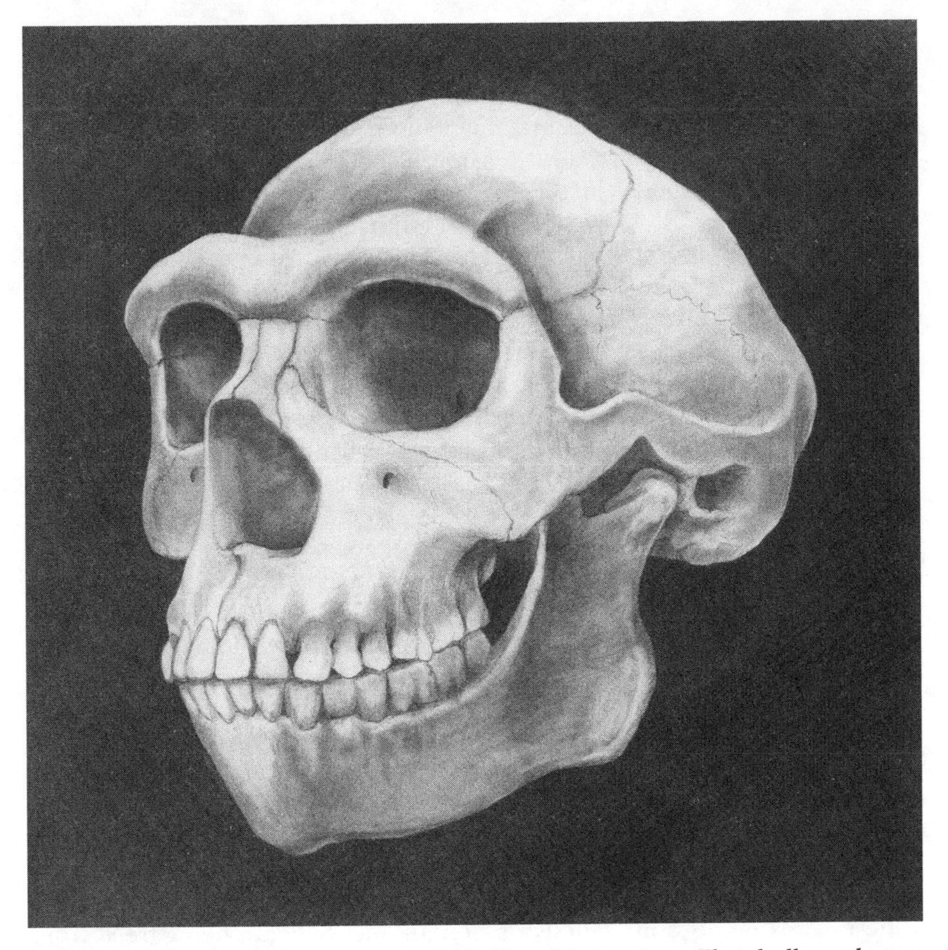

Reconstruction of a Homo erectus *skull and lower jaw. The skullcap that Dubois discovered went from the brow ridges straight back.*

teristically has a rounded or vaulted skull and a rather vertical forehead that only rarely bears prominent brow ridges, both the Javanese fossil and the living African *Anthropopithecus,* Dubois pointed out, had comparably long, low crania with markedly projecting brow ridges. But the fossil skullcap was not quite as low and did not have the same projecting brow ridges as those of living apes. Nor was the ratio of skull width to skull length of the fossil typically apelike. The upper third molar of the fossil also displayed some features that are found in modern humans and others that are seen in some apes. Like that of humans, the fossil molar had fewer roots than are found in apes, and two of the cusps were also reduced in size. Like apes, however, the third of these molar cusps was large.

But while the fossil skullcap and the molar seemed to be anatomically intermediate between living humans and living apes in the features Dubois discussed, it was the femur that put a different, and important, twist on the picture of how this extinct ape-man would have looked from the waist down. Without a doubt, the Javanese fossil femur was humanlike in its overall shape, its relative proportions, and its anatomical details.

Since the fossil femur had the same disposition of parts as a modern human femur, Dubois felt justified in concluding that this fossil individual must have been as fully upright and erect in its posture as its living human counterpart, not only in the way in which it had carried itself about on two legs but also in the relative size of its upper body and the relative length of its leg versus the size of its upper body. In apes, Dubois pointed out, the upper body is much heavier than it is in humans. For humans, a lighter upper body means that there is less of a load for an upright individual to support on its legs. Apes, however, support their relatively bulkier upper bodies with their legs as well as with their long, strong arms. Also in apes, the leg is short relative to upper-body size. By contrast, the human leg is remarkably and distinctively long relative to upper-body size. The short ape leg bears at its end a large and prehensile foot, which can grasp just as well as the hand and is of paramount importance for a more arboreal lifestyle. The long human leg is essential for fully striding, bipedal locomotion. In humans, the long femur slants in toward the knee joint, forming an angle that is not seen in any other living animal but is a mechanical necessity for a biped.

Although this fossil femur was overwhelmingly humanlike, Dubois had to admit that there were a few features—such as the more rounded shaft and the disposition of a few scars on the bone where muscles had once attached—that were more ape- than humanlike. But these features were obviously outweighed by those that resembled humans to a tee. Although all he had was a femur, Dubois was convinced that his fossil individual had not only been a fully upright and erect biped but that its body and leg had the same relative proportions as a living human and its hand, being freed from involvement in locomotion, had surely been like a human's and not an ape's. In order to drive his point home, Dubois introduced a new species name for his fossil, *erectus*.

The skullcap and molar signified to Dubois in a way that had never before been demonstrated that *Anthropopithecus erectus* represented an extinct form that bridged the gap between living humans and their living ape relatives. The humanlike femur provided incontrovertible evidence of the beginning of the transition from ape to human: With the attainment of erect posture and bipedal locomotion, our ancestors had become human first from the waist down. Dubois was ecstatic. He had validated the claims of two prominent intellectual predecessors, Lamarck and Darwin. If skull volume were any reflection of brain size, the attainment of braininess, with all the powers naturalists and philosophers invested in it, had, indeed, followed in the footsteps of the evolution of bipedal locomotion.

When Dubois compared the size, measured as a volume, that he had reconstructed for the brain of *Anthropopithecus erectus* with the brain volume of apes, he discovered that the fossil's brain had been much larger. But when he compared the fossil's brain volume with averages for brain size in modern Europeans, he found that the brain of *Anthropopithecus* was smaller, being only about two-thirds the size of a living human's. As Dubois figured it, since the brain of the largest of the living apes, the gorilla, is only about one-third the size of a European's, the brain of *Anthropopithecus erectus* had been twice as big as the ape's. The story with regard to the femur was, however, different. The length of the fossil's femur was quite comparable to that of modern Europeans, which indicated to Dubois that these hominids had been of similar height. Clearly, hominid bipedalism had evolved earlier, or at least more rapidly, than the hominid brain.

Although the brain of *Anthropopithecus erectus* had lagged significantly behind the rest of the body in evolving truly human characteristics, given that it was significantly larger than that of apes, it had clearly set off on an evolutionary course of enlargement. The impetus for the evolutionary enlargement of the brain, Dubois suggested, was the freeing of the hands from a role in locomotion, which was a direct result of assuming a fully upright posture and walking on only two legs. With the hands not otherwise occupied, they could be used to manufacture weapons for defense. Consequently, the teeth would no longer be needed for defense.

Inasmuch as the only other fossils then known were Neanderthals, and they were considered by many scholars to be very close to modern human in every way, Dubois had reason to believe that he had indeed found *the* human ancestor. Having left his homeland with the express intention of finding such an ancestor, and having chosen to look for it in Southeast Asia, Dubois felt justified in ending his article in the following manner: "And thus the factual proof is provided of what some have already conjectured, that the East Indies was the cradle of the human kind."

In 1894 Dubois published a monographic treatise on his Javanese fossil. There, not only did he announce the discovery of another tooth—this time a second upper molar—that he had found at the site of Trinil but he also argued that as apelike as the fossil species *erectus* was in various features, it was necessary and appropriate to recognize it in its own genus. Since the German embryologist and evolutionist Ernst Haeckel had already provided a genus name, *Pithecanthropus,* for a then hypothetical evolutionary transitional form between apes and humans, Dubois used it.

Haeckel's *Pithecanthropus*—which translates as "ape-man"—was perfect for Dubois's purposes. For not only had Haeckel predicted the existence of an evolutionary intermediate between apes and humans but he had also sketched out what this human ancestor would have looked like. First and foremost, Haeckel's *Pithecanthropus* was a primitive man that would have been essentially identical to *Homo sapiens* not only in its overall body structure but also, and more especially, in the proportion of its limbs. Dubois's fossil species, *erectus,* as judged by and extrapolated from the femur, certainly fit that general description. But given that it was a primitive human,

Haeckel's *Pithecanthropus* would definitely not have been as advanced as *H. sapiens* in its mental and, consequently, its linguistic development. This description, at least the mental development part of it, also fit Dubois's reconstruction of his fossil. In Dubois's publication and thereafter, the Javanese fossils from the Trinil site were no longer to be referred to as the genus *Anthropopithecus* and became known collectively as *Pithecanthropus erectus.* When other fossil hominids were discovered at other Javanese sites, the genus and species *P. erectus* was often used to embrace them as well.

Dubois's combining a seemingly apelike skull, largish molars, and a human-looking femur to make a real entity in human evolutionary history did not meet with universal acceptance. But Dubois persevered in his convictions. At the risk of subjecting himself to unbridled ridicule by the scientific community, as well as by the press, he invited many of the world's leading comparative anatomists and paleontologists to examine his specimens for themselves. To his delight, which he expressed in an article that he published two years after coining *Pithecanthropus erectus* for his fossil, Dubois received significant support for his assessment of his proposed apelike human ancestor. But in the years that followed Dubois increasingly made what he considered to be favorable comparisons between his *P. erectus* and the gibbon, which is a small, short-faced, round-headed, extraordinarily long-armed, short-legged, entirely arboreal Asian ape that is less like humans in many ways than the great apes. Nevertheless, in Dubois's hands *P. erectus* ceased to be an intermediate between the great apes and *Homo sapiens* and instead became a giant gibbon. If the discoverer of the missing link in human evolution no longer accepted his own conclusions, there was no reason for others to continue to embrace them, either.

The early twentieth century witnessed the continued discovery in Europe, particularly in France, of more, obviously Neanderthal specimens. But in Germany the lower jaw of something that was not identical to Neanderthals came to light. This latter specimen was discovered on October 21, 1907, at a site to the southeast of Heidelberg, in deposits that were identified as the Mauer sands. Although the Mauer jaw was recognizably hominid in the shapes and details of its preserved teeth and in its bony configurations, it was more apelike than a Neanderthal's jaw. The most humanlike aspect of the teeth of the German fossil—at least as it impressed the discoverer of the Mauer jaw, Otto Schoetensack—was the fact that the lower canine was neither long nor stout, and it did not project above the neighboring teeth. Rather, as in *Homo sapiens,* this tooth stood at the same level as the teeth on either side of it. In addition, the sizes of the other teeth fell within the range of living humans and were even smaller than those of some Australian Aborigines. But, as Schoetensack was quick to point out, the shape of the cross section of the front of the jaw was very similar to a gorilla's, and the posterior, vertical, and ascending part of the jaw looked as if it could have been from a large gibbon (if there were such things as large gibbons). Since Schoetensack's analysis of the associated bones of fossil animals suggested that they were quite ancient—referable "to the oldest Diluvium"—his conclusion about the Mauer jaw was that it, too, was old. As such, this specimen had to represent the oldest potential human ancestor then known. Making no mention whatsoever of Dubois's *Pithecanthropus erectus,* Schoetensack named his specimen *Homo heidelbergensis* and declared that, being apish in its jaw and at the same time humanlike in its teeth, it provided the first glimpse of what the common ancestor of humans and apes looked like.

Although Schoetensack did not delve into a total reconstruction of *Homo heidelbergensis,* or Heidelberg man, as the specimen was nicknamed, his new species of hominid did not contradict the growing impression about the sequence of important human evolutionary events: First, there was the transformation of the postcranial skeleton from that of a long-armed, short-legged arboreal ape into that of a fully functional biped; and then came the modifications of the skull, teeth, and jaws, with, of course, enlargement of the brain, development of mental faculties, and ultimately the acquisition of language. But it would be only a matter of a few years before an announcement of fossil finds in England threw a monkey wrench into this neatly packaged scenario of human evolution.

Or Was It the Brain and Then the Body?

Sometime toward the end of the nineteenth century—after Dubois had discovered and dubbed his fossil *Pithecanthropus erectus* and before Otto Schoetensack began excavating the Mauer sands—Charles Dawson, an English lawyer and antiquary, was about to begin his descent into paleontological scandal. Depending upon the source of information one chooses to cite, Dawson was on legal business at the court at Barkham Manor, at Piltdown in Sussex. Or he was strolling down a country road near Piltdown Common. By Dawson's own account, he was out on a stroll. He was walking past the commonplace stone fences around fields that were constructed of local gravels when he noticed that some recent mends in a fence had been made with brown, iron-stained flint. Brown, iron-stained flint was not known to be present in that area. Other documentation, however, portrays Dawson strolling the grounds of Barkham Manor while waiting for dinner and happening upon a pile of iron-stained flint. Although iron-stained flint was not supposed to occur in that area, it had nonetheless been dug from the fields of a neighboring farm for use in road repair.

In either case, the stories converge on the point that Dawson knew that this kind of flint should not have been found this far north because its geologic record was supposed to disappear in Kent, approximately four miles to the south of Piltdown. Eventually, whether after dinner or by diverting from his stroll along the road, Dawson was taken to the pit on the farm from which the gravels had been exhumed. The gravel pit was still being dug, the workmen told him, as it had been for years, in order to supply material for road (or fence) repair. Attuned as he was to old things, as an antiquary would be, and having worked for decades for the natural history museum in London as an honorary fossil collector, Dawson asked the laborers if they had found anything out of the ordinary while digging.* Following a negative response, Dawson charged the workmen with the task of keeping an eye out for bone as well as for anything that looked artifactual.

Shortly afterward, Dawson's request was rewarded. Upon one of his return visits to the gravel quarry, one of the workmen handed him a piece of cranial bone. It was one of the platelike bones of the cranial vault—a parietal bone, from the upper part of the side of the skull. Although Dawson recognized the bone as being human in form, he was surprised by how thick in cross section it was. Inspired by the possibility of there being more bone where this piece came from, he began a search of the gravels himself but came up empty-handed. Subsequent searches for artifacts and skeletal remains met with similar results.

Concluding that the site was totally devoid of anything fossiliferous, Dawson gave up his inquiries. However, in the fall of 1911 he made one last trip to the Piltdown gravel pit to see what he could find. Although rain was the norm for the area, it had been extraordinarily rainy, so he did not expect to find much. To his amazement, in one of the rain-washed piles of gravel lying to the side of the pit he discovered a fossil rhinoceros tooth and, more important, another piece of human-looking cranial bone. He identified the cranial bone fragment as part of the frontal bone, which is another one of the platelike skull bones. This fragment was fairly large, and much of the forehead and part of the brow region over the left eye socket had been preserved. The remarkable nature of the coincidence aside, Dawson knew that it went with the parietal bone that he had been given years earlier. Believing that whoever had had this skull in life might also have had a lower jaw like that of *Homo heidelbergensis,* Dawson wrote to the keeper of paleontology at the prestigious BM(NH), Dr. Arthur Smith Woodward.

As Dawson tells it, he immediately arranged to bring the two cranial pieces to Smith Woodward, who just as quickly embraced the potential importance of the skull these two cranial pieces represented and committed himself to working with Dawson on full-fledged excavations of the Piltdown site. Notes and letters by Dawson and others indicate, however, that Dawson had written to Smith Woodward on more than one occasion about his discoveries, first in February and then in March of 1912. But Smith Woodward must have been quite unimpressed by his longtime associate's claims, because he took himself to Europe to study dinosaur fossils instead of going to Sussex to meet with Dawson, visit the site of Piltdown, and study the purportedly ancient human remains. Since Smith Woodward would not come to

*Until its name was officially changed in the early 1990s, this museum was referred to as the British Museum (Natural History), or the BM(NH), in order to distinguish it from the other British Museum, the BM, as it is fondly called, which houses documents, archaeological artifacts, and art.

the gravel pit, then the gravel pit (or, at least its discoverer and its fossils) would come to Smith Woodward—which Dawson finally did on Friday, May 24, 1912. Apparently Smith Woodward did not leap onto the bandwagon with Dawson, because Dawson had to write again a few days later to prod the BM(NH)'s paleontologist to visit the site, which, having finally dried out, could be excavated again.

On June 2, Smith Woodward met Dawson in Sussex. Together with the young French cleric turned paleontologist Teilhard de Chardin, who had attached himself to Dawson's efforts at Piltdown, they went by motor car to Piltdown and began looking for fossils. Teilhard was thrilled when he discovered a fossil elephant tooth. But it was Dawson who, like a bee to honey, quickly zeroed in on another piece of the humanlike skull. Smith Woodward was hooked, and thereafter spent his spare time on weekends and holidays working with Dawson at Piltdown, picking over the gravel that had been dug up during the week and spread out in a single layer for the rains to wash it clean of mud and iron stain.

As the search at Piltdown continued, more parts of the skull were unearthed. First, in a pile of gravel thrown aside by a hired worker, Smith Woodward and Dawson found three pieces of the parietal bone from the right side of the same skull from which the previously discovered fragments had come. Then either Dawson or Smith Woodward discovered part of an occipital bone, which is the platelike cranial bone at the back of the skull. The occipital bone fragment fit perfectly with the first Piltdown bone discovered, the left parietal bone. And, finally, while digging at the bottom of the gravel pit, Dawson took a swing with the pointed end of his geologic hammer and the right half of a human-looking mandible went flying into the air. Although it was missing some bone from its front, the jaw nonetheless continued on to the midline. The knoblike swelling of bone at the top of the vertical part of the mandible, which would have articulated with the base of the skull, was unfortunately missing. Although the jaw fragment retained evidence of the root sockets for various teeth, it possessed only two teeth, the first and second molars, whose chewing surfaces appeared to have been worn completely flat. The bone of the skull pieces and the bone of the mandible fragment were darkly stained to the same degree. Some distance from the pit itself, Smith Woodward and Dawson found remains of fossil deer and horse.

During the time that Dawson, and then Dawson and Smith Woodward, with the help of hired diggers as well as a volunteer or two, including one of the Piltdown police, were excavating at the site, they kept their efforts shrouded in secrecy. They did so not to keep the public in the dark but in order to give themselves time to prepare their presentation of the material as well as their responses to anticipated hostility from their scientific enemies and detractors. But news of the discoveries leaked out, and on November 12, 1912, the *Manchester Guardian* ran the headlines: "THE EARLIEST MAN? A SKULL 'MILLIONS OF YEARS' OLD. ONE OF THE MOST IMPORTANT OF OUR TIME." The article, which contained various errors, proclaimed that the Piltdown skull was the oldest human fossil yet found on British soil, antedating considerably the previous holders of such claims, the Galley Hill and Ipswich men. In fact, the antiquity of Piltdown man made it a serious contender, along with the Javanese *Pithecanthropus erectus,* for being the much sought after missing link between humans and apes. Fortunately for Dawson and Smith Woodward, at least as far as their concerns of attacks from colleagues were concerned, they were scheduled to make their formal announcement of the Piltdown fossil on December 18, at the upcoming meeting of the Geological Society.

When the meeting opened, Dawson and Smith Woodward were seated at the dais with a reconstruction of the head of Piltdown man between them. Dawson spoke first, giving his account of how, while strolling along a road one day, he had come to discover the site of Piltdown, and how he and then he and Smith Woodward had discovered human and animal fossils as well as stone implements at the site. Smith Woodward gave the presentation on the human skull and mandible. He began by dismissing the possibility that these bones were pathological. He then proceeded to discuss the skull, commenting on how the bones fit together perfectly and remarking on their incredible thickness. He had even calculated that these cranial bones were thicker than those of an Australian Aborigine or a Neanderthal and much thicker than a European's.

Although the size of Piltdown man's brain may have been on the low end of that of living humans—and, as another speaker, Professor Grafton Elliot Smith, pointed out, its shape may have been more ape- than humanlike—the brain was nonetheless very large and the overall shape and most of the details of the preserved skull in which the brain had sat were definitely humanlike. In contrast to the more flat-headed and prominently browed Neanderthals and *Pithecanthropus,* Piltdown man had slight brow ridges, a fairly rounded braincase, and a very vertical forehead. Given the presumed antiquity of the Piltdown skull, one had to conclude that the shape of the modern human skull had been established well before the appearance of the more pithecoid-looking Neanderthals and *Pithecanthropus.* The mandible, on the other hand, told a less human story.

The greatest resemblance between the jaw of Piltdown man and living humans derived from Smith Woodward's reconstruction of how the fossil had chewed during life. Since the surfaces of the two preserved molars were worn very flat, as is often the case in humans, Smith Woodward suggested that the Piltdown individual had been able to move its mandible easily from side to side to the same degree that living *Homo sapiens* can. Apes, however, can move their lower jaws only a little from side to side, because their large, long, projecting upper and lower canines interlock with one another, restricting jaw movement to an up-and-down motion. As such, Smith Woodward felt justified in concluding that had a canine tooth been preserved in the Piltdown jaw, it, like the lower canine of humans, would not have risen significantly above the height of the other teeth. Although Darwin was not cited, Smith Woodward and his audience must have been conscious of the evolutionary theorist's prediction that the human ancestor would have had a small canine.

Smith Woodward's reconstruction of how all the teeth would have been spaced along the tooth row contributed, however, an unexpected detail to the picture of the low-crowned canine. Given the apparently large space available for it in the lower jaw, the root of Piltdown's lower canine would have been very stout, as in an ape. In addition, inspection of the root sockets preserved in the specimen as well as of the roots of the two molars revealed that all roots were distinctly separate and quite divergent from one another. This is what one sees in apes but not in humans. The roots of human molars are sometimes not distinct entities until near the tips. Apeness rather than humanness was also demonstrated in the greater number of cusps and elongate shapes of the two preserved molars. As for the mandible itself, everything about it spelled "ape." It was long and its ascending portion not deeply scooped out on top. Its front end was long and sloped down and back. And the sides of the jaw would have been parallel to each other. In contrast, the human mandible is short and the sides of the jaw diverge as they proceed posteriorly. And as Blumenbach had emphasized, a forwardly projecting chin lies at the front of the human lower jaw.

Given the odd combination of features of tooth and jaw, Smith Woodward was forced to confess that had the molars not been in the mandible, there would be nothing human about this jaw at all. But he carried on and speculated that although the cranial vault was essentially human, based on the apelike features of the mandible, the face of Piltdown would not have been humanlike. During his presentation on the brain of Piltdown, Professor Grafton Elliot Smith offered the opinion that, given the apelike jaw, it was entirely within reason to expect the brain to display a very simian yet humanlike configuration—a configuration that he himself had in fact reconstructed.

In recognition of the distinctive combination of human and ape features in this supposedly most ancient of fossil hominids, Smith Woodward proposed a new genus name, *Eoanthropus,* and a new species name, *dawsoni,* to receive it. As paleontologists had named the earliest horse *Eohippus,* meaning "dawn horse," so Smith Woodward meant the name *Eoanthropus* to reflect the fossil's status as "dawn man." Since *Eoanthropus,* with its rounded and high-rising cranium, looked more like living humans than did either a Neanderthal or *Pithecanthropus,* Smith Woodward did indeed have reason to believe that the more ancient fossil from Piltdown represented the true ancestor of *Homo sapiens.* But he also came to this conclusion because he subscribed to Ernst Haeckel's biogenetic law—"ontogeny recapitulates phylogeny"—complete with the (mis)understanding that the morphology of an adult ancestor is retained by the descendant. Given a choice, then, among *Eoanthropus,* Neanderthal, or *Pithecanthropus,* the logical candidate for being a human ancestor was *Eoanthropus,* who was the most similar to *H.*

sapiens, at least in cranial shape. Other, less human-looking hominids, such as Neanderthals, which were also supposedly chronologically younger than *Eoanthropus,* merely represented evolutionary dead ends, which, having evolved away from the primitive, ancestral state, became extinct.

The reception of *Eoanthropus dawsoni* by the esteemed attendees at the meeting was mixed. There were debates about the age of the fossils, the comtemporaneity of the proclaimed stone implements and the skull and the mandible, and the association of the mandible with the skull, which was impossible to assess because the mandible did not preserve its articular process. Even Smith Woodward retained doubts about uniting the apish mandible with the overwhelmingly human skull. The validity of this reconstruction, he felt, could be resolved once and for all only by finding the missing canine tooth. If this tooth had been long and projecting, it was definitively an ape's, and there would be no compelling reason to associate the mandible with the skull. If the canine turned out to have been short, then this tooth would be humanlike and the mandible representative of *Eoanthropus.*

Back to Piltdown went Dawson and Smith Woodward, but they no longer enjoyed the tranquillity of digging in relative solitude. By then, the Piltdown site had become a tourist attraction, which even saw as one of its visitors the literary father of Sherlock Holmes, Sir Arthur Conan Doyle. There were also postcards, which featured Dawson and Smith Woodward and a goose, which was apparently their adopted mascot, at the gravel pit. Teilhard de Chardin rejoined the search for more fossils, and on August 30, 1913, it was he who discovered the crucial missing link in the Piltdown controversy. When Teilhard cried out that he had discovered the lower canine, Smith Woodward and Dawson were skeptical. But when Teilhard brought them the specimen, his identification was indisputable. After Teilhard took off for the rest of the day, Dawson and Smith Woodward searched in vain for more pieces of *Eoanthropus.* The only other parts of *Eoanthropus* that were eventually discovered were a pair of tiny nasal bones.

Amazingly, the prized lower canine of *Eoanthropus* looked pretty much as Smith Woodward had predicted it would. It was also stained in the same manner and to the same degree as the other *Eoanthropus* bits and pieces. The tooth, however, was a bit smaller, more pointed, and probably would have been more upright than Smith Woodward had expected. Nevertheless, this was definitely the tooth of a primate, and a higher primate's canine tooth at that. Looking both apish and hominid but more like *Homo,* it must have belonged to *Eoanthropus,* since the latter genus was the only primate found at Piltdown. Functionally speaking, the size of the tooth would not have prevented its bearer from moving its lower jaw easily from side to side in a chewing motion that would have resulted in the kind of flat wear seen on the two lower molars.

That this tooth could have come from the same specimen whose fragments were collectively dubbed *Eoanthropus* was also indicated by its extremely worn state. The degree of wear was in keeping, so it seemed, with the severe attrition on the two lower molars of the earlier discovered mandible. From the excessive amount of wear on the inner surface of the lower canine, Smith Woodward concluded that the tooth must have been fully erupted and in use for a considerable period of time. In fact, Smith Woodward extrapolated, the length of time necessary to wear down the canine to the state it was in required this tooth to emerge quite early in the sequence of tooth eruption, probably before the second and certainly before the third molar would have broken through the gums. In apes, the canine is the last tooth to come in. In humans, however, the canine erupts prior to the second and third molars.

The case having so convincingly been made that an isolated tooth (and not just any isolated tooth but the very one that was needed to cinch the reality of *Eoanthropus* as *the* ancestor of *Homo sapiens*) really went with the jaw that supposedly went with the partial skull from Piltdown, it hardly seemed to matter that none of these pieces actually either articulated with one another or otherwise fit together. To be sure, the cranial bones did go together like contiguous pieces of a jigsaw puzzle. But the mandible was lacking its condyle, its point of articulation with the skull. And since the mandibular condyle sits in a depression of similar size at the base of the skull, just in front of the ear, no one could say whether the partial skull, which did retain this articular depression, and the mandible would have articulated together properly. As for the lower canine, not only was its root missing but it was from the right side of the jaw, whereas the piece of mandible was from the left side. Without the root of this canine, no one could assess its size relative to its predicted size or make a mirror-image cast of it to see if its length was reasonable for the depth of the preserved right side of the jaw. Any reconstruction of *Eoanthropus* that included all of the pieces that Dawson and Smith Woodward assigned to it had to have the gaps between the known parts filled in.

As history would have it, these pieces did not all go together. They did not represent the strange combination of a modern human skull with an apelike jaw. Nor were the cranial bones, partial mandible, and isolated canine ancient. But in spite of recurrent suspicions of the authenticity of *Eoanthropus,* it was decades before the truth came out, in large part because the chemical tests and the microscopes necessary to make these determinations had yet to be developed. But when, in 1953, the fraud was finally exposed, what should have seemed obvious, and in retrospect was, could no longer be denied. The skull, tooth, and mandible were quite recent in age. They had all been stained to look ancient. The cranial pieces were from a modern human skull. As the paleoanthropologist Franz Weidenreich, who would later become famous for his studies of the Peking man fossils, and others had suspected, the mandible was an orangutan's. Microscopic study of the molars revealed the scratch marks that had been made in filing down the teeth. The human skull and the orangutan mandible had been broken deliberately, and only those pieces of the skull and mandible that were necessary to carry off the hoax had been planted in the gravels at Piltdown.

Eoanthropus was a forgery of such monumental proportions as well as audacity that every aspect of the Piltdown affair continues to capture the imagination. It's quite common to read about the publication of a book or article that claims to have finally solved the mystery of "who done it." Dawson is most frequently indicted, and for apparent good reason. He had a reputation for being a legal trickster, and many of his archaeological and paleontological discoveries prior to Piltdown were eventually exposed as fraudulent. Smith Woodward, who had probably been duped by Dawson, has also been implicated in this affair, as has Teilhard de Chardin (he did find the isolated canine, after all) and even Sir Arthur Conan Doyle.

But whether the identity of the perpetrator, or perpetrators, of the Piltdown forgery will ever be resolved, it remains the case that the image of *Eoanthropus* loomed large in the field of paleoanthropology for four decades. On the one hand, *Pithecanthropus erectus* and Neanderthals, and later Peking man (*Sinanthropus pekinensis*) as well, represented a picture of human evolution in which apelike features of the skull and its brain persisted long after apelike features of the postcranial skeleton had been replaced by those of a biped. On the other hand, *Eoanthropus* reflected a course of human evolution in which the bony braincase and, especially in size, the brain within it, had become that of *Homo sapiens,* whereas, judging from the mandible, the rest of the skeleton had remained an ape's.

How were paleoanthropologists supposed to decide which picture of human evolution—which came first, the thinking or brawny parts of the human body—was correct? Until the forgery was exposed, it was impossible to choose between these opposing scenarios, given that each offered such a different set of evolutionary circumstances and outcomes. Curiously, instead of the future fossil finds helping to lead the way to questioning the centrality of Piltdown man to human evolution, the very existence of the latter made it difficult for leading paleoanthropologists to accept and think clearly about the next wave of earlier, and potentially more revealing, fossil discoveries that would be unearthed in African deposits.

Into Africa

Four years prior to Eugène Dubois's departure for Southeast Asia in search of human ancestors, one of the most important contributors to paleoanthropology, Raymond Arthur Dart, was born in a suburb of Brisbane, Australia, half a world away from the country in which he would eventually become established professionally. Dart was raised in devout Methodist and Baptist surroundings and also exposed to the fundamentalism of Plymouth Brethren. He was eighteen years old when he first encountered evolutionary ideas as a student of biology at the University of Queensland. The year, 1911, was already an important one in the Piltdown saga. In spite of his religious background,

Dart did not reject evolution, and continued his scientific studies, eventually enrolling in medical school at the University of Sydney in 1914.

Coincidentally, 1914 was also the year that the annual meeting of the British Association for the Advancement of Science was held in Sydney. Dart attended and was treated to four distinguished lectures by the eminent, world-renowned professor of anatomy, ethnology, and physical anthropology, Professor (and eventually Sir) Grafton Elliot Smith. Elliot Smith, who was then chair of anatomy at the University of Manchester in England, was himself Australian and had graduated with distinction from the medical school in Sydney. Among all of his other achievements, Elliot Smith had already become a visible and central figure in the study and promotion of *Eoanthropus.* The excitement these meetings in Sydney generated escalated to even greater heights when it was announced that the first fossil cranium ever to be unearthed from Australian soil—the Talgai skull—had been discovered. In fact, on the very night that Elliot Smith was scheduled to give a public lecture, the skull was going to be presented first to those in attendance.

During the years of the First World War, Dart was able to combine his military duties with teaching human anatomy. He became increasingly drawn by the mysteries of the nervous system and, in particular, of the brain. Determined to study with Elliot Smith after the war ended, Dart came closer, at least geographically, to fulfilling his wish when, in 1918, he and others in the Australian Army Medical Corps were posted to England. The following year, when Elliot Smith assumed the professorship and head of the anatomy department at University College, London, he hired Dart as the senior demonstrator. Since there is only one professor and head in a British academic department, this was the highest position available to Dart.

Dart's research in neurology, however, often led him to reach unusual and even strange conclusions, which eventually resulted in his being branded not only nonconformist but also antiauthoritarian. While receiving training in the study of the nervous system and comparative neurology, and seeking to understand how the brain and the nervous system developed, he quickly came to reject one of the cornerstones of neurology: that, embryologically, the cells that give rise to the body's nerve cells do not, as everyone believed, derive from the newly forming spinal cord during its early tube phase. For him, the nerve cells had an origin independent of the early-forming spinal cord. He was, however, incorrect, as studies on the fate and influence on later development of the cells that originate during the tube phase of spinal-cord formation have continually demonstrated. As for his later role in trying to decipher human origins, Dart's interest in the brain would become a central theme in his discourses on the lifestyles of early hominids.

Dart's teaching and research at University College were interrupted by a two-year fellowship in the United States, in the anatomy department of Washington University, in St. Louis. While there, Dart was greatly impressed with the comparative human skeletal collections that the head of the department, Robert J. Terry, had and continued to amass. After Dart returned to

London, Elliot Smith recommended that he apply for the position of chair of the anatomy department at the nascent University of the Witwatersrand (nicknamed "Wits") in Johannesburg, South Africa. Although Dart initially rejected this suggestion, he eventually applied for and was offered the position after the lead candidate withdrew from the competition. Dart and his American wife, whom he had met while in St. Louis, moved to Johannesburg in January 1923. Although Dart arrived as professor and head of a department that he would make internationally visible—not only in gross anatomy but also in paleoanthropology and human skeletal anatomy—he found the medical school so deep in its infancy that its buildings and grounds could easily have been thought of as works in progress that had been abandoned. His wife cried when she saw what she and her husband had moved to.

Seemingly unconnected with Dart's anatomical and neurological predilections is the story of limestone quarrying in South Africa. Limestone, which is a necessary element in the processing and extraction of pure gold, was a precious commodity in South Africa, at least for white South Africans, and the search was always on for large deposits of this mineral. As for the field of paleoanthropology, it is fortunate that a Mr. M. G. Nolan, whose brother, H.G., had established at the very end of the eighteenth century the Nolan Lime Works at the Transvaal locality of Sterkfontein, had been sent out to locate another sizable limestone deposit. Sometime after 1910, M.G. did so, quite to the south of Johannesburg, at the white rocks of Buxton, in the area known then as Taungs (which is now spelled, without the *s*, as Taung). By 1917, when the Nolan and Northern Lime Works merged, quarrying and purification of the Buxton limestone had been in full swing.

One of the frequently encountered intrusive elements that had to be removed from the limestone after it was quarried was fossilized skeletal material. Common among the fossils—the postcranial bones, jaws, and skulls of various kinds of mammals—were the remains of monkeys, specifically baboons. The first specimen of a fossil baboon was discovered at Taung in 1919. But it was not until the spring of 1924 that one of these fossil baboon skulls from Taung finally made its way into the hands of Raymond Dart. A fossil baboon skull that had been collected during a visit to the Buxton works by one of the company's directors, a Mr. E. G. Izod, went from his desk, where he thought it made an interesting paperweight, via his son Pat to fellow students at the University of the Witwatersrand, from whom a Miss Josephine Salmons borrowed it to show to her anatomy students during lab, whereupon Dart came upon it. Dart was entranced by the fossil. By way of a colleague in the geology department, Professor R. B. Young, he sent a request to the manager of the limeworks to salvage all fossils discovered thereafter. In 1924, Dart's request was rewarded through the efforts of one of the quarrymen, a Mr. De Bruyn. Having become familiar with the small baboon skulls that were frequently blasted out of the limestone, de Bruyn recognized instantly that, although its face was still encrusted in limestone matrix, another, larger, partial skull he had found was indeed something new and entirely different.

Although de Bruyn knew that this particular fossil skull was not a baboon's, he was not exactly sure what it was. But when he handed the skull over to his foreman, Mr. A. E. Spiers, he apparently referred to it as "Bushman." And it is on this point that history has two alternative scenarios. One, related by Dart's successor at Wits, Phillip Tobias, is that the sandy limestone from which the fossil had been recovered was called Bushman by the quarrymen. The other, more frequently told story is that de Bruyn referred to the skull as being Bushman because of the common but mistaken belief among white South Africans that anything representing human prehistory was Bushman, whether it was rock art, stone tools, or fossil skulls. And inasmuch as Bushmen, who are correctly referred to as the San, are very small individuals, de Bruyn might very well have thought that he had unearthed a fossil skull of one of these indigenous people. Regardless of which account is correct, it is the case that in November 1924, Spiers turned the skull over to the geologist Professor Young, who took it and two other specimens directly to Raymond Dart.

When Dart was presented with these three fossils, he knew immediately that two of them were not skulls. They were models of the inside of the braincase, called endocasts, that formed over time as water with its load of dissolved limestone dripped into the skulls and then evaporated, leaving the consolidated mineral behind as a reasonable facsimile of the shape of the once-present brain and its external features. To say that an endocast is an exact model of brain shape and external detail would, however, be inaccurate, because the brain is surrounded by three protective layers of tissue and the brain and each tissue layer are separated from one another by cerebral-spinal fluid. An endocast might approximate the volume of the braincase, but it does not reflect with precision the volume of the brain itself. In addition, the thicker the layers of tissue and of the intervening cerebral-spinal fluid, the less clearly details of the surface of the brain imprint on the bone of the inside of the braincase and, in turn, the less clear-cut are the external features of the brain that become impressed on an endocast.

Being an expert on the nervous system, Dart recognized the smaller of the two limestone endocasts as having come from a baboon or a similar kind of monkey. He also knew that the larger endocast was not a monkey's. Judging from the overall shape of the endocast, as well as from the pattern of grooves that had become recorded on the surface of the endocast, Dart was reminded more of the brains of apes than of those of monkeys. Since he was fairly certain that the larger endocast had come from the fossil partial skull with which he had been presented, Dart set to work on cleaning off the matrix that still adhered to this specimen. By December 23, 1924, he had removed all the limestone on the left side of the face, exposing most of a brow ridge, a long ovoid orbit, a relatively vertical forehead, part of a slender cheekbone, the lower face and upper jaw, and a portion of the lower jaw, or mandible. Within days, the more completely preserved right side of the face was also free of matrix.

Aside from its vertical forehead, lack of brow ridges, large ovoid orbits, slender cheekbones, small lower face, and thinly boned mandible, the Taung specimen had a small nasal opening and very humanlike teeth, including very low, non-projecting canines. In general proportions, the lower face and mandible were quite small compared with overall skull size. But, more spectacularly, the size of the preserved braincase, and of the endocast that fit perfectly into it, were huge by comparison to the size of the facial skeleton. Although most of the base of the skull was missing, a few fragments corresponding to the bone around the hole through which the spinal cord passes as it exits the braincase survived on the surface of the endocast, and they were located very far forward under the skull. By January 6, 1925, Dart had completed a manuscript on the fossil and sent the opus off to England, to the publication offices of the internationally renowned journal *Nature,* where it was published on February 7, in volume 115.

As he had done early in his research career on the nervous system, so with his publication on the Taung specimen did Dart throw down an intellectual challenge to his colleagues, and to the larger scientific community as well. With bravado, Dart began his article with the declarative statement that the importance of the Taung specimen lay in the fact that, as indicated by the features of the entire skull as well as of the skull's separate components, it represented "an extinct race of apes *intermediate between living anthropoids and man.*" The Taung specimen was clearly unlike apes, particularly the African apes of the same continent, in the overall shape of the skull as well as in the configuration of the brow (it was slight, not projecting), the forehead (it was vertical, not sloped), the nasal bones (they were short and broad, not long and thin), the cheekbones (they were slender and lightly built, not deep and robust), the orbits (they were long and ovoid and placed high on the face, not rectangular or square and low-set), the upper jaw (it was small, lightly built, and projecting forward only slightly, not deep, stout, and projecting as a snout), and the lower jaw (it was small, short, and lightly built, not deep, projecting forward, and robust). All in all, Dart was struck by the "delicate and humanoid character" of the Taung skull.

As for the teeth, Dart could study them only from the side, because he had not cleaned off the matrix that held the upper and lower jaws together. But all the milk teeth (more properly called the deciduous teeth because, like the leaves of a deciduous tree, these teeth are eventually shed) were present, and the upper and lower canines were surprisingly low-crowned. In apes, the deciduous canines, though diminutive compared with the size, height, and robusticity of the permanent canines, are definitely higher-crowned and more projecting than their counterparts in the Taung specimen. But in addition to the fossil's having a complete set of deciduous teeth—deciduous incisors, deciduous canines, and deciduous molars—its permanent teeth had begun to erupt. Specifically, all four first permanent molars were in various stages of completing their eruption into the jaws and reaching their full occlusal, or chewing, height. Since on average the first

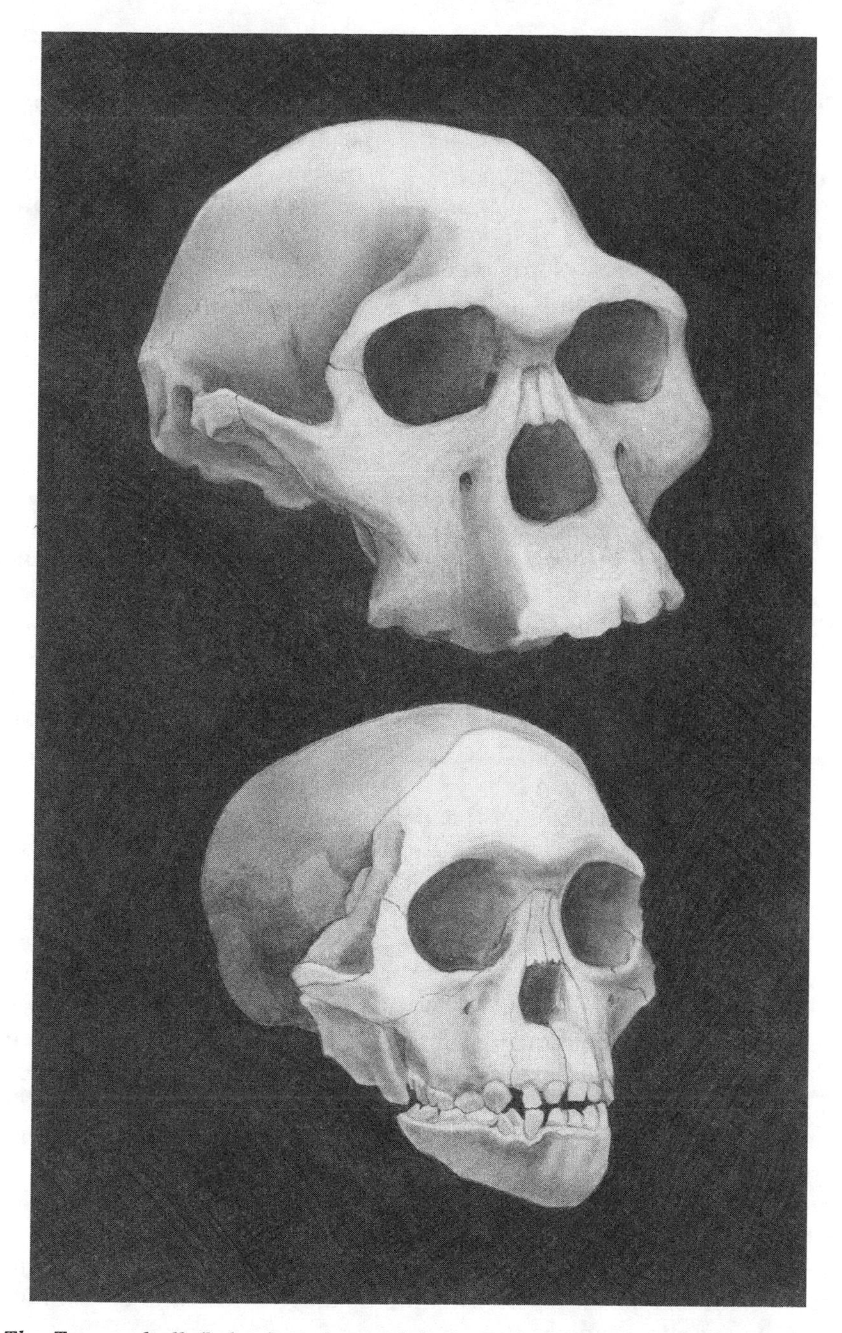

The Taung skull (below) and an adult Australopithecus africanus *from the South African site Sterkfontein (above). When Dart first worked on cleaning the specimens, he left the limestone matrix that held the face and lower jaw of the Taung child together and did not remove the limestone endocast of the brain from the fossilized bone.*

molars erupt at about the age of six in humans, Dart concluded that his humanoid must have been roughly that age at the time of its death.

But while the details of the face and of the deciduous canines were cause for Dart to embrace the Taung child as something intermediate between humans and apes, it was the brain that drove him to heights of speculation—speculation about how this individual had carried itself while locomoting, about what its brain was capable of, and about how these attributes were interrelated evolutionarily. First, because the Taung specimen retained that very informative area at the base of the skull through which the spinal cord passes, Dart could reconstruct the position in which this humanoid had held its head.

On either side of this large hole—known in anatomical jargon as the foramen magnum, which means "great hole"—lie the bony elevations by which the skull articulates with the first vertebra. In a quadrupedal animal, such as a horse or a typical monkey, these points of articulation lie almost at the back of the skull; in most quadrupeds, the head is a direct extension of the vertebral column. Since the spinal cord exits the skull and courses through a canal in the vertebral column, sending nerve fibers all the way down the tail, there would be a direct shot from brain to tail tip. In animals that do not carry their heads straight out in front of them, such as the knuckle-walking great apes, the vertebral column slopes down. Consequently, the articulations on the skull are located farther down on the skull than in a typical quadruped. The spinal column can still course directly down the vertebral canal, and the eyes can still look straight ahead.

In animals that balance their head atop a vertical vertebral column, the articulations lie at the base of the skull, often quite far forward, and they face directly down, which is the direction the spinal cord takes as it courses to the end of the vertebral column. This suite of features certainly characterizes bipeds, such as humans and kangaroos. But animals can have these features of skull-vertebral column articulation and orientation without being bipedal. An arboreal animal that brachiates and arm-swings through the forest canopy suspended below the branches, such as gibbons and siamangs and a few New World monkeys, has downwardly directed vertebral articulations situated well forward on the skull base. Although a gibbon, for example, may hang by its hands below the branch it is grasping, it still holds its torso erect and balances its head atop its vertical vertebral column. The same goes for animals that cling to vertical supports while resting and sleeping. One such animal is the tiny nocturnal prosimian primate of Southeast Asia, the tarsier, which, though not a biped or a suspensory brachiator, holds its torso erect and balances its head atop its upright vertebral column while clinging to a vertical support.

Dart did not consider all of these possibilities, even though the brachiating gibbon and the vertical-clinging tarsier had each at one time been considered a likely living model for the early evolutionary phase of human bipedalism. The brachiating gibbon and the vertical-clinging tarsier models for the evolution of human bipedalism both posited an arboreal lifestyle at the beginning. According to these models, bipedalism "arose"—or, as was often said, there was an evolutionary predisposition toward bipedalism—because the orientation of head to vertebral column had already become established. All an early human ancestor had to do, then, was drop to the ground and, after a bit of evolutionary tinkering with the legs and feet, walk.

In his scenario of the evolution of human bipedalism, however, Dart paralleled Darwin's emphasis on the African apes as being the closest living relatives of *Homo sapiens.* In *The Descent of Man,* Darwin imagined that there had been a transformation from the common type of locomotion seen among mammals, quadrupedalism (in which the torso is held horizontally), to the much more specialized form of quadrupedalism seen only in the African apes, knuckle walking (in which the torso slopes down because the arms supporting the front of the body are very long), to the unique kind of locomotion of *Homo sapiens,* bipedalism (in which the torso is held erect and only the legs support the body). For Dart, the forward position of the foramen magnum of the Taung skull, reflecting forwardly placed articulations with the vertebral column, could only mean that this humanoid had enjoyed a more erect posture than any ape.

Once Dart had imbued the Taung individual with bipedalism, he indulged in the potential evolutionary implications of such a specialized form of locomotion:

> The improved poise of the head, and the better posture of the whole body framework which accompanied this alteration in the angle at which its dominant member was supported, is of great significance. It means that a greater reliance was being placed by this group upon the feet as organs of progression, and that the hands were being freed from their more primitive function of accessory organs of locomotion. Bipedal animals, their hands were assuming a higher evolutionary rôle not only as delicate tactual, examining organs which were adding copiously to the animal's knowledge of its physical environment, but also as instruments of the growing intelligence in carrying out more elaborate, purposeful, and skilled movements, and as organs of offence and defence. The latter is rendered the more probable, in view, first of their failure to develop massive canines and hideous features, and secondly, of the fact that even living baboons and anthropoid apes can and do use sticks and stones as implements and as weapons of offence.

From here Dart went on to address the general and specific features of the brain as he thought they were reproduced on the endocast. Focusing first on the size of the Taung child's brain, Dart pointed out that it was already larger than the brain of an adult chimpanzee and almost as large as the brain of an adult gorilla. But, being the brain of a child, it would have kept enlarging and would surely have been larger than a gorilla's brain at maturity. This presumed fact was especially important to Dart because it suggested a growth curve reminiscent of the way in which the human brain grows. Although it is large at birth, the human brain continues to grow, approaching adult size by the tenth year or so but not fully reaching it until the twentieth or even the thirtieth year.

The shape of the Taung child's brain also reminded Dart more of a human than an ape. Rather than being flattened, as it is in apes, the Taung child's brain was decidedly high and rounded, as it is in *Homo sapiens.* A high and rounded brain signified to Dart that, as in humans, the faculties of associative memory and intelligent activity had been well developed and in harmony with each other in the Taung child. Inasmuch as the human brain is exalted (by humans) because of its capacity for thought, which is reflected in there being a lot of gray matter, or cerebral cortex—the "little gray cells" of Hercule Poirot—Dart was impressed by the facts that the side of the Taung child's brain was noticeably expanded and that, especially when compared with its position on an ape's brain, a particular groove in the surface of the brain, which represents the posterior boundary of the "thinking" gray matter, was located very far back and down. This particular groove, called the lunate sulcus, is located as far posteriorly and down on the human brain as seems anatomically possible.

Dart was also struck by how large, forwardly placed, and close together the bony orbits and, consequently, the eyes of the Taung child had been. There must have been, he speculated, a considerable amount of cerebral cortex tied up in vision and the processing of information about the surroundings demanded by an emphasis on sight. In fact, he thought, the cerebral cortex of the Taung child would have resounded with sensory stimulation: visual input from its eyes and tactile sensation from its hands. As such, he wrote, the hands were "being freed from their more primitive function of accessory organs of locomotion . . . were assuming a higher evolutionary role . . . as instruments of the growing intelligence." As Dart envisioned it, "the group of beings" to which the Taung child belonged could have seen, heard, and touched "with greater meaning and to fuller purpose than . . . recent apes." In short, this group "had laid down the foundations of that discriminative knowledge of the appearance, feeling, and sound of things that was a necessary milestone in the acquisition of articulate speech."

When all was said and done, Dart was convinced that the Taung child had come from a "group of beings" that indeed represented "an extinct race of apes *intermediate between living anthropoids and man.*" What about the other fossil hominids that were known at the time? Dart dismissed Dawson and Smith Woodward's *Eoanthropus dawsoni* from consideration on the basis of the configuration of its lower jaw. In *Eoanthropus,* the front of the lower jaw had sloped down and back and extended farther posteriorly as an inferior shelf of bone. Because a similar inferior shelf of bone is often found adorning the mandibles of great apes, it is referred to as a "simian shelf." In Dart's eyes, the presence of a simian shelf in *Eoanthropus,* in conjunction with the apelike slope of the front of its jaw, excluded this fossil as

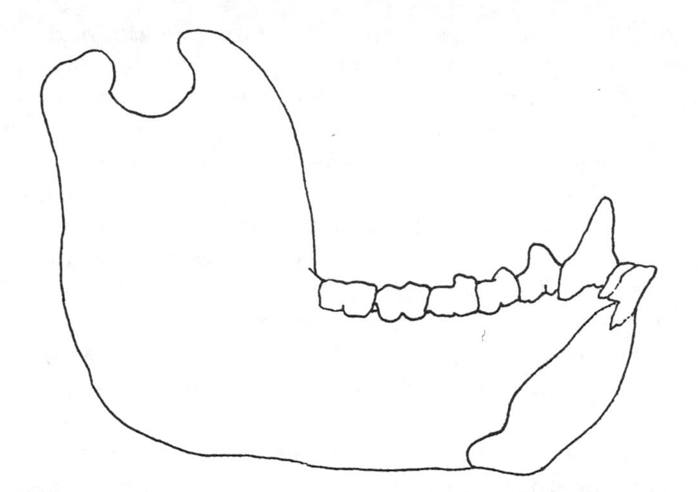

A cross section down the middle of a gorilla's lower jaw. The simian shelf is the moderate backward extension at the very bottom of the cross section.

a serious contender in human evolution. In contrast to *Eoanthropus* and apes, however, the front of the Taung child's lower jaw, like that of the Mauer jaw, was quite humanlike, being decidedly vertical in its orientation. Rather than bearing a simian shelf, the lower margin of the inside of the front of the Taung child's mandible was rounded, again as in the Mauer jaw.

As for Dubois's *Pithecanthropus erectus* being a viable fossil link between humans and their ape ancestors, Dart was even less restrained in his choice of words:

> Unlike Pithecanthropus, it [the Taung humanoid] does not represent an ape-like man, a caricature of precocious hominid failure, but a creature well advanced beyond modern anthropoids in just those characters, facial and cerebral, which are to be anticipated in an extinct link between man and his simian ancestor. At the same time, it is equally evident that a creature with anthropoid brain capacity, and lacking the distinctive, localised temporal expansions which appear to be concomitant with and necessary to articular man, is no true man. It is therefore logically regarded as a man-like ape.

In what might be regarded as a more optimistic evolutionary outlook than was reflected in Dubois's characterization of his Javanese hominid as an "ape-man" (the term itself reflects a prolonged retention of primitive simian features during human evolution), Dart referred to his Taung humanoid as a "man-like ape," in obvious reflection of the early establishment of features of human uniqueness. But in spite of acknowledging the presence of very humanlike features in the Taung humanoid, Dart continued, in the tradition of Linnaeus and others, to use language as the criterion that distinguishes a "true man" from all other contenders.

In a move that would raise the hackles of the majority of the bigwigs in the profession, Dart proposed that his humanoid—this "extinct race of apes *intermediate between living anthropoids and man*"—deserved to be placed taxonomically not only in a new genus and species but also taxonomically alone in a new family, one that subsumed this new genus and species. In recognition of this fossil humanoid's presumed intermediacy between humans and apes, Dart offered Homo-simiadae as the new family's name. As for the genus and species names, Dart coined *Australopithecus africanus* "in commemoration, first, of the extreme southern and unexpected horizon of its discovery, and secondly, of the continent in which so many new and important discoveries connected with the early history of man have recently been made, thus vindicating the Darwinian claim that Africa would prove to be the cradle of mankind."

But Dart did not stop with his proclaimed vindication of Darwin's setting human origins in Africa. He restated Darwin's selectionist argument for choosing Africa over Asia as the seat of such an evolutionary possibility:

> In anticipating the discovery of the true links between the apes and man in tropical countries, there has been a tendency to overlook the fact that, in the luxuriant forest of the tropical belts, Nature was supplying with profligate and lavish hand an easy and sluggish solution, by adaptive specialisation, of the problem of existence in creatures so well equipped mentally as living anthropoids are. For the production of man a different apprenticeship was needed to sharpen the wits and quicken the higher manifestations of intellect—a more open veldt country where competition was keener between swiftness and stealth, and where adroitness of thinking and movement played a preponderating rôle in the preservation of the species. Darwin has said, "no country in the world abounds in a greater degree with dangerous beasts than Southern Africa," and, in my opinion, Southern Africa, by providing a vast open country with occasional wooded belts and a relative scarcity of water, together with a fierce and bitter mammalian competition, furnished a laboratory such as was essential to this penultimate phase of human evolution.

Although in the years to come Dart would find himself defending his claims of and about *Australopithecus africanus,* he held firm to his convictions. In 1936, Robert Broom, a physician by training and for financial reasons but a world-renowned vertebrate paleontologist by avocation, published an article announcing the discovery of another fossil man-ape skull, this time in the Transvaal, at the limeworks at Sterkfontein. Although he referred this specimen to a new species, *transvaalensis,* Broom was sufficiently convinced of its relationships to Dart's Taung "group of beings" that he kept it in the genus *Australopithecus.* From then on, right through to the present, fossil hominids were a reality in the paleoanthropological landscape of South Africa and would eventually become so in other parts of Africa, most notably East Africa. Even if the reasons Darwin had given for why human evolution had to have occurred in Africa rather than in Asia were at best imaginary, and at worst racist, it is an indisputable fact that the earliest hominid fossils forthcoming are from the African continent.

But Dart interpreted his discovery—and for many decades to follow subsequent discoveries in South Africa would likewise be interpreted—as indicating that the driving force in human evolution had been an inextricably linked interplay between the evolution of human bipedalism and the human brain. For in the Taung child, so it seemed to Dart, these attributes, which had for so long been considered *the* bastions of humanity—the unassailable attributes of human uniqueness—were both present and necessary to each other's existence. Whereas *Eoanthropus* and *Pithecanthropus* were thrown up as alternative solutions to the question "Which evolved first, the body or the brain?" *Australopithecus africanus* provided yet a third resolution: These events happened simultaneously. For many decades after Dart's musings, this did indeed seem to have been the way the early phase of human evolution had happened.

5

Humans as Embryos

> *[T]he government which rules within the body of the human embryo proceeds along its way altogether uninfluenced by occurrences or experiences which affect the body or brain of its parents. . . . [M]an has come by his great gifts not by any effort of his own, but . . . has fallen heir to a fortune for which he has never laboured.*
>
> —Sir Arthur Keith (1928)

Taung Tied

Hindsight is perhaps the greatest of human inventions. With it one can sweep away a history of problems and inconveniences and leave in place only those aspects of, say, a person's life or ideas that are pleasant or fit one's own predispositions. For instance, Darwin and Huxley are frequently cited by paleoanthropologists as having demonstrated scientifically and by comparative anatomy that the African apes are the closest living relatives of *Homo sapiens.* But Darwin and Huxley did no such thing. Huxley sought to argue that humans are not so unique that they should not be classified with other mammals and, specifically, within a group of mammals—in this case primates. Although he concluded that humans are more similar to the tailless apes (apes) than to the tailed apes (monkeys), and that they are often quite similar to the gorilla among the tailless apes, Huxley was content to keep humans distanced from the apes by putting them in their own taxonomic family.

Darwin is another story altogether. He used a collection of different arguments, none of which was morphological, to set the stage for a totally unsupported, and unexpected, conclusion that the African apes are the closest living relatives of modern humans. In this particular case, Darwin relied on his beliefs that one would find the ancestors of closely related species in the same geographic area and that only certain conditions would provide the necessary impetus for natural selection to work. As for humans and apes, there was the orangutan in the tropical rain forests of Southeast Asia and the chimpanzee and gorilla in the tropical rain forests of Africa, and humans in both places. But it was only the harsh environment of southern Africa, with its savannas and predatory animals, that Darwin thought held the uncertainties that would allow natural selection to provoke the evolution of an animal as special as *Homo sapiens.* In addition, as Darwin saw it, the Africans were primitive humans and served as a link between his concept of an apelike human ancestor and truly civilized humans. Since apes were to be found in Africa along with primitive humans, this also satisfied Darwin's other belief, which was that closely related species will be in relatively close geographic proximity. The problem with this part of the scenario, however, is that African apes are actually found thousands of kilometers to the north of southern Africa, which is where Darwin set his scenario for human origins.

It may be true that Darwin did conclude that the African apes are more closely related to humans than any other ape. But, even if this relationship is correct, the bases upon which Darwin built his argument are not acceptable—which is a point that never comes up these days when support for one's ideas is via invoking Darwin. How scientific is it to seek validation of one's own ideas in a conclusion based on assumptions that no one today would take seriously—even if the author of that conclusion was Charles Darwin? There is a big difference between someone being of historical importance, which without a doubt Charles Darwin is, and one's ideas being forever unassailable, which no one's are.

As for Dart and his interpretation of the Taung child, texts and popular books also tend to gloss over the inconvenient parts. Like Darwin, whom he invoked in support of his contention that human evolution began in southern Africa, Dart may have come to a conclusion that later enjoyed some popularity. But, also like Darwin in this instance, Dart's conclusion was not grounded scientifically. Although many paleoanthropologists eventually came to embrace *Australopithecus africanus* as an ancestor of all other hominids, Dart's claims for this hominid derived from two arenas of speculation. One was the impact the Taung child's somewhat humanlike brain had on its state of being: its mental attributes and feelings. The other area in which Dart went from one assumption to another was in terms of the proposed intermediate evolutionary relationship of the Taung child between the apes, on the one hand, and humans, on the other. The very fact that Dart could make what he considered to be viable comparisons between the Taung child and adult modern humans highlights a particular premise—namely, that the features of the Taung child were not that dissimilar from those of the adult into which it would have grown. Consequently, the fossil child could, within reason, substitute for an adult of its species. In contrast, Dart found his comparisons between the Taung child and African apes, whether child or adult, less compelling.

It is interesting that Dart chose to imbue his Taung child with the properties of an animal whose features would not change markedly as it matured physically, because this is not a frequent happenstance among organisms. Indeed, most animals have an early phase of development, during which they look like a juvenile, with big heads, big eyes, and small faces, and then they become transformed into the configuration of the adult, often with longer snouts, smaller and beadier eyes, and relatively smaller heads. Apes, especially the African apes, are like the majority of animals in this respect. But humans are among the few animals who do not become drastically reconfigured away from the certain shapes and proportions they have as juveniles. And it is to this developmental pathway that Dart assigned his fossil child.

On Becoming Childlike: Pushing Infants into Sex

An organism's final body shape and relative proportions, and even its detailed morphology, do not always look like an adult's. Sometimes the adult looks like a child. The reason such an organism is thought of as an adult is that it is sexually mature. Biologically, reaching sexual, or reproductive, maturity has the effect of curtailing physical growth and maturation. In an ideal biological world, we would imagine that an organism would grow out of its juvenile physical phase and grow into the configuration of an adult, and that, somewhat coincident with the latter, the individual would become reproductively mature. But this is not always an ideal world, biological or otherwise.

Rates of physical maturation can be slowed or speeded up, and so can rates of sexual maturation. This means that there are two major courses by which the final morphology of the organism can be achieved. One is by speeding up or slowing down the rate at which an individual achieves sexual maturity, but doing so while keeping constant the rate at which the organism matures physically. The other way is by accelerating or decelerating the rate at which an organism matures physically but not altering the rate of sexual maturation. The only given in these possible equations is that, upon reaching reproductive maturity, an organism essentially becomes frozen in the physical state it is in at that moment in time. The interplay between rates of physical and reproductive maturation is referred to as heterochrony, which literally means "different times" or "timings." Clearly, changes in the rate of one component—physical or reproductive—will have a profound, instantaneous, and long-lasting effect on a developing organism.

Although they are not numerous in nature, there are some species of animal in which the adult retains, rather than grows away from, features characteristic of the infant. The phenomenon of remaining childlike—referred to as paedomorphosis—can be achieved in one of two ways. One way is called neoteny and the other progenesis. In progenesis, the rate of sexual maturation is accelerated relative to the rate at which the body is growing. Since the process of maturing physically will cease upon reaching sexual maturity, the individual will remain childlike in its physical features at the time reproductive maturity is reached. Amphibians, such as newts, provide a good example of the process of becoming paedomorphic by way of progenesis.

Many amphibians, including frogs, have external gills and long tails as juveniles. During this early phase of life, many amphibians live in water, where gills assist in respiration and the long tail can easily propel the individual through its aqueous habitat. For various amphibians, this juvenile phase is also referred to as the tadpole stage, and for others it is the larval stage of development. At some point in their life cycle, many amphibians, frogs included, pass from this juvenile phase into an adult phase. This transition is noted physically in the loss of the external gills and the tail. There may also be lifestyle changes. For example, as is commonly seen in frogs, at least some part of adult life is spent out of water and on dry land. With the metamorphosis from the juvenile to the adult phase, the animal also becomes sexually mature.

As it turns out, various organisms—insects and amphibians among them—are developmentally so sensitive to their environment that shifts in life-sustaining conditions can trigger changes in the timing of reaching sexual versus physical maturation. In the case of amphibians, many of whose eggs can incubate and hatch only in an aqueous environment, any threat to the persistence of their moist surroundings can greatly affect the viability of the next generation. In some salamanders—the tiger, for instance—the pace of sexual maturation will accelerate if their habitat becomes too dry too quickly for the normal rate of sexual maturation and reproduction to establish the next generation of salamanders. The potential to accelerate the rate at which sexual maturation occurs translates into individuals being able to reproduce even when they are still physically in the juvenile state, replete with external gills and a long tail. Having attained sexual maturity earlier than is typical for their species, these reproductively competent individuals will cease maturing physically, staying for the remainder of their short lives in their juvenile bodies. If the next generation of salamanders finds itself in moister, less survival-threatening conditions, individuals will revert to the norm and go through the more typical time frame of attaining sexual maturity while they pass through the phase of juvenile physical growth and then into the phase of adult physical growth. There are, however, some salamanders—such as species of the genus *Proteus,* and of the genus *Necturus*—that have become developmentally stuck in their juvenile growth phases. In *Proteus* and *Necturus,* individuals spend their entire reproductive lives with external gills and long tails, even though the environmental trigger—probably a prolonged period of dryness—that caused earlier generations to accelerate sexual maturation and truncate physical growth is no longer a threat to these animals.

The possible significance of alterations in developmental rates and patterns for understanding the questions of how evolutionary change occurs and how it is enacted was not lost on late-nineteenth- and early-twentieth-

A tiger salamander in the larval, or axolotl, stage of development (above) and a physically and sexually adult tiger salamander (below). Under normal environmental conditions, the axolotl develops physically and sexually into the adult form. But adverse environmental conditions can trigger an acceleration in the rate of reproductive maturation, which will freeze the animal in its physically juvenile state.

century embryologists. Although he was on the wrong track from the very beginning—believing that an individual's embryonic stages represented the adult forms of its ancestors—Ernst Haeckel was, nonetheless, cognizant of the possibility that a change in one part of the organism's developmental sequence, particularly if it came at the very beginning, could cause profound change by redirecting the ontogenetic trajectory. There were also a number of embryologists who believed that studying an organism's development provided clues not only to the ontogeny of the individual but also, and much more broadly, to the origin of major groups of organisms. Among the most perceptive of these scientists was Walter Garstang.

As early as the late nineteenth century, but without identifying it as such, Garstang invoked the process of progenesis to explain the evolutionary emergence of an entire class of animals, Vertebrata. The members of this diverse group are distinguished by the development of an internalized skeleton, which, in turn, is configured around a unique midline supporting struc-

An adult Necturus, *or mud puppy. This amphibian is always in the physically juvenile state.*

ture: the backbone, or vertebral column. In addition, vertebrates are clearly distinguished from invertebrates by the position of the main nerve trunk. In invertebrates, this central nervous system lies along the belly side of the body. In vertebrates, this nerve trunk—the spinal cord—is on the opposite side of the body, lying parallel to and above the bodies of the vertebral column. The question of the origin of vertebrates was a problem for paleontologists because the earliest known members of this group appear abruptly in the fossil record, without a trail of transitional intermediates leading to them from a more primitive, presumably invertebrate, ancestor.

For comparative anatomists who study only livings forms, Vertebrata constitutes a group with no easily envisioned connections to other living, complex, multicellular organisms. Even the most primitive animals with a supporting structure coursing along the midline, such as the aquatic, almost eel-like lancelet, have a stiffening rod of cartilage that extends the full length of the organism, from the tip of the snout to the end of its tail. Animals like the lancelet have been grouped with full-fledged vertebrates with vertebral columns as chordates because they all possess a stiffening rod—a notochord—at some stage during their development. The lancelet retains the notochord as an adult. Vertebrates transform the notochord into the vertebral column. But knowing this does not clarify the problem of the evolution of chordates. Where was the possible link among invertebrates to chordates? Not, Garstang pointed out in a series of articles, among any adult invertebrates or invertebrate-like animals. Rather, the best comparisons with chordates were to be found among the larval, or juvenile, stages of particular species of an invertebrate-like organism, the tunicate or sea squirt. For all intents and purposes, these larval tunicates are so morphologically and behaviorally dissimilar from the adult into which they become transformed that one would think they were different, unrelated species.

The adult tunicate is a barnacle-like animal that filter-feeds for small water-dwelling organisms. Among its other morphological configurations, its digestive tract curves around so that its anus is located near its mouth. Like a barnacle, an adult tunicate is a sessile animal, meaning that it spends its adult life anchored in one place. And, also like barnacles, these sessile adult tunicates form colonies that can sometimes reach quite a significant size. Although many sessile invertebrates live out their adulthood permanently secured to one spot, they spend their larval lives as free-swimming individuals. And such is the case with various species of tunicate.

In the free-swimming larval form, a tunicate's body looks pretty much like a tadpole without legs. The larval tunicate's body is straight, with its mouth at the front end and its anus away from it, near the base of the tail. The free-swimming larval form also has paired structures, such as right and left gill slits, and right and left muscle bundles along its flanks and tail, which makes it a bilaterally symmetrical organism, as are vertebrates. Even more amazing for a nonvertebrate is the fact that the free-swimming larval tunicate has a notochord-like stiffening rod that resides in the upper, or spinal,

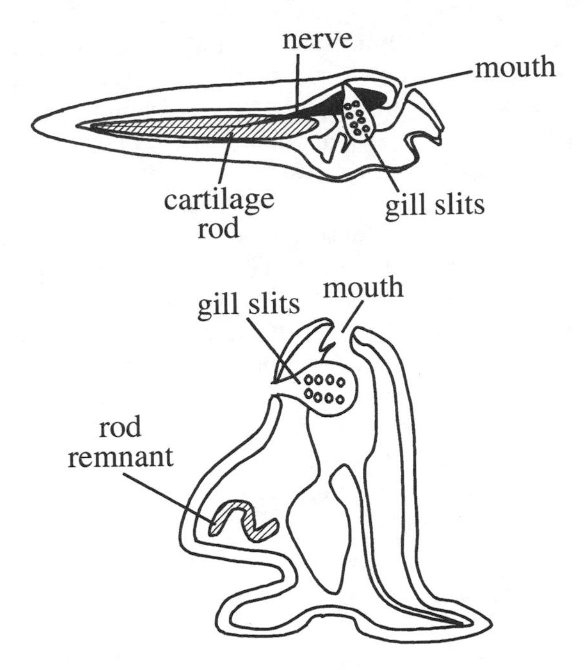

Simplified diagram of a tunicate larva (above) and an adult (below). In the larva, a cartilage rod and a major nerve trunk course along the "back" but do not maintain this position or their relationship in the adult, in which the rod becomes vestigial. In the larva, the mouth and the anus (which would exit near the tail) are separated but come to be close together in the adult.

side of its back and its main nerve trunk lies above and parallel to it. Perhaps more astonishing still, the free-swimming larval tunicate develops a groove along its back that, as Garstang argued quite elegantly, would be a necessary precursor to the development of the chordate spinal cord. When the tunicate passes from the free-swimming juvenile to the sessile adult phase, which occurs when it becomes reproductively mature, it loses the stiffening rod, becomes asymmetrical in form, and essentially doubles over, bringing its anus and mouth into close proximity and reconfiguring the relationships of its internal organs.

But what if something happened that kept the juvenile tunicate in its larval state while allowing it to be a reproductively viable organism? Clearly, Garstang suggested, such a developmentally modified organism would make a perfect ancestor for living chordates. And the mechanism by which this could occur would be progenesis: the persistence of juvenile features caused by the acceleration of the pace of sexual maturation. In short, the ancestral chordate could very easily have been a paedomorphic tunicate or tunicate-like animal, caught in a perpetual larval state by its ability to reproduce. Even a century later, Garstang's model of vertebrate origins remains a viable theory.

In contrast to progenesis, another way in which an organism can become paedomorphic is by slowing down, or decelerating, the rate of physical maturation while maintaining the normal rate at which sexual maturity is attained. This process prolongs the juvenile growth phase and, consequently, retards the onset of adult growth rates. Because, in this developmental configuration, it will take longer for an individual to go through the juvenile phase of physical development, there will be little or no chance of a transition to the adult phase prior to reaching sexual maturity. Since physical maturation ceases once reproduction is possible, the result is an individual that looks like a child physically but can, nonetheless, mate successfully. This particular mode of paedomorphosis is called neoteny. Although there is a diversity of organisms that can serve to illustrate the process of neoteny, the most frequently studied animal in this regard has been our own species, *Homo sapiens*—a species that, compared with most other mammals, does take a relatively long time to mature reproductively but also remains physically childlike in many features.

Humans as Neotenic Animals

One of the first scientists to conceptualize the potential evolutionary importance of heterochrony—of the effects of differential rates of reproductive and physical development—was the Dutch embryologist and developmental anatomist Louis L. Bolk. Although his basic, descriptive embryological studies and publications were well known and very respected through the latter decades of the nineteenth and into the early twentieth century, Bolk did not publish the first of his few but influential treatises on the subject of the interplay between developmental rates until 1915. His grand synthesis and overview appeared in print in 1926, just four years before his death at the age of sixty-four. Building on the suggestions of the early-nineteenth-century French naturalists and comparative anatomists Georges-Louis de Buffon and Étienne Geoffroy Saint-Hilaire, who had long before recognized the extreme similarities in skull shape between adult humans and juvenile apes, particularly orangutans, Bolk was determined to demonstrate the existence of such similarities in all aspects of adult human and juvenile ape anatomy.

For instance, one of Bolk's favorite examples had to do with the angle between the plane of the base of the skull, through which the spinal cord passes, and the plane of the face. As is typical of mammals in general, in human and ape fetuses the angle between the base of the skull and the face is essentially ninety degrees, a right angle. This is a pretty tight angle, which Bolk described as producing an extreme flexure between the facial skeleton and the base of the skull. In addition, because the base of the skull faces down relative to the face, the fetus's head sits atop the vertebral column and the facial skeleton lies at right angles to it.

In apes, as in most other mammals, the angle between the skull base and the facial skeleton opens up and the spatial relationship of these regions

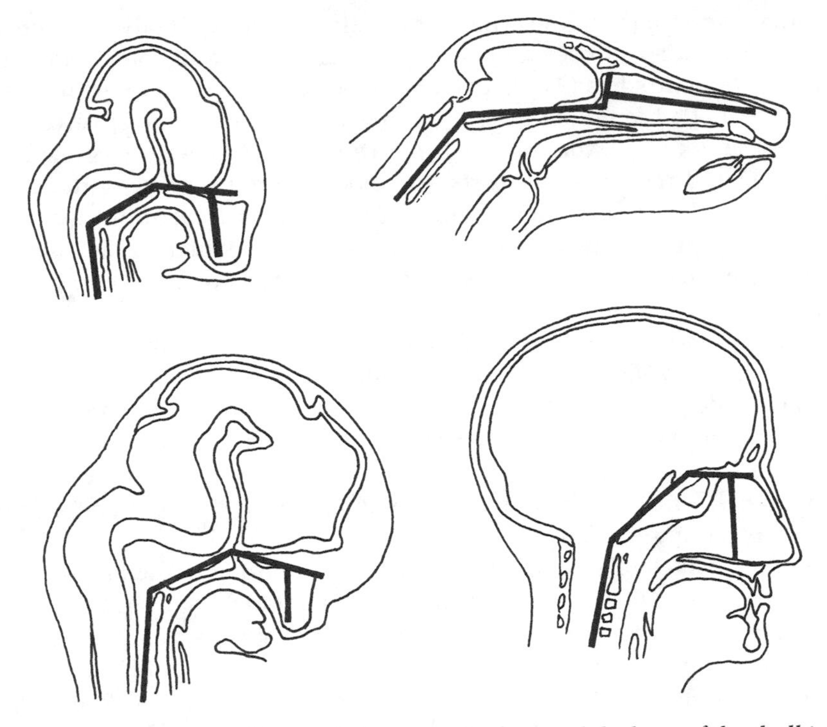

Bolk's representation of the angle between the face and the base of the skull in a dog (above) and a human (below). In the fetus (left column), the angle between is markedly flexed. In the adult dog (above right), this angle opens up. In the adult human (below right), the angle remains well flexed. (Adapted from Bolk)

changes. With growth, the base of the skull rotates away from the enlarging snout, with the result that the hole through which the spinal cord exits the skull—the foramen magnum—and the two points of articulation on the skull base for the first vertebra—the occipital condyles—point backward to varying degrees, sometimes even ending up aligned with the snout. But in humans the severe angle between the face and the base of the skull does not change with physical growth. As in the fetus, the adult human's foramen magnum and its associated occipital condyles continue to lie centrally on the base of the skull. In turn, this means both that the adult skull continues to sit right on top of the vertical vertebral column and that the face continues to lie at a right angle to the vertebral column. Of special note, of course, is that the orientation of the vertebral column in humans is associated with their bipedal form of locomotion. As Bolk envisioned it, an adult human is essentially an upright-walking embryo. And if this reflects a prolongation of juvenile growth rates into adulthood, then humans are indeed neotenic.

An anatomist, asked to point to the most striking feature of the typical vertebrate embryo, would surely note how large and dominant the head region is compared with the other parts of the developing body. Anyone who has seen newborn guppies, puppies, and even calves and colts, knows how large the neonate's head is relative to its body size. And if the head is large, so, too, must be the brain that is housed within it. But for most vertebrates, being big in skull and brain is fleeting. During their postnatal growth, the developmental rates of other parts of the body outstrip those of the cranial vault and the brain contained within it. The result is an adult whose head—the cranial vault part, not necessarily the snout, which often increases markedly in size—is now much smaller relative to body size than it was as an infant.

The exception to this generalization, at least among living land mammals, is *Homo sapiens.* Our huge brain—large compared with body size—is as good evidence as any of neoteny. In humans, the brain maintains its embryonic growth trajectory and continues to enlarge through the teens and even into the early twenties. Clearly, in humans the shift to an adult brain-growth rate, which results in the cessation of cerebral enlargement, has been greatly retarded.

From the skeletal side of things, Bolk paid attention to the joints, or contact zones, also called sutures, between the individual bones of the skull. As a general rule, the cranial sutures fuse when the brain has stopped growing. For most mammals, this occurs early in life. But typically in humans, these sutures, particularly of the bones surrounding the brain itself, remain quite patent, and often unfused, even well into adulthood. This Bolk took as evidence of human neoteny. A brain that is not only large but also takes much of an individual's lifetime to reach maximum size would certainly be put in jeopardy if the skull did not enlarge in synchrony with it. In cases where one of the cranial vault sutures closes prematurely, there is compensatory growth. The cranium becomes exaggerated in directions that are at right angles to the fused suture, with the result that head shape is altered. But when all sutures fuse prematurely, and the brain keeps growing, the result is often fatal: The brain stem gets squeezed into the foramen magnum at the base of the skull, which is the only space through which expansion would be possible.

Among the many other adult human features Bolk pointed to as being juvenile retentions, those that are particularly relevant to a history of the field of paleoanthropology and speculations about human origins include the possession of a relatively flat face, the restricted distribution of body hair, and the extraordinarily long period of time over which the teeth, especially the molars, erupt into the jaws.

According to Bolk, and as was later echoed by his intellectual successor Sir Gavin Rylands de Beer, who actually had the better grasp on the relations between heterochrony, development, and evolution, neoteny can account for humans having not only a relatively flat lower face or snout but also an upper face that is characterized by a lack of prominent brow ridges. Given the popularity of documentaries on primates, and the emphasis in zoos on breeding colonies of primates, it is simplicity itself to verify the fact that newborn primates have relatively, if not extremely, short faces, and that no known living primate—not even a gorilla or a chimpanzee—is born with any sign of brow-ridge development. If you can gain access to a museum's primate skeleton collection, take out at random the skulls of the neonates of any species of lemur, monkey, or ape, and compare them with the skull of a newborn human. You will find that in all species, regardless of how long the snout of the adult may be, the snout of the infant is remarkably much shorter; in some cases, the lower face does not even seem to project forward. As for the upper face, not only are the upper margins of the bony eye sockets of all these neonatal primates flat from side to side but they also grade smoothly into the typically vertical and often domed frontal bone. When it is viewed in profile, the large cranial vault of neonatal anthropoid primates is quite rounded, especially compared with the adult's.

But while the numerous species of juvenile anthropoid primate are so similar to one another in many features—such as in having short snouts, undistended superior orbital margins, high and domed foreheads, and relatively rounded cranial vaults—there are definite differences among the adults. To varying degrees, the forehead of an adult anthropoid primate is typically much lower and slanted backward and the skull in profile much more oblong than in the neonate. Such an ontogenetic change in cranial shape is particularly noted in the African apes as well as in many Old World monkeys, such as the mandrill and the various species of baboon. In contrast, adult orangutans and humans, the gibbons and siamangs, as well as various species of New World monkey, tend to preserve the vertical forehead and more globular cranium of the juvenile.

Also, to varying degrees, the snout of an adult anthropoid primate is longer—sometimes remarkably longer—than the juvenile's. In all large-bodied apes and in many Old and some New World monkey species, the snout of the adult is incredibly elongate compared with its size at birth. Even in the shorter-snouted anthropoids, such as the gibbons and siamangs and many New World monkeys, lower facial protrusion is relatively more pronounced in adults than in neonates. In humans, however, although the lower face lengthens vertically, it does not grow appreciably forward.

All neonatal anthropoid primates are essentially similar to one another in the general configuration of the upper face. But this is not so with the adults. Anthropoid infants lack brow-ridge development, or any supraorbital adornment, for that matter. But the superior orbital margins of adult anthropoid primates are often thickened and protrusive in various ways. Some species of Old World monkey, such as the baboon and the mandrill, have straight, barlike brow ridges coursing from side to side across their eye sockets, or orbits. In outline, the orbit itself looks like the capital letter *D* lying on its rounded side. In profile, this barlike type of brow ridge protrudes straight forward, like a mini-visor. In chimpanzees and gorillas, the thickened brow ridge follows the arc of each orbital margin, producing in some

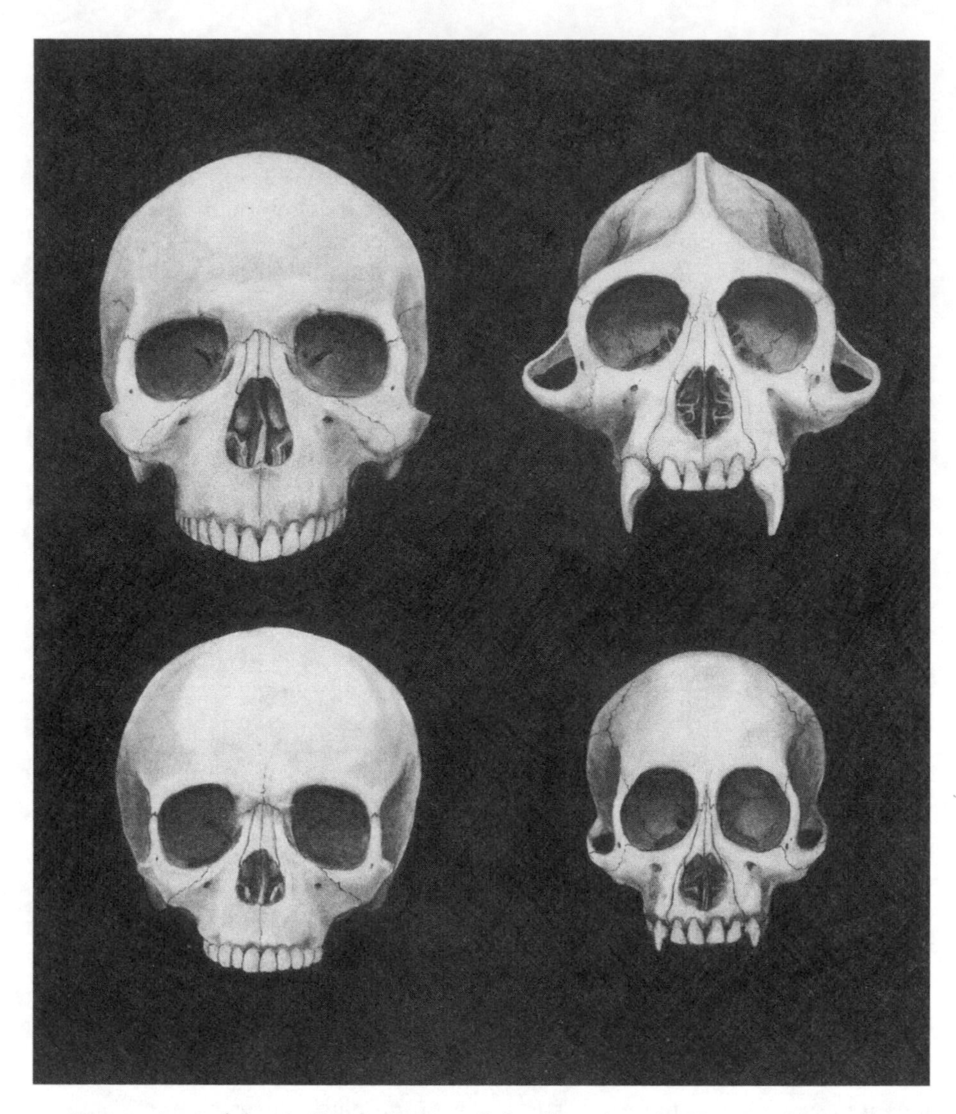

Comparison of the skulls of a human (left column) and a New World monkey (representing most anthropoid primates), the capuchin monkey (right column). The juveniles of the human and the monkey (bottom) are similar in that both have a small lower face, a relatively unprojecting snout, large eyes, and a large rounded head. The adult human typically retains this juvenile configuration, while the adults of other mammals, like the monkey, usually grow away from the juvenile configuration, developing, for instance, larger and more projecting lower faces and brow ridges.

individuals a well-defined double-arched configuration; viewed from the side, the brow ridge of African apes projects not only forward but also upward, producing a furrow between it and the sloping forehead. In the lesser apes—the gibbons and the siamangs—the entire orbital margin grows outward and forward, like a funnel, resulting in an adult animal that looks as if it were wearing swimming goggles. And in adult orangutans each orbit is ringed by a low mound of bone, which gives the adult skull the appearance of being frozen in a state of raised-eyebrow amazement. Because orangutan orbits are separated from each other by only a pair of extremely thin nasal bones, the inner portions of the growing right and left orbital mounds often become pressed against each other in the midline.

Adult *Homo sapiens* typically do not have much supraorbital enlargement. In fact, most of us develop only a moundlike swelling that looks vaguely like a pair of wings. The central and often most prominent part of this swelling occupies the region above the nose and between the eyebrows. The two "wings," which lie underneath the eyebrows, extend about halfway over the orbits. You can feel the swelling on yourself. You can also feel the transition from the end of each swelling to a flatter, somewhat backwardly angled plane of bone. In women, who tend to have higher and more vertical foreheads than males, the supraorbital regions often flow smoothly upward into the frontal bone. In men, the supraorbital swellings are often more prominent, and the forehead also often more receding, than in females; the more receding the forehead, the more pronounced any supraorbital swelling will appear to be.

If an animal is going to have brow ridges, they will begin to emerge during the transition from juvenile to adult growth rates. As such, Bolk and de Beer argued, the lack of significant supraorbital elaboration in *Homo sapiens* is further evidence of this species' neotenic tendencies. In spite of the occasional Australian Aborigine skull with enlarged supraorbital rims and a sloping forehead that a Thomas Huxley, for instance, could drag out of a museum cabinet, humans are in general still more fetuslike than not in having a relatively vertical frontal bone and minimal supraorbital elaboration.

As for human neoteny and hairlessness, Bolk offered the following observations. Although monkeys are completely hairy as adults, as newborns, they have hair only on their heads and on their backs. The hairy adult gorilla is born in an essentially hairless state, with hair adorning only its head. Newborn humans also have hair only on their heads, and, all things being relative, adult humans are not that much hairier—certainly compared with adult monkeys and apes. As Bolk saw it, the rate at which hair, representing physical features in general, spreads to cover the body has, in contrast to other primates, become markedly retarded in humans. And, as is fundamental to being a neotenic animal, the reason why humans are not totally covered with hair in old age is that their physical maturation is curtailed when they become reproductively viable.

The ultimate example of human neoteny for Bolk was the fact that, compared with other primates, it takes a long time for the teeth of humans to erupt. As in most mammals, humans and other primates have a first set of teeth, the deciduous, or milk, teeth and a second, or permanent set of teeth. The three permanent molars (the third is the so-called wisdom tooth) erupt one behind the other, beginning in the space behind the last deciduous tooth. All other permanent teeth replace deciduous teeth. In apes, the deciduous teeth break through the gums relatively soon after birth. In humans, the deciduous teeth are still erupting well into the second year of life. In apes, the molars erupt shortly after the deciduous teeth have done so, and the last teeth to come into an ape's jaws are the canines, usually the upper canines. In humans, the first permanent molar does not erupt until well into childhood, doing so on average during the sixth year (which is the criterion Dart used to age the Taung specimen). Eruption of the second human molar is delayed until the twelfth year. And the human third molar, though it sometimes erupts as early as eighteen years of age, most frequently begins to break through the gums during the twenties, well after all other permanent teeth have done so.

Although most evolutionists saw the effects of evolution as a mosaic—in an evolving lineage, some parts or features changed more rapidly or slowly than others, while some appeared to remain unchanged for aeons—Bolk believed strongly that every aspect of a human could be accounted for by a prolongation of the juvenile phase of development and, consequently, of a retardation of the onset of adult growth. He conceived of this as a process of fetalization, through which all physical parts of an individual remained in the fetal stage of development. As far as he was concerned, an adult human was nothing more than a fetus that had become sexually mature. Clearly, this is a rather extreme view of neoteny. So extreme, in fact, that Sir Gavin de Beer, who was otherwise unrestrained in his review of Bolk's ideas about human neoteny, and unabashed in embracing those suggestions he thought most valuable, did not comment at all on the latter's view of humans as reproducing fetuses. (Many years later, the evolutionary theorist Stephen Jay Gould felt compelled to put closure on this aspect of Bolk's contributions and state the obvious: A human is not a fetus that can reproduce.) De Beer did reject Bolk's term *fetalization,* preferring, instead, to use the term *neoteny* for all organisms thought to be paedomorphic by the prolongation of juvenile growth rates, because not all organisms that can be described by this process—particularly insects—have a fetal stage of development.

Pushing the Form into a Function

Not restricting themselves merely to describing the morphology, Bolk and, later, de Beer offered functional explanations for the absence of pronounced brow ridges in *Homo sapiens* and for their presence in apes and Neanderthals. Although it attracted a large and sympathetic audience, Bolk and de Beer's quest for a reason behind a feature came from D'Arcy Wentworth Thompson, who was the founder of the field of transformational growth studies. Embryologists such as Bolk and de Beer sought to understand which features did or did not change in shape or relative size during growth. But Thompson was interested in the amount, or degree, of change as well as the impact of that change. If one focused on a particular feature or attribute, the question was "How would change in its shape or size affect the total organism?" Or, if one looked at the big picture, the question was "How would individual structures change if the whole organism was transformed in some way, such as size?"

The underlying assumption in transformational growth studies is that all features have a specific purpose. When Thompsonites try to tackle the question "Why brow ridges?" it is not sufficient to acknowledge the existence of these structures and to compare their shapes and ontogenies between species. They have to try to figure out the raison d'être for brow ridges. And the explanation of purpose that became the functional morphologist's favorite, whether it be for an ape or a Neanderthal, was the presumed role of enlarged brow ridges in buttressing the upper face against the extreme bite forces that would be generated during chewing by a massive lower jaw, which, in turn, would have to be moved through its excursions by powerful chewing muscles.

As some morphologists see it, the functional argument for the presence of prominent brow ridges in the African apes is fairly straightforward. Both sexes of chimpanzees and gorillas, and especially the large male gorillas, have massive and protrusive lower jaws that are tethered to the skull by large muscles involved in chewing, or mastication. And both sexes of chimpanzees and gorillas, especially the larger males, have pronounced brow ridges. Bite forces produced at the front teeth would be great, and would have to be absorbed or deflected somewhere in the facial skeleton. A thickened strut of bone, at the otherwise weakened angle between the facial skeleton and the backwardly sloping forehead, would be functionally in the right place at the right time.

The existence of thickened orbital margins—producing what we call brow ridges—then became entwined with notions of buttressing and dissipating bite forces. But why don't orangutans, with their large and protrusive lower jaws and large chewing muscles, have prominent brow ridges? The reason given was that orangutans have a vertical forehead. Consequently, bite forces would travel up the orangutan's face, over the orbital rims, and become dispersed throughout the vertical forehead. Since an angle of vulnerability does not exist between the orangutan's lower face and the cranial vault, there is no reason for brow ridges to develop.

Brow-ridged Neanderthals are another matter altogether. They did not have large lower jaws or massive chewing muscles. Although somewhat larger than a modern human's lower jaw, the small-compared-with-an-ape's Neanderthal mandible begs the question of a functional relationship between brow-ridge development and a large lower jaw and powerful muscular chewing apparatus. Even the evidence provided by the faint scars of muscle attachment on the Neanderthal cranial vault do not support the notion that significant bite forces were generated when they chewed.

On Neanderthal skulls, chewing muscle-attachment scars are typically faintly impressed on the bone rather than being ridgelike, elevated scars. In the great apes, as well as in many monkeys, the temporal muscles originating from the top of each side of the mandible attach high up on the side of the skull, and in some individuals. If these muscles meet at the middle of the top of the skull, as they do in male gorillas, they induce the bone to produce an elevated and keel-like sagittal crest, which courses for some distance along the midline sagittal suture. But Neanderthal temporal muscles were relatively small and, as indicated by the attachments scars, confined to a low position on the side of the skull. The Neanderthal masseter muscle, which in all mammals originates at the lower back corner of the mandible and inserts on the inferior margin of the cheekbone, was probably also not very large. The Neanderthal mandible and the cheekbone are relatively thin and often bear only the slightest traces of muscle scarring. In large-jawed apes, the corresponding region of the mandible is roughly scarred and its margin distended, and the cheekbone is thick and vertically tall.

If every morphology has to have a functional reason for its being, and if there truly is a correlation between jaw use and brow-ridge development, then there had to be an explanation to accommodate Neanderthals. As might be expected, one was forthcoming: anterior tooth use. Although Neanderthals made tools of greater complexity and sophistication than any other hominid with the exception of *Homo sapiens,* paleoanthropologists clung to the notion that they were like apes and used their jaws and teeth as implements. And as various Neanderthal specimens do have very heavily worn front teeth, one could still construct a functional explanation for the existence of brow ridges in Neanderthals—in spite of the contradictory evidence of the muscle markings.

As for *Homo sapiens,* the characteristic *absence* of pronounced brow ridges also had to be explained. Two such proposals became central to interpretations of human evolutionary change.

One explanation for the lack of brow ridges in *Homo sapiens* came from the observation that, especially in contrast to the great apes, the human lower jaw is very small. As de Beer put it, with a small mandible, bite forces were minimal. There was simply no need for buttressing brow ridges. The second argument was that the buttressing function had been taken over by the forehead, which rises vertically up from the orbital margins. Bite forces could just go straight up into the steep frontal bone, where they would become dispersed and eventually dissipate. But lest we forget where this discussion began, the *developmental* reason for *Homo sapiens*' vertical forehead, as well as for the lack of brow ridges, was neoteny.

Thoughts on the evolutionary significance of human neoteny as evidenced by the skull have had a profound effect on thoughts about human evolution throughout most of the twentieth century. Bolstered by a sense of security provided by the developmental and functional arguments for human neoteny, de Beer boldly stated that "[i]t is because it has no brow-ridges that the fossil Piltdown man is regarded as close to the line of modern man's descent." Similarly, backed by the power of human neoteny, Sir Grafton Elliot Smith was able to declare without reservation that the back and base of the Piltdown skull was in shape and detail comparable not to an adult but to a juvenile ape. In this respect, Piltdown was just like *Homo sapiens.*

Whether articulated or not, the implications of neoteny have loomed large in the history of who among fossil hominids was the real human ancestor.

Neoteny Enters Human Evolution

Bolk's influence on the scientific world was not limited to growth and the effects of altering developmental rates. Bolk believed that his ideas on human neoteny (his particular theory of fetalization) were profoundly relevant to an understanding of human evolution. Grounded in a version of Haeckel's biogenetic law—"ontogeny recapitulates phylogeny"—Bolk proposed that the course of human evolution was one of progressively retarded physical maturation. As he saw it, in the earliest of a series of ancestors, the phase of fetalization was merely a brief moment in the overall growth span of the individual. However, with each successive descendant, the phase of fetalization became protracted, consuming ever larger chunks of an individual's life. Eventually, the result was an organism—*Homo sapiens*—that developed so slowly that even prolonging death would not provide the time necessary for the adult phase to be introduced, much less to become expressed.

Although Haeckel's biogenetic law would be discredited soundly by luminaries such as de Beer, who also rejected Bolk's specific application of his fetalization theory to human origins, the potential evolutionary significance of comparing the morphology of juvenile apes and humans was not lost on the paleoanthropological community. When the occasional specimen of juvenile fossil was found, its scientific value soared accordingly, precisely because it could potentially provide additional clues to an understanding of the role of neoteny in human evolution. One of the few illustrations Dart included in his announcement in the journal *Nature* of the discovery of the Taung child was a line-drawing comparison of the profiles of this fossil and a chimpanzee and gorilla of presumed similar age. Visually, the effect says it all. These juvenile individuals may look generally similar, but, even at this young age, the African apes already had longer snouts than the relatively shorter-faced Taung child. The Taung child, on the other hand, had a much larger and more rounded cranial vault than did the juvenile chimpanzee apes, particularly the juvenile gorilla. Dart's published measure of cerebral length relative to facial length confirmed this observation: On a scale of 100, the chimpanzee came out at 88, the gorilla at 80, and the Taung individual at 70, with the higher figures indicating a larger facial skeleton relative to

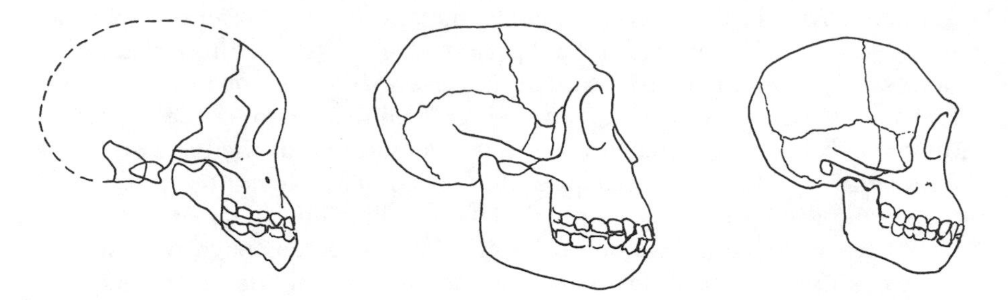

Dart's comparison of the Taung child (left) with the skull of a comparably aged gorilla (middle) and a chimpanzee (right). He wanted to point out that even though they were still quite young, the apes had begun to grow away from the juvenile state, toward having flatter and proportionately smaller brain cases and prominent brow ridges and jaws. The Taung specimen had small jaws, no brow ridges, and a very large brain. (Adapted from Dart)

cranial length. Dart's assumption that specific as well as general aspects of the size and shape of the Taung partial cranium were representative of the adult state of this species, *Australopithecus africanus,* is obvious from his rejection of Dubois's *Pithecanthropus erectus* as being in the direct line of modern human origins:

> Unlike Pithecanthropus, it [the Taung child] does not represent an ape-like man, a caricature of precocious hominid failure, but a creature well advanced beyond modern anthropoids in just those characters, facial and cerebral, which are to be anticipated in an extinct link between man and his simian ancestor.

Since the only meaningful comparison with the *Pithecanthropus* skullcap, which was from an adult individual, is with another adult specimen, Dart had to have assumed that the Taung child would have matured into an adult individual whose juvenile features were retained largely intact. There is no reasonable explanation other than neoteny for Dart's replacing the adult *Pithecanthropus erectus* specimen with the juvenile *Australopithecus africanus* specimen in the hypothesized transition from ape to human. In this context, it is enlightening to turn to the writings of Sir Arthur Keith, who was himself a proponent of the role of neoteny (although he did not refer to it as such) in human evolution. In one of many presentations of his belief in the importance of neoteny in higher primate and especially human evolution, Keith stated:

> In some of the higher primates there is a tendency to prolong the characters of infancy into childhood and the characters of childhood into adolescence and adolescence into the years of maturity. In man this tendency has been carried . . . to an extreme degree. By the working of this evolutionary process man has come by many of his most distinctive features. The more bestial traits of body—and also of mind—which have come to him by inheritance are thereby shed.

But although Keith admitted that "[i]n the Taungs skull we see a tendency to delay the onset of those growth movements which produce prominent supra-orbital ridges, muscular crests on the skull and massiveness of jaw"—as seen in the African apes, and particularly in the gorilla—he was unwilling to embrace Dart's interpretation of the fossil as being a man-ape, a form leading from an ape ancestry to the emergence of full-fledged hominids. Rather, Keith was firm in his belief that the Taung child was nothing more than a fossil ape. Curiously, his argument against Dart's interpretation of the place of the Taung child in human evolution relied heavily on development and human neoteny. Being the generally acknowledged expert at the time on the comparative anatomy of the brain, Keith devoted an entire chapter in his book *New Discoveries Relating to the Antiquity of Man* to dismantling Dart's assertion that in general proportions as well as in specific details of the preserved surface of the endocranial cast of the braincase the Taung child was more human- than apelike.

One of Dart's points was that the brain of the Taung child was high-rising from top to bottom and relatively narrow from side to side. In contrast, the brains of adult apes, and especially adult gorillas, are lower in profile and wider from side to side. It is only in the fetal stages of development that the ape brain is about as high as it is wide. Keith acknowledged these facts and observations. But he then proceeded to point out that some researchers (Bolk among them), having been provoked by Dart's fossil, had located specimens of gorilla in which the brain was compressed from side to side, as in the Taung child. In addition, Keith observed that in Neanderthals, especially in the Gibraltar skull, the brain tended to be low from top to bottom and wide from side to side, as in ape brains. As for Dart's conclusion that the fissures and elevations of the convolutions preserved on his fossil's endocast were more human- than apelike in pattern and arrangement, Keith could not disagree more. As far as he was concerned, "the recognizable markings of the Taungs brain . . . were of an anthropoid character and in nowise lifted that fossil from above the condition seen in gorillas and chimpanzee with well-developed brains." Keith's conclusion on the "mental status of Australopithecus" was blunt: "[W]e may infer that it [the Taung child] also resembled these living forms [apes] in its habits and faculties." Being a brain expert, Keith felt that the case on *Australopithecus africanus* was closed: This fossil had nothing to do with human ancestry. Nevertheless, perhaps to further seal the Taung child's coffin, Keith continued his assault on Dart with his overview of the teeth and the face.

Choosing the skull of an Australian Aborigine of approximately the same age as the Taung child for comparison, Keith proceeded to evaluate similarities and differences between the two specimens in the size and shape of the cranial vault and brain, the facial skeleton, the upper and lower jaws, the orbits, the cheekbones, and even the tiny nasal bones. In the Taung child, the nasal bones were more depressed and concave in profile than in the human child, in which the nasal bones projected forward over the lower face. The cheekbone of the Taung child was very robust compared with that of the human child. But in other features Keith looked at the Taung child was very similar to the Australian Aborigine. When he compared the Taung child with a juvenile chimpanzee of even younger age, he found that the two were very much alike in the concaveness of the nasal-bone region and in the forward projection of the jaws. Although the chimpanzee specimen was much smaller in size than that of the Taung child, Keith felt that the two were quite similar in overall form. In contrast, although the Taung child and a juvenile gorilla skull with the same teeth in the same states of eruption as the fossil were similar in size, they were quite dissimilar in many ways. The forehead of the ape was more receding, but its nasal region was much more forward-projecting than that of the Taung child. Also, the mandible of the gorilla was more massive, particularly in height.

The Taung child was similar both to the juvenile chimpanzee and to the juvenile gorilla in the persistence of the suture that courses completely between the maxillary bone and the premaxillary bone in front, in which the upper incisors are rooted. Although Goethe had sought to demonstrate that humans were like other mammals in the possession of a distinct premaxillary bone, it is still the case that the suture between these two bones of the upper jaw is never fully visible in humans and what there is of it—coursing only a short distance down from the lower margin of the orbit—rapidly disappears after birth. In having an indisputably complete premaxillary-maxillary suture, the Taung child was clearly similar to the juvenile specimens of African ape, and equally as obviously dissimilar to the human condition. In deciduous molar size, the Taung child also fell in with the African apes rather than with humans, although the reverse was true of relative anterior tooth size.

As for Dart's contention that the foramen magnum of the Taung child was as forwardly placed as in *Homo sapiens,* and that its central position at the base of the skull indicated that the fossil species had walked upright and bipedally, Keith rejected the whole package on the grounds that the comparison Dart had made between the fossil and an adult human was entirely inappropriate. Keith's comparisons between the Taung child, a human child, and a juvenile chimpanzee—the *proper* comparisons—yielded a totally different picture. In the human child, the foramen magnum is even more centrally placed than in the adult and, consequently, more centrally placed than in the Taung child. The better comparison in foramen-magnum position, Keith believed, lay between the Taung child and the juvenile chimpanzee. Although he did not state it outright, it is obvious that Keith was implying that, as in the chimpanzee, with growth, the position of the foramen magnum in the Taung individual would have shifted backward, widening the angle of flexure of the cranial base between the axis of the snout and the foramen magnum. At the end of his long discussion about the Taung child's head and neck region, and the attachment and orientation of neck muscles, Keith concluded that rather than walking upright and bipedally, the Taung individual had moved about in a manner similar to gorillas and chimpanzees—apes that locomote by "supporting their bodies in a semi-upright posture by the aid of their arms."

But there was one particular feature in which the Taung child differed from both the chimpanzee and the gorilla in the same way in which *Homo sapiens* differs from the African apes. In very young African apes, even younger then the Taung child, the brow ridges are already well developed. In the Taung child, the supraorbital regions were smooth and undistended, as in juvenile and many adult humans. In further similarity to humans, the palate of the Taung child was relatively shorter than that of the two African apes, and the spacing along the jaws and the relative sizes of the deciduous incisor and the canine teeth more human- than apelike. This latter observation notwithstanding, Keith felt that he could demonstrate far more favorable comparisons between the Taung child and the juvenile chimpanzee, and even between the fossil and juveniles of both African apes, than between *Australopithecus* and a juvenile *H. sapiens.*

When Keith began his methodical deconstruction of the Taung child's status as being humanoid, he admitted that "[in] the Taungs skull we see a tendency to delay the onset of those growth movements which produce prominent supra-orbital ridges, muscular crests on the skull and massiveness of jaw." Later in the same passage, he continues:

> The growth transmutations of the gorilla are enormous; his skull, smooth and rounded in infancy, becomes crowned ultimately with great bony crests. The chimpanzee, on the other hand, clings to more youthful features. In Australopithecus the same tendency to retention of early stages in growth was more marked than in the chimpanzee; the same tendency has become extreme in man. We must give the effects produced by the working of this law due weight when we assess the claims of Australopithecus to a place in the human phylum.

Although one might predict from this that Keith would conclude that *Australopithecus* was more closely related to humans than to the African apes, he actually came to the opposite conclusion. Because he was convinced from the outset that the Taung child was an ape, his final assessment of *Australopithecus* was predetermined. As such, the most positive statement about Dart's child that Keith could make was that "the discovery at Taungs has . . . provided those who seek man's origin in the ancestry of the great anthropoids with strong support." As for dealing with the differences he recorded between the chimpanzee, the gorilla, and the Taung child, Keith chose the chimpanzee to represent "the older and more primitive form" from which "the gorilla has evolved in the direction of increased brutalization, while Australopithecus has branched in an opposite direction, thus assuming many human traits." In spite of its inconsistencies and labored and contrite argumentation, Keith's dismissal of the Taung child as merely a fossil ape that happened to have a few humanlike tendencies held sway with much of the paleoanthropological community, especially in Great Britain. It also had a devastating effect on Dart. Thank goodness Dart had his Huxley in Robert Broom.

Putting *Australopithecus* Back into the Human Family Tree

Sir Arthur Keith's status in the scientific community did not secure his interpretation of the Taung child as being ape- rather than human-related. In South Africa, the physician and internationally recognized paleontologist Robert Broom rose to the challenge and became an integral player in the search for fossil hominids. Broom was actually interested in much older geologic deposits than those that would yield fossil hominids and had devoted his paleontological researches primarily to reptiles rather than to mammals. However, he was sufficiently intrigued by Dart's 1925 article in *Nature* on the Taung child, and by the outraged letters to the same journal the following week by the likes of such notables as Sir Arthur Keith, Professor Grafton Elliot Smith, and Sir Arthur Smith-Woodward, that he arranged to go to Johannesburg to see the specimen in question for himself. As recounted in his autobiography, Broom made clear his differences with the outspoken Sir Arthur Keith: "As a paleontologist I did not greatly worry about the size and shape of the brain or the convolutions, but I was convinced from the structure of the teeth that the Taungs child was not allied to either the chimpanzee or the gorilla, and that it was closely allied to man."

Inasmuch as teeth are often the only surviving fossilized remains of an extinct vertebrate, it is not surprising to see Broom's own anatomical bias come so clearly to the fore. But he was so convinced—by the morphology of the teeth, as well as by features of the skull—that the Taung child was humanlike that he immediately sent a detailed description accompanied by a good set of illustrations (better than those that Dart had published in his *Nature* article) to his longtime colleague at Oxford University, Professor William Johnson Sollas. Sollas, who was famous for his theory that European Neanderthals had been driven to extinction by modern humans invading from the east, received Broom's missive and interpretation with open arms. While admitting that he, like many of his colleagues in England, had thought on the basis of Dart's article alone that the Taung child was nothing more than a chimpanzee- or gorilla-related ape, Sollas replied to Broom in late March 1925 that he could see from the description and illustrations he had been sent that this might not be the case. And, by the way, could Broom send some additional information on the position of various cranial landmarks typically used for estimating skull shape and volume?

Broom did as Sollas requested. Shortly thereafter, in late May 1925, he received a reply that not only Sollas but also the earlier disbelieving Elliot Smith were now convinced that Dart had been correct. Sollas had compared the details of the Taung child with a very large sample of juvenile chimpanzee skulls, with the result that, as he wrote to Broom, "the more I compare them with the Taungs the more difference I see, and the more human

Taungs appears. . . . I should have named it *Homunculus!*" In fact, Sollas was so convinced of Dart's correctness that he sent a note to *Nature* arguing that the Taung child clearly differed from a chimpanzee of similar age and should be considered humanlike, as Dart had originally claimed. This note was published in 1925 (certain things happened more quickly then than they do now). Sollas followed up the next year with a longer publication based on the cross-sectional information Broom had earlier sent to him.

Although Broom was a confirmed supporter of Dart's interpretation of the Taung skull, he was unable to do any fieldwork relating to human origins because his medical practice was in the township of Maquassi, well away from the Transvaal region that would prove to be the source of most of South Africa's earliest hominid fossils. But colleagues of Broom's who had connections managed to get a position at the Transvaal Museum in Pretoria for him, and he moved there in August 1934. He devoted his first year and a half at the Transvaal Museum to attending to his long-term interest—fossil reptiles—but then dived full-force into the business of searching for human fossil relatives. If he failed, it would not have mattered. In contrast to the majority of those who seek human ancestors, Broom would have been just as happy to have found a good deposit of fossil anything.

In his first foray into the field of fossil-hominid hunting, Broom went to a limestone works not too far from Pretoria, where he did find fossils, including a small saber-toothed tiger, a giant baboon, and a slew of species of ancient relatives of rats and moles. He delighted in being able to describe the new species of fossil rat and mole from tooth morphology alone. But it was the discovery of the giant baboon that prompted two of Dart's students to visit and inform him of their discovery of bones at another limework, Sterkfontein, which is located about forty miles from Pretoria and thirty miles from Johannesburg. Mining at Sterkfontein had begun in 1895, and from the beginning it was clear that the limestone was rife with fossils of all sorts of mammals, large and small, including primates. There also had to be specimens of hominids. And some of these must have escaped the limestone kilns, because Broom stumbled upon a blurb in a guidebook that read, "Come to Sterkfontein and find the missing link," which Mr. R. Cooper, the proprietor of the works at that time, had himself written. In fact, fossil selling seemed to be a second occupation for some of the miners, including the quarry foreman, a Mr. G. W. Barlow. Nevertheless, Broom managed to enlist Barlow's aid in keeping an eye out for, and placing in safekeeping, any humanlike fossils that might emerge during mining operations.

On Broom's third visit to Sterkfontein, in mid-August 1936, his request was rewarded. Barlow presented him with two-thirds of an endocast of the brain of something that was either ape- or humanlike. Upon searching the quarry, Broom and Barlow found a cast in the limestone of the top of the skull that could have yielded the endocast. The base of the skull, into which the endocast fit, was found the next day, as were parts of bones of the cranial vault. All of these pieces had to be cleaned, but some of the limestone matrix was left underneath the skull as support for the fragile bone. When these fragments were assembled, they constituted much of the base of the skull, brow, and orbital regions. On the right side, the second premolar and the first two molars were preserved. Unfortunately, some crushing and displacement had occurred during fossilization. Nevertheless, Broom was astute enough to recognize that the pieces had come from the same individual. He also believed that this specimen was an adult counterpart of the juvenile found at Taung, although he considered that the latter was possibly much older geologically than the Sterkfontein specimen. He published this conclusion in *Nature* in 1936.

The new specimen differed, of course, from the Taung child in that it was an adult and, consequently, much larger. The Sterkfontein specimen had brow ridges of moderate size that were penetrated by large frontal sinuses. Since the facial region of the Sterkfontein specimen had been distorted, it was not possible to make comparisons with the Taung child. Broom published a sketched reconstruction showing the specimen with a long, slightly sloping facial profile that had a hint of nasal bone elevation. The base of the skull could not be studied. The endocast and preserved teeth could have been compared in detail with the Taung child's, but they were not. Instead, Broom, the inveterate "tooth man," compared the upper first molar of the Sterkfontein specimen with the upper first molar of a fossil apelike hominid, *Dryopithecus rhenanus,* which some paleontologists thought could have been the earliest ancestor of apes and humans. He found the two specimens to be quite similar in cusp disposition. As for the endocasts, the only comment Broom made was that the brain of the new specimen would have been considerably wider, especially toward the front, than that of the Taung specimen.

Given that Broom cited nothing that indicated a special relationship between the Sterkfontein specimen and the Taung child, it is somewhat surprising that toward the end of the article in *Nature* he declared that "[t]his newly-found primate [from Sterkfontein] probably agrees fairly closely with the Taungs ape" (which earlier in the article he had affirmed was not an ape). Because of the morphological differences, whether stated or implied, between the Sterkfontein specimen and the Taung child, and because of the presumed geologic age difference between them, Broom made the even less well-supported pronouncement, "I therefore think that it is advisable to place the new form in a distinct species, though provisionally it may be put in the same genus as the Taungs ape." As a consequence, and in spite of the lack of any substantive basis for the claim, the paleoanthropological world now had a new species of *Australopithecus, A. transvaalensis.* Later, Broom would place the species, *transvaalensis,* in a new genus, which he called *Plesianthropus.*

The next major "discovery" of Broom's—which he purchased from Barlow on Wednesday, June 8, 1938, for a couple of pounds—was part of an upper jaw with the first molar preserved and evidence that four others had recently been broken off the specimen. Broom recognized that this specimen represented something like the Sterkfontein ape-man, although it was larger, and that the matrix that adhered to it was different from the limestone at Sterkfontein. When Barlow was pressed for the location of the site that yielded this new specimen, he was at first unresponsive. But Broom's persistence led him to a boy, Gert Terblanche, who had actually found the specimen. Gert was carrying the four missing teeth around in his pocket. Broom bought these teeth from the boy, who took the physician-cum-paleontologist to his fossil hiding place, which was just up the valley from Sterkfontein. There Gert had secreted a relatively intact lower jaw preserving two teeth. Broom scrounged around on the ground of the area—which we know as the site Kromdraai—and came up with a number of cranial fragments. When these fragments were cleaned up and put together, they produced much of the left side of the lower part of the skull, as well as the lower face and the upper jaw. There was also a fairly complete lower jaw, with all of its premolars and molars, which was complementary to the upper jaw.

In Broom's mind, this was no African ape. The Kromdraai hominid was humanlike in two significant ways. One humanlike feature of the fossil, Broom pointed out, was that the bony tube that leads to the fleshy ear, and houses the eardrum and inner ear bones, lay below the plane of the cheekbone. In apes, this bony tube lies level with the plane of the cheekbone. The occipital condyles, which articulate with the first vertebra, were in line with this bony tube, which was an indication of how far forward these articular surfaces were on the cranial base. For Broom, this was clear evidence that this fossil hominid had had a more erect posture than had been seen in the chimpanzee and the gorilla.

But the Kromdraai individual was not quite the same as the *Australopithecus transvaalensis* specimen from Sterkfontein. For one thing, it was larger. It had larger teeth, which differed in morphological detail, as well, and the jaw was more massive. The Kromdraai face was also flatter. Clearly, this was a totally different hominid, one that deserved to have its own genus and species names. Broom also suggested that this specimen was more similar to modern humans than was *A. transvaalensis* because he thought the fossil may have had a smaller canine than did the Sterkfontein specimen.

Broom set to writing up an announcement of the Kromdraai discovery for *Nature*. It was published in August 1938. But before introducing the fossil from this site, Broom discussed the hominid species from Sterkfontein, *transvaalensis,* and reminded his readers that he had only provisionally placed it in the genus *Australopithecus*. Now, however, he had evidence that validated his uneasiness about subsuming the Sterkfontein hominid in that genus. He had the front part of the lower jaw of a young male, of perhaps, he would later suggest on the basis of human-growth criteria, nine or ten years of age at death. Broom believed that the outline of the midline of this youth's mandible was sufficiently different from that of the Taung child to warrant placing all Sterkfontein hominid material in a different genus, for which he created the name *Plesianthropus,* which literally means "near man."

In 1946, with G. W. H. Schepers, who had been one of the two Dart students who brought the elder paleontologist to Sterkfontein, Broom published an opus on all known South African fossil hominids. In that work, he pursued the argument that these ape-men were true links between the great apes and modern humans. He pointed to the fact that the outline of the midline cross section of the lower jaws of both the Taung child and the Kromdraai youth were remarkably similar to a chimpanzee's.

Although he was primarily responsible for demonstrating that there had been early, primitive human relatives in southern Africa, Broom, like others in the profession, often made assertions on the flimsiest grounds. Eventually, all fossil-hominid specimens from Sterkfontein, as well as very similar fossils later discovered at the South African site farther to the north, Makapansgat, became subsumed in Dart's *Australopithecus africanus.*

After rather casually naming *Plesianthropus,* Broom proceeded to summarize how Gert Terblanche had found the site of Kromdraai and the first fossil-hominid specimens from it, and how, since it was a larger and yet in some ways more humanlike type than *Plesianthropus transvaalensis,* this new ape-man deserved to be recognized in its own genus and species. Since this new find was larger and more massive than *P. transvaalensis,* Broom proposed the species name *robustus* for it. Because he thought that his Kromdraai hominid was more on a par with humans than with apes, he coined the new genus name *Paranthropus* for it. As more specimens of the *Australopithecus-Plesianthropus* and *Paranthropus* types of early hominid were discovered at South African cave sites, it became commonplace to refer to the larger and more massive *Paranthropus* as a "robust australopithecine," and the smaller and more lightly built *Australopithecus* as a "gracile australopithecine"—Australopithecine being the colloquial name for the subfamily Australopithecinae, into which they were all placed.

New specimens of South African australopithecine did come to light. In 1947, on April 18, the crew under Broom's direction, blasting away at the Sterkfontein limestones, where the type specimen of *Australopithecus transvaalensis* had been discovered, unexpectedly exposed a ring of bone and, within its walls, the cavity for the brain. The piece that had been blasted away was found, and it contained the skullcap. When the matrix of limestone was removed, there, but for the teeth, which were all missing, was an essentially complete and perfectly preserved skull. It had a gently concave or dished-out lower face that terminated below in a somewhat protruding upper jaw. Its brow was only slightly thickened and distended. Its cranial vault was domed and rather high-rising. Its cheekbone was slender. In shape and detail, the base was more reminiscent of humans than of apes, including having a more forward position of the foramen magnum. In general, its bone was thin and bore only minimal evidence of muscle-attachment scarring. In the May 17 issue of *Nature,* Broom announced the discovery of this remarkable skull, which had been cataloged as Sterkfontein 5, and which he attributed to the genus *Plesianthropus.*

In August of the same year, a nearly complete pelvis was discovered at Sterkfontein. Although it differed in certain features from the pelvis of *Homo sapiens,* the Sterkfontein pelvis was clearly a closer match with it than with any ape, even, as Broom comments, the orangutan. Most striking was that the bladelike upper portion of the fossil pelvis—the ilium—was very short from top to bottom, very deep from front to back, and quite expanded posteriorly, as in humans. In apes and most land mammals, the ilium is very tall vertically, rather narrow, almost straight up and down along the posterior border, and expanded anteriorly at the top. In conjunction with the position of the foramen magnum, the configuration of the pelvic region gave clear evidence of the human features of this early fossil relative.

As for the robust australopithecine type of early hominid, Broom expanded its known fossil record. In late 1948, he was asked to open a new site, Swartkrans, which lies within view of the site of Sterkfontein. With the luck rarely enjoyed by any paleontologist, Broom soon found the left side of a lower jaw in which the premolars and molars were preserved. In overall shape, the teeth impressed Broom as being humanlike. In size, however, these teeth even surpassed those of the *Paranthropus robustus* specimen from Kromdraai. After finding a few isolated teeth that he thought belonged to the same hominid as the mandible, Broom sent a quick note to *Nature* to announce the discovery of yet another robust ape-man. Because of its even larger teeth, he placed it in a new species, *P. crassidens.*

Excavation at Swartkrans during the next eight or so months yielded part of a lower face and an upper jaw, an even more massive mandible, and, eventually, three skulls. Broom attributed all of these specimens to *Paranthropus crassidens.* One of the skulls was quite crushed. Another skull, missing the back and right side, was otherwise in pretty good shape, although the lower face had turned up a bit during fossilization. But the third skull, which was only slightly distorted, was essentially complete. It primarily lacked some of the temporal muscle-induced sagittal crest, the very front of the upper jaw and the teeth that would have been there, and bits of the base of the skull. This skull contrasted markedly with the Sterkfontein 5 skull of *Plesianthropus.*

The Swartkrans skull was clearly a more robust individual, with a much more massive jaw, brow, and facial skeleton. The cheekbones were tall and thick. The face was broad from side to side, essentially flat, and almost vertical in profile. The cranial vault, although somewhat crushed downward, would certainly never have been as domed or as high-rising as that of Sterkfontein 5. Not only was the whole appearance of this *Paranthropus* individual more brutish and robust but it also bore evidence of larger, more powerful muscles than the *Plesianthropus* skull, especially in the development of a sagittal crest for the attachment of what must have been the huge temporal muscles that would have been necessary to move the kind of massive lower jaw also found at Swartkrans.

Excavations at Swartkrans through the remainder of 1949 as well as the following year yielded many more specimens of fossil hominids. Although there were more cranial, mandibular, and dental pieces to add to the puzzle, one discovery that deserves special mention is most of a right pelvis bone (this bone itself is technically referred to by anatomists as the os coxa). This os coxa was larger than the ones found at Sterkfontein. But, as Broom noted, the Swartkrans os coxa was as humanlike as those attributed to *Plesianthropus* and differed from apes in the same ways, too.

The place of the australopithecines in human evolution was secure, the diminishing ranks of stalwart detractors notwithstanding. Broom, increasingly buoyed by the growing acceptance of his discoveries of ape-men, and with the validation of Dart's claims about the Taung child, took pride in quoting a comment of praise offered by the eminent embryologist and comparative anatomist Sir Gavin de Beer: "It is difficult to disagree with Broom's conclusion that 'man arose from a Pliocene member of the Australopithecines, probably very near to *Australopithecus* itself.' "

Neoteny in Human Evolution Comes of Age

It is understandable that embryologists such as Sir Gavin de Beer should take such an obvious interest in discoveries of fossil hominids. On the basis of the comparative anatomy and embryology of present-day vertebrates, one could argue that humans were neotenic. But if there was a connection between ontogeny and phylogeny, and one wanted to document neoteny in human evolution, one needed more and older fossils than those that were known in the early decades of the twentieth century. But by the mid-1900s, there was a sufficient number of juvenile as well as adult specimens of hominid fossils known that embryologically based theories of heterochrony—of which the developmentally altered states of neoteny and progenesis were but two manifestations—could be brought to bear on questions of human origins.

The leader of the new wave of proponents of human neoteny was the Australian physical anthropologist Andrew Arthur Abbie. Although he was aware of the term *neoteny,* Abbey preferred to use *paedomorphism* instead. In parallel with de Beer and others in their rejection of Haeckel's belief that ontogeny revealed phylogeny by a recapitulation of the adult versions of previous ancestors, Abbie was quite adamant about the shortsightedness of physical anthropologists and paleoanthropologists in focusing their studies primarily on adult specimens. Echoing Arthur Keith's emphasis on understanding evolution and the adult form by way of studying ontogeny, Abbie argued that scientists had to pay more attention to the fetal state of the individual if they hoped to have any chance of deciphering human origins. As Huxley had used comparative embryology to establish "man's place in nature," so Abbie suggested that the only possibility of discovering the course of human evolution was by studying the embryological stages of primates. For only then could one speculate with conviction on what the generalized ancestral primate fetal form had been. From this hypothetical state, it would then be possible to reconstruct hypothetical ontogenies of fossil species by analogy to ontogenetic changes that could be observed in living species. In 1952, Abbie presented this idea in the form of a series of simple line drawings of skulls in profile going from the neonatal form, to the child (standardized at the time of eruption of the first permanent molars), to the adult female, to a paedomorphic male, to a more brutish-looking, less paedomorphic, and more differentiated adult male.

Old World monkeys were represented by the macaque. This sequence of skulls very clearly shows a change from the round and large cranial vault, small, flat face, and nonprotrusive jaws of the neonate, to the less highly vaulted skull with larger and more protrusive jaws of the child, to the even less highly vaulted skull with larger and more protrusive jaws of the adult female, and so on through the less paedomorphic adult male. The upper canines also become increasingly larger and longer from the adult female to the more differentiated male. Curiously, though, only the adult female is drawn with brow ridges, although they are of moderate size.

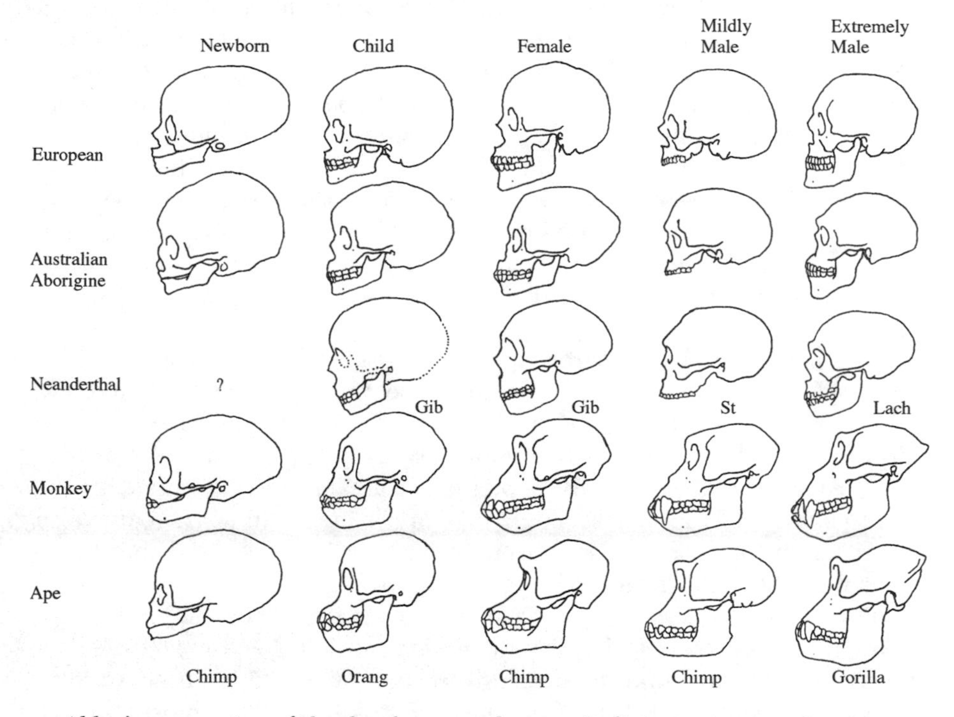

Abbie's conception of the developmental stages in living primates—humans (upper two rows), a monkey, and an ape—and in Neanderthals. The monkey he chose was the macaque, an Old World monkey. His apes were mixed. He did not have a newborn Neanderthal to begin that sequence, but he did have a child (Gib is the Gibraltar child). The second Gib is the Gibraltar adult, St is the Steinheim skull (which is not a Neanderthal), and Lach is the La Chapelle-aux-Saints skull. (Adapted from Abbie)

The ape skull sequence is interesting in that it mixes up all three great apes. The neonate is a chimpanzee, but at this age neonatal cranial shape and jaw development are essentially the same in all great apes. The child stage is represented by a juvenile orangutan, which retains the dome-shaped cranial vault of the neonate but has larger and more protrusive, as well as slightly upturned, jaws. (The adult orangutan, male and female, retains the high, rounded skull shape of the neonate.) In the adult-female column is a funny-looking chimpanzee with strangely thickened, forwardly oriented, and very upwardly protruding brow ridges, a modestly domed cranial vault, and large and quite protrusive, but not upturned, jaws. A more normal-looking chimpanzee, with brow ridges that project equally forward and upward, a relatively flat and somewhat elongate cranial vault, and large, protrusive, and somewhat downwardly deflected jaws is the "paedomorphic male." A gorilla, with more pronounced brow ridges, some cranial crests, more massive jaws but a surprisingly more rounded cranial vault, represents the hyper-male.

Abbie provided two sequences of skulls for *Homo sapiens:* European and Australian Aborigine. As might be expected, given the historical precedent, Europeans, who occupied the top row in the illustration, deviated less from the juvenile shape than the Australian Aborigines. Even an Australian Aborigine child's skull had a less domed cranial vault and more protrusive jaws than did the European child's skull. The trend of deviating from the paedomorphic shape was exaggerated further in the Australian Aborigine female, more so in the supposed paedomorphic male, and even more so in the hyper-male. Within the European sequence, it is barely possible to see a difference between the female, the paedomorphic male, and the more differentiated male.

The message is clear: Europeans (that is, some human races) are supposed to be more neotenic than Australian Aborigines (other races). In rebuttal, the British-by-birth, but ultimately Princeton University–based physical anthropologist Montagu F. Ashley Montagu argued that the real evidence actually indicated that the most neotenic of humans are to be found among Asian people and that black populations are neotenic in some features in which Europeans are not. Not surprisingly, Bolk had preferred that Europeans were the most "fetalized" race of *H. sapiens.*

For Abbie, the less completely neotenic, or paedomorphic, Australian Aborigine served as a good comparison between living *Homo sapiens* and the then-known hominid fossil record. In the early 1950s, in addition to undoubted recent fossil *H. sapiens,* the other fossil hominids most paleoanthropologists accepted were *Australopithecus, Pithecanthropus,* Piltdown, and Neanderthals.

When most paleoanthropologists surveyed this fossil assemblage, they saw the diversity that was reflected in placing different specimens in different genera and species. When Abbie came to look at the same assortment of hominids, he could not understand how his colleagues saw what they did.

He characterized their problem as being stuck on the fact that the fossil forms seemed to show "an apparently incongruous assemblage of physical characters." If his fellow paleoanthropologists had just not been so confounded by this presumed incongruity, they certainly wouldn't have thought that the extinct hominids represented distinct species. Abbie was dead set on demonstrating his colleagues' incorrectness. He used development to make his case, starting first with the argument behind the series of skulls: *H. sapiens* is a paedomorphic (read "neotenic") species within which there is great variability in the expression of physical features.

Within Abbie's conceptual framework, in which minor differences in developmental timing can be invoked to account for every difference seen between adults, there is an infinite number of possible combinations of different representations of physical features. Because of this, if one looked long and hard enough through the skeletal collections of the world's natural-history museums, one could find the occasional present-day human with a Neanderthal, *Pithecanthropus,* or even an *Australopithecus* feature. Indeed, the internationally renowned and domineering Swiss comparative primate anatomist Adolf Schultz had found a supposed *Pithecanthropus*-like gap between the upper permanent second incisor and the upper permanent canine in a "modern negress," and Abbie himself had stumbled upon something similar in a "living Aborigine." This mind-set also worked to explain why the fossil forms could seemingly have contradictory combinations of morphological features: *Australopithecus* being an ape with humanoid teeth and limbs; the Piltdown skull having a human cranium but apelike jaws and teeth; *Pithecanthropus* having an apelike skull but a human femur; and Neanderthals having a less apelike skull but a (presumably) more apelike femur.

Echoing Thomas Huxley's sentiments of a century earlier, Abbie felt justified in proclaiming that since so-called Neanderthal features could also be found in living *H. sapiens,* the extinct hominid could easily be incorporated into the latter species. As uniquely different as Neanderthals might otherwise be—for instance, in facial protrusion, brow-ridge configuration, nasal-region size and shape, as well as in many postcranial features—Abbie was convinced that, "even in his extreme form, [the Neanderthal] is no more than a local specialization within the normal range of human variation." To illustrate this point, Abbie put his Neanderthal skulls right below the Australian Aborigines, which had been placed below the Europeans. The Neanderthals went from a child's skull with rounded, small face and jaws found at the site of Devil's Tower, Gibraltar, to the less cranially domed but large-faced presumed female Neanderthal from Forbes's Quarry, Gibraltar, to the longer, lower-skulled, and largish-faced Steinheim skull from Germany, representing the paedomorphic male in the series, to the very long- and low-skulled and large-faced and -jawed La Chapelle-aux-Saints skull from France, representing the more differentiated male. Since the macaque and ape skull series lie immediately below the fossils in Abbie's illustration, it is obvious that Neanderthals look more like *H. sapiens*—which is sufficient for Abbie to conclude that, at the species level, the fossils are nothing more than variants of *H. sapiens.*

Since he believed that human morphological variability is unbounded in the possibilities of expression—so that virtually any combination of features can end up in the same individual—Abbie would have subsumed all known fossil hominids within *H. sapiens.* The only thing that prevented him from doing so was his belief that an absolutely large brain size and a large brain/body ratio together define the domain of being human. It was not sufficient that an animal have a large brain, because that would include whales and elephants, or have a large brain/body ratio, because that would include at least one of the New World marmosets. Consequently, while *Australopithecus* was hominid-like in its skeleton, it was still something apart from "humanity," because it did not meet the criteria of absolute *and* relative brain size. On the other hand, "[w]ith that possible exception," Abbie declared, "we have no right to consider any recognised hominid as being anything else but human, without any specific—much less generic—distinction."

Without stating it in taxonomic terms, what Abbie was saying was that all non-*Australopithecus* fossil hominids, by virtue of their showing some *H. sapiens* features and by *H. sapiens'* occasionally displaying one of their features, should be classified as *H. sapiens.* The morphological differences that emerge within the now unnaturally widely variable species *H. sapiens* Abbie explained in terms of differences in developmental timing. Differences were due only to the degree to which any individual is paedomorphic (little differentiated from a fetal standard) or gerontomorphic (more differentiated from a fetal standard). For Abbie, paedomorphism was the slowing down of differentiation from the fetal state (again, read "neoteny"). Gerontomorphism was the acceleration of differentiation from the fetal state.

In a broader comparative context, since most primate fetuses have smooth and rounded crania and small jaws, and this combination of features is often retained in adult *Homo sapiens,* Abbie concluded that what we see in humans—paedomorphism—must be the ancestral, or primitive, condition for primates. As such, gerontomorphism must be the more evolved, derived, or specialized state, which is expressed either as a suite of characters, as in most features of the skull and jaw of a gorilla, or in only a few features. Although *H. sapiens* is generally paedomorphic, Abbie saw the species as gerontomorphic in having a highly arched nose and long legs. But regardless of the adult state, Abbie was most interested in the fetal state because he believed it was the most revealing in terms of evolution: "If a common generalized foetal form could be discovered the problem of man's ancestry would be much closer to solution than it is now."

Abbie conceived of this "common generalized foetal form" as being similar to the human embryo at about the seventh week of gestation, which is still well within the first trimester. The reason he chose such an early stage of development is that the fetus has not yet begun to display features specific to being a particular kind of primate. Most important for Abbie, although the digits of the hand of a human fetus at seven weeks have begun to differentiate, the digits of the foot have not. The foot, of course, going back to Aristotle, had long been recognized as the singular clue to distinguishing humans from other primates. In most primates, the big toe is well separated from the other toes and is functionally like a thumb in its ability to grasp in conjunction with the other toes. In humans, the big toe is in line with the other toes and the cleft separating it from the others is not deeply invaginated. For Abbie, gerontomorphism was at work in creating the opposable big toe of nonhuman primates by accelerating into early development the process of clefting that would free the big toe from the others. We are now more familiar with the details of development, which are different from what Abbie understood them to be.

Abbie also proposed an intriguing developmental model for how humans, among primates, came to have such relatively long legs. His argument relied on the fact that, starting with the head region, the different parts of a developing embryo become distinct entities from the top down. Of particular importance for his theory was the observation that the limb buds for the arms begin to emerge prior to those for the legs. The difference, Abbie suggested, between an organism with short legs and one with long legs was due merely to a change in the time at which these limb buds appeared and the length of time during which they grew. As such, short-leggedness, as seen in chimpanzees, would be paedomorphic, the result of delaying the growth of the legs, which are embryonically late-emergent structures to begin with. Humans, on the other hand, must have evolved long-leggedness through a process that involved advancing the time at which the limb buds of the legs start to grow and continuing this accelerated growth rate until well after birth.

To be sure, Abbie was a victim of many of the biases about races and interpreting fossils that beset his predecessors and contemporaries. But his attempts to explain morphological differences between fossil and living hominids, and to interpret these differences evolutionarily in terms of differential rates of growth, deserve some recognition. He was dead wrong, however, on many aspects of development.

Although he conceived of the development of the grasping big toe as well as of the long toes of apes and other primates as resulting from an acceleration of the rate of clefting between digits, it is actually the case that the grasping thumb and long fingers of humans, apes, and other primates differentiate embryonically in the precise configuration they will have in the adult: The digits develop within paddle-shaped ends of the arms and the legs in the arrangement they will have throughout life. Cell death between digits frees them from their embryonic sock.

Abbie was also dead wrong about the absence of a tail in humans being a manifestation of paedomorphism. All vertebrate embryos have a long tail.

But during development, some species reduce or resorb the length of this axial structure, others keep it about the same length, and still others elongate it. Reduction or loss of a structure is altogether a developmentally different phenomenon than the pure realization of paedomorphism or gerontomorphism.

As for paedomorphism's being a primitive primate condition, it is probably the other way around. Since most vertebrates, including most primates, develop away from the fetal configuration, the common condition would be gerontomorphism. As such, this would be the primitive condition. As the less frequently encountered developmental solution among vertebrates, primates included, paedomorphism—specifically neoteny—would be the more restricted, the more specialized, the more derived developmental phenomenon.

But, though he was off the mark on various counts, Abbie arrived at some conclusions that have too long been ignored. To begin with, he made an enlightening analogy between development and a movie: "Development can be looked upon as a cinematographic film which comprises the whole of differentiation and can be run fast or slow as desired, or at different speeds at different times." The entire film, or only parts of it, can be speeded up or slowed down. When speeded up, more film can be played; when slowed down, less. As Abbie put it,

> We have no idea of the proper speed for the film, or even its length . . . but it would be interesting to speculate on the results if the speeds were changed. . . . When looking for some common ancestor for man and other primates it is necessary to seek among embryos, not adults. . . . Conceivably, a minor embryological twist in any primate stock could introduce the changes in timing necessary to produce the human stem and "selection" would do the rest.
>
> While it is true that the ultimate fate of [the] embryo is already determined at conception, it is equally true that a minor shift in emphasis could direct differentiation into any of the lines that end up each in its own specific kind of primate.

Abbie used the example of the differences between humans and apes in the disposition of the big toe to prove the point that a small change early in development could effect a significant morphological difference between species. Darwin sought to find examples of intermediate stages in the presumed transition from an apelike foot with its grasping big toe to a fully modern human foot with its short, forward-facing, and non-grasping big toe. Darwin even pointed to the foot of some "savages" as still retaining some of the prehensility characteristic of the ape foot. In this century, a physician and human anatomist who specialized in foot anatomy, Dudley Joy Morton, suggested that the big toe in one of the subspecies of gorilla, the mountain gorilla, lies closer to the other toes than it does in the lowland gorilla, and that the former configuration may represent an intermediate condition in the evolution of the human foot. But Darwin and Morton were comparing adults, the end products of growth. As far as Abbie was concerned, Morton had it all wrong. An ape foot is an ape foot, and a human foot is a human foot, regardless of the presumed variations or minor differences among individuals of the same species. The source of the differences had to be sought in the fetus. For it is there that the fundamental developmental differences between the foot of an ape and the foot of a human can be clearly seen.

On a more general level of applicability to evolutionary thought, Abbie's ideas as summarized above are not, of course, uniquely his. Bolk, Garstang, and de Beer, among others, had been serious proponents of heterochrony. Historically, it is interesting that Abbie tried to think in terms of heterochrony, as it relates to not just the morphology of living humans but also as it relates to fossil hominids—even though he went beyond the biological reality of just how variable a species can be. But in spite of the fact that he erased species boundaries through an invocation of heterochronic processes, Abbie was aware of the possibility of seeking clues to the origins of morphological novelty in pathological and developmental disorders. Without mentioning individuals by name, Abbie was resurrecting the ideas of an important but unfairly ignored group of paleontologists, geneticists, and developmental anatomists from the earlier part of the twentieth century.

6

Development, Inheritance, and Evolutionary Change

> *When a child resembles a parent or grandparent in some striking particular, we say it* inherits *such-and-such a characteristic from the parent or grandparent in question.* By heredity, then, we mean organic resemblance based on descent.
>
> —William Castle (1913)

The Myth

This is how the story goes. In the late nineteenth century, Gregor Mendel, an Austrian monk, labored day in and day out breeding, crossbreeding, and back-breeding different strains of the simple garden sweet pea and eventually figured out two major principles of heredity. First, that inheritance occurred via discrete units (that is, in a particulate manner by means of what we know as genes) rather than through a process of blending. And, second, that some gene states were dominant over their recessive counterparts. Mendel even calculated that the morphology determined by the dominant gene would be expressed in a ratio of 3 to 1 over the morphology determined by the recessive gene in a population of individuals carrying one of each. But, sadly, Mendel's discoveries went unnoticed until they were "rediscovered" independently at the turn of the century by a handful of geneticists who went on to develop the field of population genetics and apply its principles to the Darwinian theory of evolution by natural selection. Through the work of these geneticists, not only the mechanism of inheritance but also that of speciation could now be explained from a scientific base that was lacking when Darwin formulated his evolutionary theory. The grand evolutionary synthesis of the 1930s, which brought together the diverse fields of paleontology, zoology, botany, and genetics under the common language of evolution by natural selection, was the triumphant realization of this natural history of events.

This, in a nutshell, is the gist of the post-Darwin evolution story. But the story is not entirely true.

The World Before Genes

In order to more fully appreciate the history of the field of genetics, it is important to understand how early biologists and evolutionists thought about inheritance and its consequences. In light of this, it is not unreasonable to begin with Darwin, for even though he offered various dubious suggestions on how evolution and heredity might work, he also served the very crucial role of focusing public and scholarly attention on questions that remain important today. Although Darwin's ideas on natural selection and evolution survived an initial period of skepticism and public ridicule and eventually received the intellectual support of many scientists of diverse interests, they almost as quickly sank into oblivion forever. The demise of Darwinism was due in part to the work of the early geneticists, who viewed the fundamentals of genetics as incompatible with theories of natural selection and adaptation. It was also abetted by the failure of the two fields of investigation that were most likely to provide validation, if not vindication, of Darwinism to do so—paleontology and embryology.

Paleontology would seem to be the discipline that could demonstrate Darwin's picture of evolution as the gradual accumulation of minute transformations through the action of natural selection. This scenario demands the existence of a continuum of intermediate forms, as the following quote makes clear: "Natural selection can act only by the preservation and accumulation of infinitesimally small inherited modifications, each profitable to the preserved being . . . [and] if it be a true principle, [it will] banish the belief of the continued creation of new organic beings, or of any great and sudden modification in their structure." With species changing slowly and continually, rather than by the leaps and bounds espoused by such saltationists as Thomas Huxley, the fossil record should be jam-packed with examples of evolutionary change through an inexorable series of intermediates forming an unbroken chain of ancestors and their descendants. All, it seemed, that paleontologists had to do in order to document these successions of forms was go into the field, find a promising fossil locality, and dig. But although fossils were indeed found, quite often in supreme abundance, the continuum of gradual evolution was rarely convincingly documented. Rather, new species seemed to appear as abruptly in the fossil record as older ones disappeared into the oblivion of extinction.

If one could not reconstruct a history of gradual evolutionary change from a compilation of fossil sequences, one might at least get a sense of it from the study of the ontogeny—the developmental unfolding—of the individuals of a species. Earlier in the nineteenth century, the German embryologist Karl Ernst von Baer—who had become famous for many embryological insights, including the discovery of the human ovum—had put the final touches on a seemingly airtight case against interpreting the stages of the development of an individual as the adults of the animals that were supposedly reflected in the individual's ontogeny. Accordingly, a human embryo at a stage of development where it presumably had gill slits was not going through a fish stage; but even if, by some chance, it were, a human embryo was not the equivalent of an adult fish. Basing his conclusion on better evidence than his predecessors had, von Baer argued that the only common ground between the ontogenies of organisms was their shared embryonic stages. But at any particular common embryonic stage the developing organism was not in the form it would attain as an adult. Humans and fish might share certain developmental stages, but these shared stages are not the specific features of being a human or a fish that would later develop. The features that would specify "human" or "fish" would be acquired after the developing organisms had deviated from the shared path of ontogeny.

Although the evidence for von Baer's "laws" was accessible to any embryologist, it would be swamped by an alternative explanation: that the shared stages of ontogeny that exist between different kinds of organisms represent the adult stages of organisms as they are arranged in a sequence of increasing morphological complexity. Now, however, there was also an evolutionary twist to this idea: During growth, each individual of each species passes through its own evolutionary history. Thanks to Ernst Haeckel, who began publishing on the subject in 1866, the adult was put firmly back into ontogeny. Since one could supposedly view the course of evolution of a species by studying the unfolding ontogeny of individuals of that species, the adult also became the reflection of that phylogeny.

For Haeckel, the presumed gill-slit stage in human ontogeny was the equivalent of an adult fish. (In reality, such a stage does not occur; there are only the folds of the gill arches, which, among other structures, develop into our hyoid bone, inner ear bones, and jaws.) In his eyes, evolutionary change often came about by tacking on to the end of an individual's ontogenetic sequence a new form and squeezing the rest of the series of adult forms farther back into the ontogenetic sequence. An ontogeny was simply an evolutionary sequence of a series of adult ancestors. Consequently, by studying the ontogeny of an organism, one was actually observing a mini-version of the evolutionary past of that individual and that individual's species. Haeckel referred to his scheme of "ontogeny recapitulates phylogeny" as "the first principle of Biogeny," which became more popularly known as the biogenetic law.

In proposing "the first principle of Biogeny," Haeckel proclaimed his firm support of Darwin's general themes. First, he wholeheartedly embraced "the simple fundamental idea that the Struggle for Existence in Nature modifies organisms." Second, he was in full agreement with Darwin on the point that just as "man produces new domesticated varieties of animals and plants," so in nature there is a "constant preference or selection of the individuals most suitable for propagation." And third, he accepted the theory "that the interaction of Heredity and Adaptation acts as a modifying cause."

As he saw it, Darwin's ideas could be melded with the biogenetic law because the struggle for existence occurred between adults, and natural selection acted on adults, picking and choosing the more fit individuals from each generation.

Darwin had been relatively successful in making his case for a struggle for existence in nature. He had laid the groundwork for this argument by pointing out two obvious facts, which anyone could verify: Most organisms can and will strive to reproduce as often as possible; and in order to survive so that they can reproduce, all organisms are dependent on the availability of the essential natural resources of air, water, and sources of energy.

Darwin then reminded his readers that the natural resources an organism might need in order to survive are neither globally ubiquitous in their distribution nor constant in their availability. In addition, it is a given that variation exists among individuals in both physical and behavioral attributes. Consequently, putting all of this together, one is forced to conclude that organisms must compete for access to natural resources—that there is a struggle for existence—and that variation among individuals in physical and behavioral attributes probably derives from such a struggle for existence. Simply put, the existence of finite resources leads to struggle and competition among individuals for these resources. In contrast, an increase in the availability of resources, or a reduction in reproductive potential, can lead to a decrease in the intensity of struggle and competition that would normally exist among individuals.

According to Darwin, the only way in which evolutionary change occurs is through struggle and competition. He firmly believed that it was only in such an antagonistic situation that natural selection could act, culling the less fit to survive under those circumstance from the more fit of the species. His relentless adherence to this scenario forced him to bypass the rain forests of southern Asia in favor of the harsh savannas of southern Africa as the possible seat of human origins. Because of this, he also had to reject the orangutan as a possible close human relative. His preference for Africa as a place harsh enough for natural selection to evolve humans then directed his choice of close human relatives: the African apes. This in spite of the fact that these apes are not only found in forested areas of Africa but these forests lie thousands of kilometers to the north of the savannas of southern Africa.

Darwin developed the case for natural selection by first delineating three principles, which could be verified by anyone. First, offspring are similar to their parents. Second, individuals who have certain kinds of physical or behavior attributes will be able to obtain more resources than individuals who do not have these attributes; consequently, some traits are more advantageous than others. And third, individuals who have more advantageous attributes will have higher survival rates; in turn, these individuals will have greater reproductive success than those who are endowed with less advantageous attributes. To these three principles Darwin added three observations: There is a struggle for existence; there is variation among individuals in physical and behavioral attributes; and physical and/or behavioral attributes are either advantageous, disadvantageous, or of no consequence whatsoever in allowing the individual who possesses them to gain access to scarce resources. The conclusion Darwin drew from these observations and higher-level principles was that there must be a selection from among the available variations of physical and behavioral attributes, that selection will tend to favor individuals with the more advantageous attributes, and that these favorable attributes will increase in the species or population because the individuals who possess these more advantageous attributes will have greater reproductive success. These individuals will be able to produce more offspring who have these attributes.

The stumbling block for Darwin, as for all those who were struggling with variations on the evolution theme and with the problems of heredity, was the question of how features were transmitted from the individuals of one generation to those of the next. There were various notions in the intellectual air as to how features might change, and everyone who had any pretensions to evolutionary ideas invoked them at one time or another. For instance, there was the ever-popular concept—formalized in the phrase "the inheritance of acquired characteristics"—that an individual's conscious or unconscious desires affected the use or disuse of an organ and this, in turn, engendered change in that organ. Change so acquired would be passed on to the offspring. Although the notion of use and disuse was the brainchild of Lamarck's generation, it was also a major theme in Darwin's writings.

But how do features get passed from parent to offspring? In spite of von Baer's having confirmed very early in the nineteenth century that humans were just like other mammals in producing eggs, or ova, that would eventually produce offspring, the riddle of inheritance remained shrouded in mystery. In the eighteenth century, early embryologists believed that there were germ cells and that each germ cell contained all the parts of an organism. Each part existed in a preformed but miniaturized state, and in its proper anatomical position; it had only to enlarge during growth. Inasmuch as embryonic growth constituted an unfolding of parts, and the original meaning of the word *evolution* came from its use in an ontogenetic context, this particular theory was referred to as the theory of unfolding, or the theory of evolution.

The embryological theory of evolution gave rise to the theory of encasement. In its first formulation, the theory of encasement postulated that the female germ cell contained the germ cells of all subsequent generations of offspring. Each female germ cell contained the germ cells of the males and the females of the next generation, the germ cells of the latter contained the germ cells of the next generation, and so on ad infinitum. Given that science in the eighteenth and early nineteenth centuries was conducted according to the rules of the Great Chain of Being, people believed that the progenitors of all human germ cells had originated in the ovaries of Eve, just as the first female of every other species had held in her ovaries the germ cells of every generation that would follow. Until 1827, when von Baer discovered from his studies on dogs that the mammalian egg lay within the ovary, the ovary itself was believed to be the germ cell.

But the theory of encasement took a different twist when, sometime in the latter part of the seventeenth century, the Dutch linen merchant and father of the microscope, Antoni van Leeuwenhoek, discovered male sex cells, the spermatozoa, in humans. Looking for all the world like minuscule but real animals thrashing about, with a filamentous tail propelling an often ovoid head, sperm were promoted by some embryologists as the source of the future generations of individuals. In fact, another Dutch scientist, Nicolas Hartsoeker, believed he could see tiny preformed men, which he referred to as homunculi, in the sperm that he scrutinized through his microscope. In opposition to the Ovulists, or Ovists, who defined themselves as "believers in the egg," arose the Animalculists, or the "believers in the sperm." Animalculists maintained that it was the first male of the species created—and in terms of humans, Adam—whose germ cells contained all future generations of that species. The egg merely served the purpose of nurturing the homunculi-bearing sperm.

Upon being confronted by the Animalculists, the Ovulists regrouped and altered their position so as to engulf the discovery of sperm. In reality, the Ovulists reckoned, the sperm only provided the signal for the unfolding of the preformed individual contained within the egg to begin. The egg still contained the future generations of the species.

The Ovulists seemed to have won the day once and for all when, in 1740, the Swiss embryologist Charles Bonnet discovered that the unfertilized eggs of female plant lice could produce offspring. Bonnet interpreted this as direct proof that the female carried in her germ cells the males and females of future generations. Until the reality of "virginal generation," or parthenogenesis, was clearly understood—that it is commonplace in various insects, such as ants in the production of worker males or drones, for offspring to develop from unfertilized eggs—Bonnet's discovery remained a hard egg for the Animalculists to crack.

The demise of preformationist theories began with the work of the German embryologist Caspar Friedrich Wolff. Wolff was a student of natural science and medicine first in Berlin and then in Halle. In his doctoral dissertation, which he defended in late November 1759, he proposed a new theory, which he called the theory of epigenesis, in which he argued that the embryo developed from undifferentiated material. Like many of his colleagues and contemporaries, Wolff used as his major source of study the chicken egg, which was an easy subject for embryological research because of its size and availability. Although he ascribed physical growth to the presence of an omnipresent vital force in nature acting on the undifferentiated fluid of the egg—the idea of a vital force in nature being an accepted idea at the time—Wolff argued theoretically, and demonstrated by studying the ontogeny of the chick, that the embryo was not a miniature adult. In the case of the chicken, the embryo passed through definite stages of development in which juvenile features were transformed into those of the adult. A chick embryo's limbs were not merely tiny versions of an adult chicken's wings and legs. The complex system of blood vessels, and the long, complex intestinal tract of a chicken, to which the lungs and liver, as well as a number of important glands, are tethered, are not hidden in the early egg in a miniaturized form. Rather, both the vascular system and these organs begin to differentiate as the originally flat, undifferentiated embryonic body itself differentiates.

Wolff also set the stage for von Baer by discovering, and realizing, that it is only in their early stages that the embryos of different animals are similar to one another, and that animals become different as they mature. His investigations revealed a general plan among higher animals: that the embryos of all multicellular animals develop through a process of differentiation of cellular layers, and that the different organs as well as the vascular and nervous systems develop from these different cellular layers. More specifically, Wolff recognized that the various systems appear in a particular order, with, for example, the nervous system being laid down prior to the appearance of the vascular system. Although various claims by Wolff, as well as others, were discarded or revised with advances in technology, many of his insights into embryonic differentiation were groundbreaking. A generality that is still taught in anatomy classes is that the arrangement of nerves is more constant than that of arteries, which, in turn, is more constant than that of veins, and the reason for this is that this is the sequence in which these systems are laid down.

In addition to bringing his discoveries of normal development to bear in his arguments against the preformationist theory, Wolff was profoundly struck by the evidence of abnormal, or pathological, development. How, he wondered, can one reconcile the appearance of malformed, or "monstrous," individuals with the relative uniformity of form from one generation to the next predicted by the preformationist theory?

Similar to the fate of so many insightful scholars, Wolff's theory of epigenesis, as well as his concrete discoveries, went unnoticed by many in the field. Those whose attention Wolff's ideas did attract were swayed by the counterclaims of an influential professor of medicine at the University of Göttingen, Albrecht von Haller, a Swiss by birth. Although he did not embrace the preformationist theory as popularly portrayed—with the individual being present in the germ cell in a miniaturized form—Haller did maintain a version of it. He proclaimed that from his study of chick embryo development it was evident that the chick was already present in the unfertilized egg. What could not be ignored, however—and this is why so many embryologists before had difficulty—was the fact that the embryo's structures were invisible in the prefertilized state. Upon fertilization, the process of development begins, the heart starts to beat, and, eventually, the blood vessels and other organs become visible. Haller's notion of the invisible-but-nevertheless-present-individual-within-the-egg was applicable to all animals, including mammals. In mammals, the presence of the preformed individual was even more difficult to discern because of the small size of the ovum. But, invisible prior to fertilization or not, if female, this embryonic individual contained within her eggs all the future males and females of the species.

Haller's stand against Wolff's theory of epigenesis resulted in the latter scholar's being unable to secure a position in Germany or to present his ideas in public lectures. When, by supreme good fortune, he was invited to St. Petersburg by Catherine the Great of Russia in 1766, he went and stayed there for the remaining twenty-eight years of his life. During these years he continued his studies, which the St. Petersburg Academy of Sciences eagerly published. In 1812 the German embryologist Johann Meckel, who taught and conducted his research at the very university in Halle at which Wolff had earlier received his degree, translated one of Wolff's pre-Russia articles from Latin into German. As a result, the latter scholar's place in history was restored.

Meckel, who is perhaps best known to fledgling anatomists because the cartilaginous precursor of the bony lower jaw and a persistent part of the embryonic intestinal duct are named after him, also left his mark on the field of embryology. He set the stage for von Baer and Haeckel by suggesting that the ontogeny of an organism passed through the stages of the organisms that were less developed than it, with, perhaps, a new form or species "evolving" from the last in the sequence. In conjunction with the research that led him to this idea, Meckel was also attracted to the study of pathologically and abnormally formed—"monstrous"—embryos, fetuses, and newborns. Having supposedly found in these human monsters the organs of lower animals, Meckel explained this phenomenon as the result of a disruption in the normal course of the growth and development of the embryo. If, in its ontogeny, a "higher" animal passed through a series of stages of "lower" animals, then, Meckel concluded, the reason a human embryo, for instance, had the organs of, say, a fish, was that it had been prematurely derailed at that stage from its normal course of development up the ascending scale of organismal complexity.

The intellectual path from Wolff (belatedly) to Meckel and his contemporaries and then to von Baer and his colleagues led to an understanding of the fundamentals of ontogeny. But in spite of these insights into developmental processes and the differentiation of the organism after fertilization, by the time Darwin published *On the Origin of Species,* the scientific world was still in the dark as to how features were passed on from parent to offspring.

In Search of Heredity

Sometime during the years 1837–1838, Darwin scratched his pen in the pages of the notebook devoted primarily to thoughts on "the transmutation of the species," which is identified as Notebook B, and wrote: "The condition of every animal is partly due to direct adaptation & partly to hereditary taint." Possibly without knowing it, he was close to embarking on an attempt to explain how traits, and variants of traits, were passed on from one generation to the next. In 1838, in Notebook D, which he filled in only three months with his idiosyncratic shorthand approach to jotting down thoughts, arguments, counterarguments, and things to remember, Darwin formulated his theory of inheritance. Basic to this theory was the idea that the features a parent passes on to its offspring depend upon how long the former has lived and how much or what experiences it has lived through. Darwin's adherence to the so-called Lamarckian notions of use and disuse were very much a part of his theory of evolution. For the older a potential parent is, the greater the opportunity to acquire new characteristics or to modify others. In his notes about what kinds of features might be passed on as acquired features, Darwin distinguished between a mutilation, which would not be hereditary, and a true use-disuse situation, such as altering the size of one's muscles.

Darwin also speculated on hybrids and developmental monsters, or, as they were also then called, "sports of nature." In his mind a mule, which is the infertile offspring of a horse and an ass, and a person born with extra fingers or toes or a defective palate were similarly inconsequential examples of inheritance and evolutionary change. Although people born with malformations—and even the occasional mule—could produce offspring, they were oddities. For Darwin, monsters and sports of nature, as well as their spawn, were only fleeting and ephemeral examples of epiphenomena that were totally irrelevant to the course of evolution. On the other hand, Darwin strongly believed that "[w]hat has long been in blood, will remain in blood." As he explained it, what this meant was that "an animal . . . is <only> able to transmit . . . those peculiarities, to its offspring, which have been *gained slowly.*" Since mules can be created by breeders in one generation, and developmental malformations can appear in one generation and disappear just as quickly in the next, cases like these cannot serve as real examples of the course of inheritance and evolutionary change—which, according to Darwin, is slow and gradual.

Darwin's global picture of both gradual inheritance and evolutionary change is clearly expressed in a passage (D17) from Notebook D, which includes the shorthand notations he used when reworking his earlier penned thoughts (the sideways darts enclose his insertions):

> When two animals cross, each sends his own likeness, & the union makes hybrid, in fact the parents beget child like themselves. expression of countenances, organic diseases, mental disposition, stature, are slowly obtained & hereditary; <but if> if the change be congenital (that is most slowly obtained with respect to that individual) it is more easily inherited.—<<but if change be in blood long, it becomes part of animal &>> by a succession of <such changes> generations, these small changes become multiplied, & great change be effected.

As for how animals may pass their attributes on to their offspring, Darwin hinted at his solution when he stated in Notebook D that the "whole parent [must be] imbued with the change." As such, every part of an organism contributes to the stuff of heredity as passed from parent to offspring. This idea became the core of Darwin's theory of inheritance, which he called pangenesis.

While discussing whether "[w]ith respect to future destinies of mankind, some of species or varieties are becoming extinct," Darwin provided an additional clue to his ideas on inheritance, as well as on speciation. He used an imaginary example based on the consequence of the slave trade in Africa in general, and of the colonization of South Africa in particular. Both of these activities resulted in separate indigenous African tribes being placed in contact with one another. If, Darwin speculated, these tribes had remained apart, their distinctiveness would have been further accentuated by natural selection. Instead, he noted, "the tribes become blended & prevent that strong separation which otherwise would have taken place . . . otherwise in 10,000 years [the] Negro probably [would become] a distinct species."

Darwin's concept of "blending" in this quote may have been proposed only in reference to the mixing of the various tribes that became pushed together because of the slave trade and colonization. But he was probably also thinking of inheritance, because, at the time, features inherited from both parents were thought to combine in the offspring. If, however, these African groups were to remain separate from one another for a long enough period of time, there would be no blending, and changes that accrued via the action of natural selection would eventually lead to the creation of separate species. Although Darwin did not pursue models of speciation in his published works, in Notebook D he did approach formulating the mechanism that became popular in the twentieth century as the major source of organismal diversity: speciation by geographic separation, which is formally referred to as allopatric speciation.

The undertone of allopatric speciation in Darwin's staccato of thoughts that are recorded in Notebook D is interesting in light of the fact that although the title of his most well-known opus is *On the Origin of Species*, he did not articulate within its pages a model for producing diversity of life. All of Darwin's examples of evolutionary change, whether in *On the Origin of Species* or, later, in *The Descent of Man*, are those of linear transformation: change of a particular species over time brought about by the accumulation from generation to generation of myriad minuscule changes. As he put it in one of his entries in Notebook D, "by a succession of . . . generations, these small changes become multiplied, & great change be effected." Although according to this model of evolutionary change a species 100,000 years from now may be very different in traits that are typical of its collective membership from that species today (and whether you care to refer something 100,000 years from now to the same species as its present-day ancestor is another question altogether), there will still be only one species.

To today's evolutionary biologists, one of the ways in which organismal diversity can be produced via allopatric speciation is by disrupting genetic exchange between subgroups of a species (most easily by means of a physical barrier) and then letting natural selection take its course, picking and choosing among the advantageous and disadvantageous traits of each isolated subgroup. Over time, usually perceived as lots of time, each subgroup will become sufficiently different from the others that the members of a subgroup will no longer be able to interbreed successfully with the members of the other subgroups. Each subgroup would now be a new species, and diversity of life would have been achieved.

Although most schoolchildren today know something about genes and inheritance, none of the naturalists of the mid-nineteenth century who were trying to figure out evolutionary change, diversification of life, and the inheritance of traits did. Darwin may have struggled with notions of allopatric speciation in Notebook D, but, ultimately, he did not articulate a model—which might explain why various Darwin specialists, such as Peter Bowler, have argued that Darwin did not embrace geographic isolation as an important feature in the formation of new species. Given that Darwin was ignorant of genetics, it is no wonder that he proposed various models, including the notion that struggle and competition, in conjunction with natural selection, were sufficient to cause divergence between subgroups.

Among Darwin's theories was one on inheritance, which he put forth in a two-volume work entitled *The Variation of Animals and Plants under Domestication,* which he completed after *On the Origin.* According to letters and other documents, Darwin began working on *Variation of Animals and Plants* in early 1860 and completed it in late 1867. The work was published the following January. He referred to his theory of inheritance as the theory of pangenesis. In his journal, Darwin recorded completing the section on pangenesis in late November 1866. As indicated by his correspondence, Darwin first used the word *pangenesis* publicly in a letter that he wrote to Joseph Dalton Hooker, England's most eminent botanist, on April 4, 1867. There Darwin thanked Hooker for sending him an article in which a Mr. Traill reported having successfully joined the eyes of two kinds of potatoes to form a "mottled mongrel." For Darwin "[i]t would prove . . . to demonstrate that propagation by buds and by the sexual elements are essentially the same process, as pangenesis in the most solemn manner declares to be the case."

Reminiscent of the quote I cited earlier from Notebook D—the "whole parent [must be] imbued with the change"—Darwin's theory of pangenesis depicted inheritance as the sum total of contributions from every part of the parent. As he envisioned it, although all body parts go into making a complete individual, each body part is also an independent entity. At every stage of an organism's development, each body part, each organ, throws off minute gemmules, which disperse throughout the body and are capable of reproducing themselves. Most gemmules are active, but some are dormant and can be inherited from generations past. Dormant gemmules can be expressed in future generations, causing what appears to be a reversion, or atavism. Active gemmules can be suppressed, which must be the reason why a particular feature of one but not both parents is present in an offspring. The ability of gemmules to multiply and to express from every stage of development the organ or bodily part from which they were dispersed explains how a wound can heal, as well as how an amphibian can regenerate a lost tail or limb. Almost every cell of an organism contains the gemmules of every part of the organism—which is why, Darwin speculated further, botanists could believe that every cell of a plant is potentially capable of producing an entire plant.

Except for the unfertilized eggs of ants or other parthenogenetic animals or plants that can produce viable offspring, the cells that do not contain a complete representation of the body's gemmules are the sex cells. Upon fertilization, the gemmules from both parents combine or blend to produce the traits of the offspring. Blending also explains why the features of offspring often appear as intermediate in expression between the character states seen in the parents. Since gemmules come from every phase of growth of a particular body unit, a child does not grow into an adult per se. Rather, according to Darwin, it is the succession of gemmules representing different stages and ages that creates the adult. Since different gemmules represent different stages of life—old as well as young—avenues are available for passing on to the offspring characteristics that were acquired as an adult.

A theory of heredity based in part on the notion that gemmules represent features of later life also allowed Darwin to explain sexual dimorphism: how males and females of the same species can develop different characteristics. For example, consider the extremes in plumage size and coloration between the male peacock and the female peahen. In the male, the tail feathers, in particular, are far longer than the body of the individual and are brightly colored and conspicuously patterned. In contrast, the feathers of a peahen are remarkable neither in size nor coloration. In fact, the peahen is drab in comparison with the male.

When Darwin attempted to explain sexually dimorphic differences in the peacock in *On the Origin of Species,* he invoked a selection argument. He suggested that, in the female, natural selection had checked the development of conspicuous coloration and a long tail because the former trait would be dangerous for the female to have (vis-à-vis predation) and the latter an inconvenience while she was incubating her eggs. In *The Descent of Man,* however, which he completed in 1870, well after *Variation in Animals and Plants,* Darwin retracted his earlier explanation for the evolution of sexual dimorphism, stating that "after mature reflection on all the facts which I have been able to collect, I am now inclined to believe that when the sexes differ, the successive variations have generally been from the first limited in their transmission to the same sex in which they first arose."

How did Darwin envision the transmission of features to offspring of the same sex as the parent? Let us look at the peacock example again in the light of his mature reflection. The excessive and highly colored plumage of the male appears late, rather than early, in development. Indeed, Darwin had observed in a variety of different species of bird and mammal that features specific to one or the other sex that have to do with color, size, physical adornment or elaboration—the secondary sexual characteristics, as opposed to the primary features directly related to reproduction—tend to appear later rather than earlier in the course of physical maturation. Since males pass on to male offspring male secondary characteristics, and females do likewise with female offspring, it follows, Darwin argued, that features that develop later in life will be passed on to offspring of the same sex.

The same day, January 30, 1868, on which *Variation of Animals and Plants* was published, Darwin wrote to his friend, colleague, and public advocate, Thomas Henry Huxley:

> Most sincere thanks for your kind congratulations. I never received a note from you in my life without pleasure; but whether this will be so after you have read pangenesis, I am very doubtful. Oh Lord, what a blowing up I may receive. I write now partly to say that you must not think of looking at my book till the summer, when I hope you will read pangenesis, for I care for your opinion on such a subject more than for that of any other man in Europe. You are so terribly sharp-sighted and so confoundedly honest!

And on April 22, 1868, Darwin wrote to the British explorer and botanist George Bentham:

> I have been extremely much pleased by your letter, and I take it as a very great compliment that you should have written to me at such length. . . . I am not at all surprised that you cannot digest pangenesis: it is enough to give any one an indigestion; but to my mind the idea has been an immense relief, as I could not endure to keep so many large classes of facts all floating loose in my mind without some thread of connection to tie them together in a tangible method.

What Darwin did not know (shades of Alfred Russel Wallace coming up with a theory of evolution by means of natural selection before Darwin had published his version) when he was formulating his theory of pangenesis was that approximately nineteen hundred years earlier the renowned Greek physician Hippocrates, the father of medicine, had outlined the essence of this theory in two of his writings, *On the Sacred Disease* and *On Airs, Waters, and Places.* The major difference was that Hippocrates spoke of a "seed," whereas Darwin had his gemmule. But, like Darwin's gemmule, Hippocrates' seed derived from all parts of the body. Hippocrates had even speculated, using changes in head shape as an example, that artificially induced, acquired characteristics could be incorporated into the seed and subsequently passed on to the next generation. But what was probably more devastating to Darwin than his belated discovery of Hippocrates' priority in proposing a theory similar to his was the hurt he felt at what he took to be a misinformed public attack on his theory of pangenesis by his cousin Francis Galton.

Like Darwin, Galton had inherited sufficient family money that he did not have to work in order to support himself, his family, or his scientific and musical pursuits. Also like Darwin, Galton was interested in just about anything intellectual that crossed his path. He was an explorer, a geographer, a biologist, an evolutionist, an anthropologist, one of the first biometricians, and also claimed the now-unflattering title of the father of eugenics. Galton, who was similarly deeply invested in deciphering the mysteries of heredity, was one of the few scholars who took Darwin's theory of pangenesis seriously—seriously enough that he set out to conduct experiments to document it. Even while Galton was conducting these experiments, which did not support what he understood to be his cousin's theory of pangenesis, he kept Darwin informed of what he was doing.

Galton believed that when Darwin wrote about gemmules being dispersed throughout the body, he meant that they were carried via the blood. Consequently, Galton set about injecting "alien blood" into purebred strains of animals and then breeding the transfused animals to see what the offspring would look like. Theoretically, as Galton understood Darwin's theory of pangenesis, purebred strains transfused with the blood of mongrels should produce offspring of which at least some would display features of the donor of the alien blood. The animals Galton chose to "mongrelize" were rabbits, and the particular strain was the silver-gray.

Silver-gray rabbits seemed the ideal experimental animal for a variety of reasons. First, they show changes with age. With the occasional exception of a white streak down the face, they are born with a completely black coat, which becomes gray a few weeks later. The white streak, if present, eventually disappears. Second, silver-grays never have lop ears. And third, if variants are born, they are both infrequent and their peculiar characteristics sufficiently well known that anyone could tell that these individuals were born of a silver-gray stock.

In one experiment, Galton took two ounces of blood from the animals to be transfused and replaced it with the same amount of blood from a donor. As for the donors, which he sacrificed, Galton took any rabbit he could get: a yellow, common gray, or black-and-white rabbit. In another experiment, he took out three ounces of blood and replaced it with the same amount from donors. In both of these experiments, Galton stirred or even whipped up the blood to remove what he called fibrin. And in the third experiment, in which the silver-gray and the donor were anesthetized and their carotid arteries connected by tubes, Galton replaced all of the whole blood of one individual with that of the donor. Following these experiments, Galton, ever the mathematician, calculated how much donor, or "alienized," blood each silver-gray had in order to determine the percentage of non-silver-gray gemmules the offspring of the experimental silver-grays would inherit. Prior to mating, he let the males in the experiment get rid of "old reproductive material" by putting them in cages with nonexperimental females.

The results of breeding the alienized silver-grays were disastrous. Some males and females would or could not breed at all. Of the litters produced, there was one individual with a white leg. As for the males and females that could not breed after being injected with donor blood, Galton had this to say: "My results thus far came to this, viz. that by injecting . . . blood I had produced no other effect than temporary sterility. . . . I had injected alien corpuscles but not alien gemmules." Galton would have accepted pangenesis had the total transfusions, which replaced whole blood with whole blood, worked out. But they did not. He had to conclude that there was no proof of pangenesis, and that "if the reproductive elements do not depend on the body and blood together, they must reside either in the solid structure of the gland, when they are set free by an ordinary process of growth, the blood merely affording nutriment to that growth, or else that they [are] . . . temporary inhabitants of [the blood], given off by existing cells, either in a fully developed state or else in one so rudimentary that we could only ascertain their existence by inference."

With Galton reading this paper first on March 30, 1871, at a meeting of the Royal Society of London, and then its being published within the next few weeks, the criticism of Darwin was aired twice publicly and in a short period of time. Because Galton had kept him apprised of his experiments, Darwin should not have been surprised by his cousin's results and conclusions, or by the next logical step of presenting these results in a broader scientific forum. However, Darwin was sufficiently upset, and offended, to respond in a letter to *Nature*. He vehemently, and at great length, denied even leaving the possibility open for someone to think that gemmules were circulated through the body via the blood system. It is difficult to understand why Darwin did not stop Galton before his experiments were well under way, or at least let the latter scientist know that he did not believe he had meant that gemmules were transported through the blood. But in his letter to *Nature* Darwin did allow that "when I first heard of Mr. Galton's experiments, I did not sufficiently reflect on the subject, and saw not the difficulty of believing in the presence of gemmules in the blood." In an obvious tone of sarcasm, Darwin closed his letter with the backhanded compliment that although Galton had been totally off base, "every one will admit that his experiments are extremely curious, and that he deserves the highest credit for his ingenuity and perseverance."

Now that Galton was also publicly embarrassed, he, too, drafted a response to *Nature*. His case was based on Darwin's use in *Variations in Animals and Plants* of three words—*circulate, freely,* and *diffused*—in reference to the dispersion of gemmules throughout the body. What else but "blood circulation" could be implied by these words and the phrases in which they were used—"circulating freely" and "the steady circulation of fluids"? But Galton's hurt was palpable. He believed that he had been betrayed by Darwin's ambiguity both in language and in action. In his long conclusion, Galton put an ironic twist on an analogy to a scene Darwin had employed while arguing how language could have evolved, a scene in which an ancient human predecessor had imitated the sound of a predator in order to inform his comrades of potential danger:

> I feel as if . . . having heard my trusted leader utter a cry, not particularly well articulated, but to my ears more like that of a hyena than any other animal, and seeing none of my companions stir a step, I had, like a loyal member of the flock, dashed down a path of which I had happily caught sight, into the plain below, followed by the approving nods and kindly grunts of my wise and most-respected chief. And I now feel, after returning from my hard expedition, full of information that the suspected danger was a mistake, for there was no sign of a hyena anywhere in the neighbourhood. I am given to understand for the first time that my leader's cry had no reference to a hyena down in the plain, but to a leopard somewhere up in the trees; his throat had been a little out of order—that was all. Well, my labour has not been in vain; it is something to have established the fact that there are no hyenas in the plain, and I think I see my way to a good position for a look out for leopards among the branches of the trees. In the meantime, *Vive* Pangenesis.

In spite of these exchanges, and the fact that Darwin chose to make negative reference to Galton's transfusion experiments in the second edition of *Variation of Animals and Plants,* the two cousins remained close friends, colleagues, and intellectual sparring partners. In 1875, the year it was published, Galton lectured on his new theory of heredity at the Anthropological Institute in London. In November, and again in December of the same year, Darwin sent his cousin correspondence that was both congratulatory and critical of the new theory.

Galton's theory differed significantly from Darwin's theory of pangenesis in that it envisioned the stuff of heredity as being derived from the reproductive organs themselves, rather than being a concentration of products from all parts of the body. Galton coined a new term for the unit of heredity, "stirp," which he took from the Latin word for "root" (*stirpes*). He proposed that the germs of each body part are contained in the stirp. Continuity from generation to generation is maintained by the contributions of the parental stirps to the offspring's stirp. Stirps that do not go into the development and formation of the offspring serve as the source of that individual's stirp, which, in turn, is passed on to the next generation. In contrast to the general views on heredity at the time—that inheritance provided an identity or sameness from one generation to the next—Galton expressed the belief that "the personal structure of the child is no more than an imperfect representation of his own stirp, and the personal structure of each of the parents is no more than an imperfect representation of each of their own stirps." As such, while stirp was passed from parent to child, individuality was maintained.

Inasmuch as the stirp was produced in the sexual organs, there was no allowance in Galton's theory of heredity for the effects of use and disuse of organs and the inheritance of acquired characteristics. The only possibility Galton conceived of in which acquired characteristics could be passed on to the next generation was that some external force might impact the sex organs. Even so, Galton considered the probability of inheriting acquired characteristics to be slim.

As for the generally held notion that characters blend, Galton again took a different position. First, somewhat like Darwin, he broke away from the idea that the primary focus of heredity was between parent and offspring. This idea of heredity made no allowance for reversals, or atavisms. Instead, Galton envisioned heredity as including the stuff of ancestors as well as of parents. But since each parent can contribute to the next generation only half of its stirp and the germs contained therein, there will be a competition that will allow only some of the germs to be passed on. Second, and here is where he was unique, Galton proposed that the germs acted independently of one another. They were not, as was commonly believed, inherited as factors that blended or hybridized. Instead, they were "particles" that maintained their own identities as units. Galton was able to see that this type of inheritance made the best sense of how a child could have certain features of one parent and other features of the other parent. In one fell swoop, Galton's theory of heredity was able to explain from the same naturally occurring mechanism both how there could be physical similarity between offspring and parent and how individuals could differ from one another.

But Galton put another twist on his theory of heredity, one that eventually led to his proposing a program of eugenics, in which the occasional "great" individual would be bred to create more like himself. Whereas Darwin perceived the differences between individuals of the same species as constituting a continuum of variation, Galton, who thought that "primitive" African tribes were racially distinct, believed that differences between groups within a species were discrete and discontinuous. For Darwin, evolutionary change occurs slowly over time and through the gradual shift in the representation of traits within a species. As Galton saw it, discontinuous traits meant that evolutionary change comes about through a leap, or saltation, from one plateau or average representation of traits to another. Natural selection and sexual selection had no place in Galton's view of evolution.

Galton sent Darwin a copy of his article in November 1875. On November 7, Darwin wrote that "[u]nless you can make several parts clearer I believe . . . that only a few will endeavour or succeed in fathoming your meaning." Starting with *stirp* and *germ,* Darwin proceeded to enumerate his questions and calls for clarification. But whether others found Galton's presentation of his theory of heredity as difficult to understand as Darwin did, it is the case that his theory had little impact on the contemporary scientific community. Indeed, it was fifteen years later that Alfred Russel Wallace, the "other" author of evolution by natural selection, would admit in a footnote deep in the bowels of one of his books that he had only recently come upon Galton's theory of heredity. But, having done so, he offered the opinion that if this theory were to be "established," it "may be considered . . . the most important contribution to the evolution theory since the appearance of the *Origin of Species.*"

Unfortunately for Galton, few scientists shared Wallace's sentiments, in spite of the fact that Galton did provide, at least in some form, ideas that would later be more fully explored by others. Indeed, in addition to broadly theoretical notions—such as those that would be important to understanding the discrete particulate nature of inheritance—he also developed various mathematical formulas to express hereditary relations. Because he thought in terms of heredity across many generations, rather than only from parent to child, Galton was able to calculate the percent of contribution of heritable material not only from parents but also from grandparents, great-grandparents, and so on into the generations of the past. These percentages are fundamental to an understanding of simple population genetics and are basic to the tenets of the now-popular field of sociobiology.

The Discovery of Inheritance

Darwin's theory of pangenesis and Galton's theory of heredity may have differed in where the stuff of inheritance came from, and in how it was expressed in offspring, but it is unlikely that either of the theories would have been conceived were it not for the research on cells that took place during the first half of the eighteenth century. Of course, studies at the cellular level would not have been possible were it not for the invention of the microscope. Thanks to the work of the German botanist Matthias Jakob Schleiden and those he either recruited to work with him, such as the zoologist Theodor Schwann, or those whose work he encouraged, such as the botanist Karl Wilhelm Nägeli, the role of the cell and its products as the basis of organismal development and life itself began to be understood. But while these giants in the field of cytology—the study of the cell—opened the door to the function of the cell, they remained in the dark regarding a cell's origin. It was only when, after the mid-century, the German physician and anthropologist Rudolf Virchow, among others, discovered that cells derive from other cells, and do not, as was commonly believed, arise through a process of concentration of materials, that advances in biology, and particularly in the study of heredity, could be made. As Virchow stated so clearly on February 17, 1858, in the second of a series of lectures presented at the Pathological Institute of Berlin: "[*N*]*o development of any kind begins* de novo. . . . Where a cell arises, there a cell must have previously existed . . . just as an animal can spring only from an animal, a plant only from a plant."

Armed with a basic, working knowledge of the cell, August Friedrich Leopold Weismann, the German physician turned university professor of comparative anatomy and zoology, conducted research on microscopic pond-dwelling invertebrates, flealike daphnia and tentacled medusae (also called hydras), which eventually led to his formulation of a theory of heredity based on the continuity of what he called germ plasm. His driving question was how features are passed on to offspring with such obvious precision. He was also concerned with the matter of how, over geologic time, structures can persist essentially unchanged in species of plant and animal. The answer, it was much later discovered, was that the information necessary to produce an entire organism is contained in the sex, or germ, cells. Although the great disparity in size between the egg and the sperm of an animal, or their equivalents in a plant, had proved troublesome to earlier cytologists who attempted to understand which cell and which part of that cell contributed most to the formation of the offspring, Weismann's own studies contributed to finding the answer.

Weismann identified the hereditary material as the germ plasm or, at times, as the germ nucleoplasm. He referred to the properties of the germ plasm as "the tendencies of heredity." He further argued that the germ plasm is contained within the center, or nucleus, of each germ cell, not in the fluid encased within the shell of the cell, within which the nucleus in its membranous covering resides. Upon fertilization, the tail of the sperm remains outside the egg, and only the nuclei of the sperm and the ovum fuse. In turn, each nucleus contributes the 50 percent of its germ plasm and the half of its chromosomes (which at the time were called loops) that are necessary for the formation of the first cell of the embryo. After fertilization, the egg has the full complement of loops and germ nucleoplasm of an individual of that species.

Weismann astutely recognized that it was the germ plasm that provided the information necessary for the subsequent development and formation of the individual. Reminiscent of Galton's depiction of the stirp, Weismann suggested that, while some of the germ plasm is involved in the development of the individual, a part of it is preserved intact to produce the germ cells of that individual. The picture is simple and elegant. The germ plasm resides in the nucleus of the sex, or germ, cell. It gives rise to an individual's body, or somatic, cells and their nucleoplasm, as well as to the sex, or germ, cells and their germ plasm. Extending the idea of this process back in time, Weismann suggested that an individual inherits part of the germ plasm not only of its parents but of their parents, their grandparents, their great-grandparents, and so on. This aspect of inheritance explains how a feature, presumably long lost in an earlier ancestor, can reappear as an atavism.

As for how an individual, which is made up of different cells that form different structures and perform different functions, develops, Weismann proposed an equally elegant solution. As everyone in the profession knew, from the fertilized egg, the individual enlarges and grows by cell division and multiplication. But Weismann suggested that, with cell division, changes arise in the nucleoplasm that cause differences to appear in the next generation of cells. The differentiation of cell layers, tissues, and organs, as was first noted by Wolff, is a direct result of changes in the nucleoplasm that occur during the development of the individual. In multicellular animals, for example, cell divisions give rise to two layers of cells, an inner (endoderm) and an outer (ectoderm) layer. Later, ectoderm cells divide and differentiate into cells that give rise to the nervous system and to cells that will become the outer covering of skin. As the cells of the nervous system divide, they, in turn, become differentiated into cells that produce the sense organs and those that produce other organs. Simply put, the process of differentiation proceeds from the generalized to the specific. For Weismann, the key to this orderly progression of development and differentiation lay in the fact that although a mother cell divided in half to produce two externally identical daughter cells, the division was unequal in quality. Although the details of cellular layers and differentiation were yet to be worked out correctly, Weismann provided a model of heredity that explained the process of cellular specificity.

In addition to formulating a theory of heredity that recognized the significance of the contents of the nucleus of sex cells, Weismann used the information at hand to argue convincingly that Darwin's theory of pangenesis was invalid. How could it be otherwise? If the "tendencies of heredity" are dictated by the germ plasm, and the germ plasm is contained within the nucleus of the sex cells, then the concept of gemmules cannot be true. If development begins with the fertilization of the egg, and differentiation of the parts and subparts of the developing individual results from information specified by the nucleoplasm of the cells themselves, then acquired characteristics could never be inherited. Although the botanist Nägeli would try to rejuvenate the pangenesis version of the inheritance of acquired characteristics in the 1880s by way of idioplasm, which he thought resided in the substance of the cell lying outside the nucleus, Darwin's theory did not stand up to the realities of biology. Heredity and the formation of parts were inextricably tied to the cell and the contents of the nucleus. But it would be some time before the basis of how cells become organized to produce a complete, fully functional, but minutely differentiated organism would be understood.

On the Road to Inheritance

Many advances in medicine have arisen through the sheer need to deal with the conditions and atrocities of war rather than in the pristine environment of a medical laboratory. Similarly, many discoveries in biology came about for economic and practical reasons rather than as a result of pure intellectual curiosity. So it is that early experiments in animal husbandry and plant hybridization, which eventually led to an understanding of the mechanisms of inheritance, were originally directed toward providing food, or plants and animals of beauty or curiosity, with a known and hopefully minimized risk to the farmer.

In 1822 Johann Mendel was born to a peasant farm family in Silesia (now in the northern part of the Czech Republic). Although he could very well have ended up spending his life tending to the family farm, first his father and then his sister supported his education. By the age of twenty-one, he had succeeded in his studies of physics and mathematics at the Olmütz Philosophical Institute and had been recommended by his physics professor, Friedrich Franz, to the Altbrünn Monastery in Brno (then called Brünn), not with the primary intent of becoming a priest but in order to continue his studies. In 1847 Mendel, who took the name Gregor upon becoming a novice at this Augustinian monastery, was ordained into the priesthood, which gave him protection from daily wants and also afforded him entrée into the cultural and scientific communities. He taught mathematics and eventually came to tend the gardens of the monastery and experiment with plant hybrids, particularly those of the garden pea. Today, every schoolchild has probably read about Gregor Mendel's discoveries of the basic laws of inheritance: that characters segregate, that some features are dominant over others, and that breeding hybrids derived from known stock produces descendants in a specific ratio. But, as the historian of science Robert Olby has remarked in his review of Mendel's ideas, given the general interest of society at large in plant breeding and hybridization, the monk's studies were neither unique nor extraordinary.

Mendel's personal papers, and along with them the history of how he came to begin his studies in genetics, were destroyed. But we do have his scientific papers and official documents. According to Olby's sleuthing, Mendel had probably been schooled not only in matters of the cell, fertilization, and organismal growth, especially plant growth and the work of Nägeli, but also in a language of evolution. And it was apparently an interest in the latter subject that led Mendel to pursue his experiments in plant hybridization.

The primary question with which Mendel struggled was whether species were stable in nature, or whether variation arose naturally, with higher-level differences being those that distinguished species from one another and lower-level differences being those of the varieties or subspecies within a single species. Inasmuch as plant breeders in pursuit of new and stable hybrids would have to propagate generation after generation of plants in order to reach their goals, Mendel knew that the task at hand would be daunting. But he had to go through these rigors of breeding, crossbreeding, and back-breeding if he were to have even a chance of formulating "a general law governing the formation and development of hybrids," from which it would be "possible to determine the number of different forms under which the offspring of hybrids appear, or to arrange these forms with certainty according to their separate generations, or definitely to ascertain their statistical relations." It is here that Mendel's background in science, particularly in physics, came into play. For it was the calculation of the statistical relations of the offspring of hybrids that would distinguish his experiments from the work of the typical dedicated plant breeder.

It was fortunate that Mendel used the common edible pea as his experimental plant. In this particular plant, the characters he focused on—such as seed shape or seed color—are not linked to one another, which means that the expression or lack of expression of each feature would be independent of the others. Because of this stroke of luck, Mendel could follow the course of inheritance of each feature separately. Before he began his experiments, he had to make certain that the strains, or varieties, of the three different species of pea he chose to work with were as pure in their features as possible—that each variety would breed true to its distinguishing characteristics. Beginning in 1856, he spent two years following every generation of each variety of pea to ensure that each variety bred true. He artificially pollinated each parental generation to protect against his plants being fertilized by foreign pollen. Having satisfied himself at the end of these two years that his chosen varieties remained "constant" in their features, Mendel set about his experiment in hybridization.

It was known among plant breeders that the features of two parental plants were passed on intact to the hybrid generation, and also that the expression of these features in the post-hybrid generations was variable. But no one knew if this was a random event or whether a law governed the expression of features in the offspring of hybrids. This was the problem Mendel set out to tackle. In order to determine this law statistically, Mendel had to perform "as many separate experiments as there are constantly differentiating characters presented in the experimental plants." The categories of characters, and the different character states, that Mendel followed through in his experiments were differences in form of ripe seed (round or wrinkled); color of the food supply within the seed (yellowish or greenish); color of seed coat (whitish or grayish); form of ripe pod (inflated or wrinkled); color of unripe pod (green or yellow); position of flowers (along the stem or at the top); and length of stem (although variable, constant enough for each variety). Mendel bred more than ten thousand plants, the vast majority of which were planted in the monastery gardens, with a few grown in pots in the greenhouse as controls.

Although it occasionally happened in some plants that some features in the hybrid generation appeared to be intermediate between the features of the parents, this, fortunately, was not the case with Mendel's peas. Rather, as other plant breeders had also found quite frequently in their experiments, hybrid peas more closely resembled one or the other parent. To reflect the differential inheritance of parental features, Mendel coined the terms "dominant" and "recessive." "Dominant" and "recessive," respectively, were synonymous with similar terms—"active" and "latent"—other plant breeders, such as the Dutchman Hugo de Vries, used to describe similar phenomena. Dominant characters, Mendel wrote, are those "which are transmitted entire, or almost unchanged in the hybridisation, and therefore in themselves constitute the characters of the hybrid." Recessive characters are those features "which become latent in the process . . . [and] withdraw or entirely disappear in the hybrids, but nevertheless reappear unchanged in their progeny." In his notes, Mendel used a capital letter (such as *A* or *B*) to denote the dominant character state, and a small letter (*a* or *b*) for the recessive character state. Since Mendel knew which parent contributed which character state for a particular character (such as round versus wrinkled seed shape), he could figure out which character state was the dominant and which the recessive one by seeing which was expressed in the hybrid.

The test of what happened to the recessive character states came with the breeding of hybrid with hybrid. Again, Mendel was fortunate in that none of the offspring of this mating were intermediate in their characters. In the cross between hybrids there were some individuals that had the dominant character state and others that had the recessive character state. This was no surprise. Plant breeders had long known that features that were absent in one generation might appear in another. But what Mendel did succeed in demonstrating with his hybrid crosses was that dominant and recessive character states were expressed within the population in a particular ratio, 3 to 1. The dominant character state was represented in three times as many individuals as the recessive character state.

In one experiment, for example, involving seed shape, out of 7,324 seeds retrieved, 5,474 were round and 1,850 were wrinkled, yielding a ratio of 2.96 to 1, or essentially 3 to 1. In another experiment, involving color of nutrient matter within the seed, 6,022 of 8,023 seeds were yellow and the remaining 2,001 were green, yielding a ratio of 3.01 to 1, or essentially 3 to 1. In every other cross-hybrid experiment, for every other category of character under scrutiny, Mendel found ratios approximating the magical 3 to 1. The recessive character states reemerged intact and unaltered from the first parental generation.

In the next step, Mendel bred the second hybrid generation of offspring with one another. The crosses between plants with the recessive character state resulted in offspring in which the feature of the recessive character state was expressed. But the crosses between hybird plants with the dominant character state produced some offspring with the imprint of the dominant character state and others with that of the recessive character state. Again, recessive character states reemerged. When he totaled his figures, Mendel found that the ratio of dominant versus recessive-based morphologies was 3 to 1, with the ratio of hybrids to the pure dominant forms being essentially 2 to 1. The total proportion of pure dominant, versus hybrid, versus pure recessive offspring was then 1 to 2 to 1. Just as the ratio 3 to 1 was a constant in terms of how many individuals of a cross between hybrids bred from pure strains expressed the dominant character state and how many expressed the recessive character state, so, too, the ratio 1 to 2 to 1 consistently reflected the nature of the cross between individuals of the 3 to 1 generation. In terms of the kinds of characters Mendel studied, these ratios still hold true today.

Using capital versus lower-case letters to denote dominant and recessive character states, Mendel proposed the following expression for the 1 to 2 to 1 generation:

$$A + 2Aa + a.$$

This formula allows us to understand the background of the character that will be featured in the adult that develops from a particular seed. Of the three seeds that grow into plants expressing the dominant character state—which is what we see as the physical attributes—only one individual is hereditarily pure for that trait. The other two individuals, which on the surface bear the stamp of the dominant character state, are carrying an unexpressed recessive character state. They are hybrids. And as only one individual is pure for the dominant character state, only one individual is pure for the recessive character state, which is the only way that an individual can express the recessive character state. In the twentieth century, this formula would be represented as:

$$AA + 2Aa + aa.$$

Olby has argued that Mendel was not Mendelian if part of the definition is that traits are inherited in pairs. Olby's interpretation certainly follows from the $A + 2Aa + a$ formula that Mendel used repeatedly. However, in one section of his article, Mendel did detail the possible combination of A's and a's coming from pollen and egg cells, and from this he derived the following formula:

$$A/A + A/a + a/A + a/a.$$

He then compressed this formula into the more familiar:

$$A + 2Aa + a.$$

The first of these two formulas appears to represent what is actually inherited from each parent. The second of the two formulas, which Mendel used more frequently, reflects the morphology of the individuals and at the same time recognizes the presence of an unexpressed recessive in the hybrid. Just how completely or incompletely a Mendelian Mendel really was will always be a matter of interpretation.

Now, it is often portrayed in general textbooks on the subject, and is indeed part of the lore, that Mendel not only discovered particulate inheritance but that he also identified the process in these terms. This, as Olby has clearly argued, does not appear to be the case. However, the other claim about Mendel—that he recognized that independent, unlinked characters can segregate and be inherited in different combinations—is true, and is obvious from the following passage:

> There is therefore no doubt that for the whole of the characters involved in the experiments the principle applies that *the offspring of the hybrids in which several essentially different characters are combined exhibit the terms of a series of combinations, in which the developmental series for each pair of differentiating characters are united . . . that the relation of each pair of different characters in hybrid union is independent of the other differences in the two original parental stocks.*

Mendel discussed his experiments and reported his results at two different meetings, February 8 and March 8, 1865, of the Natural History Society of Brno. The response from the audience was not even underwhelming. There was no response, no taking the challenge to pursue additional experiments along the lines of inquiry that he had established. Nonetheless, he was asked to submit a manuscript for publication in the *Proceedings* of the society, which he did. It was published the following year, 1866, and Mendel was given forty reprints of his article to send to colleagues.

One of the people to whom Mendel sent a reprint was Nägeli. This would seem to be a natural choice, since Nägeli was the premier botanist of the time. But Mendel might as well have kept the reprint for himself for all the impact it had on Nägeli. Nägeli was not impressed with Mendel's demonstration of constant forms. He was convinced that, eventually, the constancy that Mendel described for his hybrids would break down, and variation and intermediacy—the supposed norm—would prevail. It is also the case that Nägeli's own theory of the idioplasm could not accommodate Mendel's conclusions. Without Nägeli's support, and from it the public recognition he would have needed to promote his ideas, Mendel retreated into the world of his monastery. There he spent the rest of his life, breeding hybrids of different sizes and tolerances that would enable his plants to grow well in the available space and light of the monastery's gardens.

7

Genetics and the Demise of Darwinism

Of course, with the single steps of evolution [natural selection] has nothing to do. Only after the step has been taken, the sieve acts, eliminating the unfit. The problem, as to the manner in which the individual steps are brought about, is quite another side of the question.

—Hugo de Vries (1906)

Hugo de Vries, the Rediscovery of Mendel, and the Questioning of Darwinism

The year 1900 is generally accorded the distinction of being when Mendel's results were rediscovered independently by three plant physiologists and hybrid experimenters, the German Karl Erich Correns, the Austrian Erich Tschermak von Seysenegg, and the Dutchman Hugo de Vries. The degree to which each of these scientists actually rediscovered and truly understood the significance of Mendel's results is still a matter of debate.

Perhaps the most historically visible of these three botanists was the Dutchman, Hugo de Vries. While he, Correns, and Tschermak von Seysenegg had been working independently during the late nineteenth century on hybrid experiments that would eventually lead each of them to the same results Mendel achieved—the law of segregation and the sacred 3 to 1 ratio of dominant versus recessive character-state expression in the hybrid generation of two pure strains—de Vries was the first to publish a version of this formula.

The problem with rediscovery, however, is that it is difficult to know just how much inside information the rediscoverer had beforehand. Although Mendel's publication of 1866 was never widely known, it did fall into the hands of a few plant breeders who then passed it on to others, including Tschermak von Seysenegg, de Vries, and Correns. Tschermak von Seysenegg, who apparently understood the implications of Mendel's laws the least, obtained his results from large-scale crossings between various varieties of a particular species of pea, *Pisum sativum.* Although Tschermak von Seysenegg ended up corroborating Mendel's ratio, his research goal had actually been to investigate the viability of cross-fertilization in this species of pea. Correns, on the other hand, who had studied under the famous and influential botanist Nägeli and had even married the senior scientist's niece, was probably most in tune with the implications of the segregation of characters and the 3 to 1 ratio. His first forays into the realm of Mendel's experiments, which he had hoped to replicate, had also been with peas, crossing plants with different seed colors.

Although de Vries may have inadvertently stumbled onto the 3 to 1 ratio and the law of segregation, he saw that these basic tenets of inheritance might have some significance not only for plant breeders but also and especially for his own theory of inheritance. More important, from the point of view of the history of rediscovery, de Vries managed to get his ideas into print before the others did. Coincidentally, de Vries happened to send Correns a reprint of his article on the subject—"On the Laws of Segregation in Hybrids" (in the original French, "Sur la Loi de Disjonction des Hybrides"), which he published in the journal of the French Academy of Sciences, *Comptes Rendus,* during the very period when Correns was putting the finishing touches on his own manuscript on the very same subject. (Here, again, are shades of Alfred Russel Wallace sending Darwin a letter spelling out the very ideas of evolution by means of natural selection that the elder naturalist himself had been working on for years.) Correns received de Vries's article on April 21, 1900, and worked like a fiend to complete his manuscript, which he did within twenty-four hours. He then immediately sent it to the German Botanical Society, which published it shortly thereafter as a report in its journal series.

With obvious bitterness at having been upstaged by de Vries, Correns made a point in this article of acknowledging Mendel as the true discoverer of the law of segregation and of the 3 to 1 ratio. Although de Vries had the essentials of Mendel's discoveries down pat in his *Comptes Rendus* article—including his use of the terms *dominant* and *recessive* instead of his favorite words for the same concepts, *active* and *latent*—he had made no mention whatsoever of Mendel. The scientific detective work of the historian Onno G. Meijer suggests, however, that, as early as 1897, de Vries had in his possession a copy of Mendel's 1886 article. But, judging by the articles, notes, and classroom presentations that are preserved in de Vries's archives, he did not understand the significance of Mendel's discoveries at the time. As Meijer determined from the evidence, however, by 1900, not only had de Vries reread Mendel's article but, this time, he also understood Mendel's law of segregation and the 3 to 1 ratio. If this story is true, it is of particular interest that de Vries failed to mention Mendel in his *Comptes Rendus* article. No wonder Correns, who had cited Mendel the previous year in an article he wrote on the observation that a hybrid will express the feature of only one of its parents, was miffed that de Vries had stolen his own thunder. It is also not surprising that Correns would go out of his way to propose in his German Botanical Society article that de Vries must have been as surprised as he when he finally figured out that Mendel had actually discovered the law of segregation and the 3 to 1 ratio some thirty-five years earlier.

Whether Correns had planned to claim independent discovery of these laws of inheritance had de Vries not beaten him to the punch is unclear. But, given the event of April 21, 1900—receiving a copy of de Vries's rediscovery article in the mail—the most that Correns could do was try to undercut de Vries's priority of discovery by giving full credit to Mendel. Unbeknownst to Correns, while he was madly reworking his manuscript on April 22 in order to deal with de Vries's article, was that more than a month earlier de Vries had sent another manuscript on the subject to the journal of the German Botanical Society in which he did credit Mendel with having earlier discovered the law of segregation and the 3 to 1 ratio. In this second manuscript, which appeared in print only a few months later, although de Vries claimed independent discovery of Mendel's laws, he tried to diminish them in importance by relegating them to the status of a special case with peas. Nevertheless, de Vries did acknowledge Mendel's discovery of the principles of inheritance, which he later rediscovered, fleshed out, and proved valid for true hybrids.

But while for Correns the major accomplishment of hybrid breeding experiments was the rediscovery of Mendel's laws, this was not the be-all and end-all for de Vries. As early as the late 1880s, de Vries had begun developing theories of inheritance and of evolution. And it was toward a demonstration and explication of these theories that he pursued the hybridization experiments that eventually led him to intersect with Mendel.

In 1889 de Vries published his theory of inheritance, which he called intracellular pangenesis. The "pangenesis" part of his theory was not coincidental. De Vries had read Darwin's theory of pangenesis. Having been intrigued by some of Darwin's propositions, he sought to develop a research plan that would elaborate upon these parts of the Englishman's theory that seemed viable. As de Vries saw it, there were two major aspects to Darwin's theory of pangenesis. The first was that characters are inherited through real, material particles (Darwin's gemmules), and that these particles are capable of reproducing themselves. The second was that these particles derive from every cell of the body during every stage of the life of an individual, and this ongoing process is what allows for the inheritance of acquired characteristics. It was in the first part of Darwin's theory of pangenesis (particles of inheritance), not the second (the inheritance of acquired characteristics), that de Vries found potential import. After Weismann's rejection of Darwin's basis for the inheritance of acquired characteristics, no one, including de Vries, could embrace it. But one could accept the possibility that the inheritance of real biological characters occurred through the inheritance of particles.

De Vries called the particles that were responsible for the inheritance of characters "pangenes," or "pangens." With pangens being the basic units of inheritance, he could explain what happened in hybridization and crossbreeding experiments in terms of the expression of the pangens that were passed on from each parent. Although de Vries was not Mendelian in the sense of envisioning the inheritance of pairs of pangens—with one of each pair of pangens being inherited from each parent—he did equate each pangen with the expression of a character. Although many pangens were to exist side by side and concurrently in the same individual, for de Vries they were discrete and independent units. This explained how, for example, male and female flowers could develop on the same plant.

De Vries's concept of pangens (and, consequently, the morphologies for which they are responsible) as independent entities also led him to break away from Darwin's view of how characters are distributed among individuals of the same species. Darwin saw variation among individuals as being continuous: constituting an essentially unbroken continuum of minute differences. Such a perception of continuous variation among individuals would make sense to Darwin because he also believed in blending inheritance. In order for blending inheritance to take place, however, features could not be discrete entities. For de Vries, who envisioned pangens as separate, independent, nonblending units of inheritance, features could be discontinuous. In contrast to continuous variation, with characters grading insensibly among individuals of the same species, discontinuous variation is characterized by discrete characteristics, which are distinguished by tangible morphological breaks among individuals of the same species. And the reason why there could be discontinuous variation is that the pangen behind each feature is also a discrete unit. But how does variability arise? De Vries proposed two possible mechanisms: Variation is produced either by a change in the number of pangens or by the advent of a new pangen.

Because of the lackluster reception of his theory of intracellular pangenesis, de Vries sought to demonstrate his predictions through experiments in plant hybridization and crossbreeding. As a result of his discovery, as early as 1886, of the chance production of novel features in the evening primrose—"sports," or "monstrosities," as they were called—throughout his scientific life he focused on the generation of novel features in nature. Contrary to what plant breeders commonly believed, de Vries demonstrated that these sudden and supposedly randomly appearing novel characteristics were hereditary. He also showed how by selecting for a particular monstrosity or novel feature, such as a twisted stem or an excessively broadened leaf, which might be infrequently expressed in one generation of primrose, he could manipulate breeding so that the feature occurred with a much higher frequency among individuals of the next generation. Eventually, in 1900, the year of the rediscovery of Mendel's pea experiments, de Vries published the first volume of the culmination of his life's work, which he called his mutation theory.

De Vries's Mutation Theory, Evolution, and Human Origins

De Vries believed that the number of pangens could change as a result of changing environmental conditions. In turn, a change in the number of pangens was responsible for the variability in the expression of the physical characteristics one sees among individuals within a species. If a shift in the expression of a particular number of features within a species was sufficient to create a race, or subspecies—a group of individuals who are characterized by this cluster of features—this was only a transient phenomenon. Another shift in environmental conditions could just as easily submerge this race, or subspecies, into the general pool of individuals of the species, with all traces of the existence of this group of individuals being erased forever.

While for Darwin individual variation was the fodder of evolution and the source of new species—the source of change from which natural selection could pick and choose the better from the less well adapted—this was not the case for de Vries. As de Vries conceived of it, individual variation was just that: minor differences among individuals of the same species that had nothing to do with evolution, especially the evolution of new species. In de Vries's mutation theory, the emergence of a new species derived from changes to pangens—from mutations—that occurred while these pangens were in the process of replication. De Vries believed that it was upon these new pangens, or mutations, that natural selection acted. Because he had seen the same novel feature, or monstrosity, appear simultaneously in several individuals of the same generation of evening primrose, de Vries hypothesized that, in general, the same mutations can arise in more than one individual of the same generation. Again in direct opposition to Darwin's adaptationist point of view, de Vries also argued that since mutations in pangens occur by chance, they have no adaptive significance.

According to de Vries, the following principles were the essential elements of heredity and evolution: Pangens are the units of inheritance. One pangen is responsible for one character. Most characteristics are discontinuous—there are discrete breaks between individuals in the features they possess. A low-level form of variability among individuals is due to changes in the number or frequency of unaltered pangens. Another level of variability, of the sort that distinguishes species from one another, results from the introduction of new pangens, or mutations, which arise during the process of replication. The novel features that consequently arise from the new pangens, by the very nature of their rapid and random appearance, are not necessarily adaptive—they cannot be viewed as adaptations crucial to the existence of the individuals possessing them. Natural selection acts on the novel features (the sports, the monstrosities) and, by quickly spreading the mutations in subsequent generations, can create a new species. Although there is continuity of life from one generation to the next, the evolution of novelty and, eventually, of new species is discontinuous in nature. It is saltational.

Darwin's evolution was gradual. De Vries's was rapid. Darwin's evolution was utilitarian, with every feature having an adaptive significance. De Vries's was random. Darwin's species arose through the winnowing effects of natural selection on minute differences in what was, overall, a morphological continuum. De Vries's species arose through the advent of mutations that created sports and monstrosities. Consequently, individuals could be distinctly different from other individuals.

In his review of Hugo de Vries's work, the historian Peter Bowler made the point that the mutation theory served to provide alternatives to a strictly orthodox Darwinian view of evolution. Physicists, led by the English baron Lord William Thomson Kelvin, had a difficult time embracing Darwin's theory of evolution by means of natural selection because the latter scientist kept invoking longer and longer periods of time over which the transmutation of species by the gradual accumulation of infinitesimally small differences would have to occur. With every edition of the *Origin,* Darwin demanded an older and older earth in order to accommodate the amount of time he believed was necessary for evolutionary change to accrue. But eventually Darwin pushed this point beyond the calculations that Lord Kelvin and his colleagues could allow as being possible for the age of the earth.

De Vries's theory of evolution was much more palatable than Darwin's to many scientists of different disciplines. In de Vries's view of life, evolutionary change could be rapid. Consequently, there was no cause for special pleas for an unrealistically ancient earth. This satisfied Lord Kelvin and his band of physicists and also found support in the saltationists among the evolutionists. By arguing that not all characters have to be adaptive, de Vries provided a way out for Darwinians who were mired in the utilitarianism demanded by Darwin's brand of evolution, in which all features are viewed as adaptations, with natural selection choosing only the more favorable adaptations. Curiously, although de Vries was overtly non-Darwinian in many of his propositions, he was also actually a staunch selectionist. But by arguing a position in which it doesn't matter whether some features are more advantageous or favorable to their possessors than others might be, or whether there is an inextricable interplay between natural selection and adaptation, de Vries was also rejecting the primary premise of Darwin's theory of evolution by natural selection: that there is always a struggle in nature, that in this struggle some features will be more advantageous than others, and that it is because of this struggle that natural selection can cull the more advantageous features and at the same time both bestow greater reproductive success upon their possessors and chart the course of evolution. For de Vries, selection meant not necessarily the selection of a favorable feature but primarily the selection against unfavorable characteristics.

De Vries's adherence to a belief in discontinuous features and to the abrupt appearance of mutations that could give rise to new species could have led him to suggest that there really was a major gap between *Homo sapiens* and the rest of the animal world, and that the races of the human species might themselves be viewed as separate species. Obviously, Darwin's insistence on a slow, gradual evolutionary process, in which early humans originated from an apelike ancestor and then progressed, via an evolutionary continuum, from the lowliest of savages to the most civilized members of the human species, left no room for the speculations of polygenecists. But de Vries could also be in agreement with Darwin on the evolutionary closeness of humans and other animals and of the unity of the races of *H. sapiens* without also joining ranks with him theoretically. In fact, de Vries's view of nature and evolution made his argument for the origin of humans and human races less racist than Darwin's.

Darwin's notion of gradual evolution over a long period of time could lead to the conclusion that features that purportedly distinguished one race from another were stable in nature. For, according to his theory of evolution by means of natural selection, differences arise only as the result of a struggle among individuals, whose traits are expressed differentially. In contrast, de Vries's idea that variability within a species is both transient and inconsequential for evolution allowed for a more fluid interpretation of potential racial differences: Variability and, consequently, any possibility of race formation, were fleeting in terms of evolutionary time. For de Vries, struggle was not a factor in the evolutionary development or expression of what some might identify as racial difference. Because of this theoretical position, de Vries's writings had the effect of countering the notions of social Darwinism, such as those espoused by Herbert Spencer, in which the Darwinian emphases on struggle and the survival of the fittest were invoked to justify a racial hierarchy, colonization, and the subjugation of the "savage" by the "civilized" races. In de Vries's world, only species are the result of evolutionarily significant processes. And species, including *Homo sapiens,* arise as a result of mutation, with natural selection acting on the novel features.

William Bateson, Discontinuous Variation, and Repetition of Parts

In the formulation of a case for separate units of heredity and, consequently, for the recognition of the reality of discontinuous variability in nature, de Vries sought the support of the British zoologist William Bateson. Bateson is perhaps best remembered among present-day geneticists for being the translator and champion of Mendel's work in the English-speaking scientific world. But well before his discovery of Mendel's experiments through the articles of de Vries, Correns, and Tschermak von Seysenegg, Bateson believed in discontinuous variation. In terms of the advance of evolutionary science, Bateson was an important link in the theoretical and experimental endeavors to combine ideas of inheritance with those of evolution. Until then, such attempts had remained solely within the domain of botanists.

As detailed by his wife, Beatrice, in her memoir, Will "found" himself academically at St. John's College, Cambridge University. Initially through the support of a more senior St. John's student in comparative embryology and biology, Walter Frank Raphael Weldon, who would eventually ascend to professorship at Oxford University, and then with the sponsorship of the eminent American zoologist William Keith Brooks, Bateson took on the study of the poorly known wormlike creature of the genus *Balanoglossus*. Following a brilliant embryological study, Bateson was able to argue convincingly that *Balanoglossus* was indeed a chordate, albeit a primitive and unsegmented chordate at best. After the *Balanoglossus* study, Bateson became interested in questions of variation and evolution. What, exactly, was variation? How did individual variation come about? And how did variation fit into a workable model of evolutionary change?

Darwin had based his theory of evolution by means of natural selection in large part on both the availability of variation and on the correlation between variation and the effects of the environment. But no one, not even Darwin, had actually provided data demonstrating the link between organismal change and environmental change: that, indeed, organisms do track their environment, the former changing in the wake of shifts in the latter. Being an admirer of Darwin's meticulousness in amassing huge amounts of detailed information to support his case that individuals of a species do vary to some degree from one another, Bateson set out to perform a similar Herculean task of data collection. He sought to find support for the proposal that the production of variability is intimately connected to the environment in which organisms find themselves.

The research plan Bateson outlined for himself to investigate this question took him to Western Central Asia, to the lakes and drying-up lake basins of that region of Russia. There, from the spring of 1886 through the fall of 1887, he studied the fauna of the lakes and dying lakes, recording everything, from locality data to local environmental conditions, water density, water salinity, and lake depth, ad infinitum. While there, he also became fluent enough in Russian and Kirghiz to engage in colloquially accurate conversation. But at the end of his studies and arduous fieldwork, Bateson returned to England not full of examples of organisms varying in synchrony with their environments, as Darwin had claimed was the case, but with a conviction that, to the contrary, a tracking of the environment by the organism was not the rule in nature. There appeared to be no correlation whatsoever between change in environmental conditions—in his particular study, differences in salinity between viable versus dying lakes—and organismal change. Sometimes many features seemed to change in a number of individuals. Sometimes there was no obvious pattern. Just as certain shifts in environmental conditions might alter the average representation of certain character states of a population in one direction, so, too, could a change in the opposite direction just as easily return individuals to the previous averages of expressed character states. As far as Bateson could tell, there was little here that supported the Darwinian idea that slight shifts in character states from within the set of character states already present within a species would lead to evolutionary change.

Bateson's experiences in Western Central Asia, and a similar but shorter study of brackish-water organisms in northern Egypt, although utter failures in his mind, prompted him to pursue an in-depth research program on variability in plants and animals. This he accomplished largely through worldwide correspondence, the ransacking (to paraphrase his wife, Beatrice) of museums, libraries and private collections, and extensive travel in order to see for himself purported examples of variation and abnormality. His excitement about this research was palpable in his correspondence. He had not been engaged in this study for even a year when, in September 1888, he wrote a letter to his sister, Anna, in which his intellectual state almost burst from the page: "My brain boils with Evolution. . . . It is becoming a perfect nightmare to me."

In 1890, three years into this enormous task, Bateson applied for the position of Deputy of the Linacre Professor at Oxford University. He did so even though he knew perfectly well that the candidate of choice was Professor Ray Lankester, who was already a formidable figure in comparative zoology and England's leading proponent of the central role of natural selection in evolution. Bateson lost to Lankester, and although this thwarted for many years his plans to pursue a career that included teaching, it also freed him not only of time-demanding academic duties but also of any feelings of obligation that might have prevented him from airing publicly his opinions on variation and selection.

Eventually, through a series of short publications, and then in his huge and vastly detailed volume on variation, *Materials for the Study of Variation,* which was published in 1894—in which he compiled 886 examples of discontinuous variation—Bateson came out clearly and solidly on the side of discontinuous variation among organisms. His work on lake salinity and variation in lake-dwelling organisms had convinced him that "most of the elements of the physical environment are continuous in their gradations, while, as a rule, the forms of life are discontinuous." Not only did he question the nature of organismal variation and its importance in evolution but he also rejected the utilitarian Darwinian notion of adaptation and downplayed the role of natural selection in generating evolutionary change. Like Francis Galton and Hugo de Vries, Bateson relegated the minor variability among individuals that Darwin had promoted as the stuff of evolution to just that: the minor ways in which individuals differ from one another in one feature or another. For Bateson, it was in the recognition of discontinuity between characteristics that one gained insight into the workings of evolution: Evolutionary change is discontinuous in the sense that new features arise rapidly—a proposition that was wholeheartedly embraced by two of the staunchest advocates of saltation in evolution, Francis Galton and Thomas Henry Huxley.

Even though Bateson set himself in theoretical opposition to Darwin's most fundamental premises, he never lost sight of the importance of the earlier champion of evolution, whom he never ceased to admire as a historical figure of extreme significance. As he wrote in the opening paragraphs of *Materials for the Study of Variation:*

> To solve the problem of the forms of living things is the aim with which the naturalist of to-day comes to his work. How have living things become what they are, and what are the laws that govern their forms? These are the questions which the naturalist has set himself to answer.
>
> It is more than thirty years since the *Origin of Species* was written, but for many these questions are in no sense answered yet. In owning that it is so, we shall not honour Darwin's memory the less; for whatever may be the part which shall be finally assigned to Natural Selection, it will always be remembered that it was through Darwin's work that men saw for the first time that the problem is one which man may reasonably hope to solve. If Darwin did not solve the problem himself, he first gave us the hope of a solution, perhaps a greater thing. How great a feat this was, we who have heard it all from childhood can scarcely know.
>
> In the present work an attempt is made to find a way of attacking parts of the problem afresh.

By the fall of 1891, Bateson was well on his way to becoming forever convinced that a large part of the answer to evolutionary change lay in the study of repeated, or meristic, body parts, from the perspective of both normal and abnormal development. In a letter to Anna, dated September 14 of that year, he summarized his new idea. He called it the "vibratory theory of repetition of parts," which he would also refer to as the "Undulatory Hypothesis," and which he believed applied equally to plants and animals:

> It is the best idea I ever had or am likely to have. . . . Divisions between segments, petals etc. are *internodal* lines like those in sand figures made by sound, i.e. lines of maximum vibratory strain, while the mid-segmental lines and the petals, etc. are the *nodal* lines, or places of minimum movement. Hence all the *patterns* and *recurrence of patterns* in animals and plants—hence the perfection of symmetry—hence bilaterally symmetrical variation, and the *completeness* of repetition whether of a part repeated in a radial or linear series etc. etc. I am, as you see, in a great fluster. I have been talking to F. D. [Francis Darwin, one of Charles's sons] about it—and he thinks "Really is very neat, upon my word."
>
> P.S. Of course Heredity becomes quite a simple phenomenon in the light of this.

Spurred on by the encouragement of Francis Darwin, Bateson continued his research into the normal and abnormal development of repeated parts and included examples from animals in *Materials for the Study of Variation* and from both plants and animals in almost every article, book, and lecture he wrote thereafter.

He described repetition of parts in animals such as starfish, sea anemones, sea urchins, hydras, octopuses, and squid, whose bodies are essentially configured like spoked wheels: They have a midpoint around which multiple and virtually identical appendages are arranged radially. He also focused on repetition of parts in animals that are bilaterally symmetrical—

having a midline axis and right and left sides—such as worms, insects, and shellfish, as well as vertebrates. Many species of worm and their kin are segmented along their bodies by muscular rings. In arthropod invertebrates, such as insects and crustaceans, repeated parts are seen in the segments of the antennae, in the segments of the legs, and in the series of articulating plates that form the hingelike shell in which the animal is encased. In vertebrates, repeated parts are found in the series of teeth along the jaws, in the series of vertebrae that course from the articulation with the base of the skull to the tip of the tail, in the ribs that articulate with vertebrae, in the bones of arms, legs, hands, and feet, in the rays that support fins, and in the scales that cover the bodies of fish and reptiles. In nonhuman mammals, the parallel rows of mammary glands are an obvious example of a series of repeated parts. Repetition of parts in plants follows two basic patterns. In ferns, the leaves on either side of the stem are arranged serially along the frond, as is often true of the segments of the leaves of deciduous trees. In many flowering plants, the petals are arranged radially around the eye of the flower.

Bateson was not interested solely in the development of the normal arrangement of teeth, vertebrae, tentacles, and petals. He derived most of his insights into the development of repeated parts from studies on the "sports," the "monstrosities," the "chimeras" of nature—when animals would have more or fewer of one or another feature than is typical for the species. Common cases of individuals having more or fewer elements could be found in the study of teeth, vertebrae, ribs, foot or hand bones, and claws. Sometimes a mammal has more mammary glands than is typical for its species. Or an animal may be born with two heads. Examples from the plant world included differences in petal number, which happens to occur as a multiple of the expected norm and is either halved or doubled.

Bateson was also provoked by studies on the development of sports either directly or as a consequence of addressing a larger question, such as the regeneration of lost body parts. He was particularly struck by experiments on the tiny aquatic hydra, which, with all its tentacles arranged around the top of its body stalk, looks like a medusa's head on a stilt. How and why is it that in radially configured animals, such as hydras, which can regenerate lost parts, one tentacle will replace two that have been cut off? Although Bateson found it more difficult to generate an overarching theory to accommodate the patterning of a radially arranged organism, throughout his life he continued to invoke the idea of wave motion to explain serial segmentation.

Whether it be in the rhythmic patterning of banding along an organism's body, of barring along an organ, such as a feather, or of rippling, such as the surface of a goat's horn, Bateson saw elegant analogy in the propagation of wave motion. An example of wave motion that he used time and again was the encroaching and subsiding tide, which creates in its wake a regular series of troughs and blunt ridges of sand. And, like the ripple effect of the tide on sand, anomalies in number of serially repeated elements—whether they be in number of hand or foot bones, of vertebrae, or of ribs—were typically discrete and discontinuous in the sense that a single developmental event, rather than an accumulation of minute variants over time, was responsible for the novel form. As a wave of greater or lesser intensity courses up and then back along the beach, creating a different number of sandy ridges, so a developmental wave of a certain intensity will determine the expression and number of elements in a series of repeated structures.

Bateson Becomes Mendelized

In 1894, the year in which *Materials for the Study of Variation* was published, the Royal Society acted on Francis Galton's suggestion that a committee that would study measurable characteristics in plants and animals using statistical techniques should be established. Three years later, Bateson was asked to join this committee. Shortly thereafter, this committee was reconstituted as the Evolution Committee and its domain of research concerns was expanded considerably. Bateson eventually became the secretary of the Evolution Committee. Backed by modest financing from this committee, he began inquiries into plant and animal, specifically poultry, breeding. Within a short period of time, the grounds of Bateson's house were filled to overflowing with students conducting breeding and crossbreeding experiments on varieties of all sorts of plants and animals. There was, however, always the problem of maintaining even the minimal funding necessary to keep this prodigious research effort afloat.

When in 1899 Bateson was approached—for the second time—about his opinion on whether the Royal Society should acquire Darwin's house in Down, Kent, for use as a biological research station, he responded with the outline of a research plan for determining the laws of inheritance in animals and plants through experimentation and statistical methods. He was not, however, convinced that Down House would be suitable for such a biological research station; for one thing, its library was severely wanting, and, for another, it was too far away for Bateson to get to easily. But it is obvious from the content of his letter that, even if his wife's claim of his approaching Mendel's methods of work is not accurate, he would certainly have been intellectually ready to embrace Mendel's conclusions.

According to his wife's account—which the historian of science Robert Olby has suggested may not be true—Bateson was on his way by train to lecture before the Royal Horticultural Society on May 8, 1900, when he happened to read the copy of Mendel's article that he had acquired by way of having first come across the rediscovery articles of de Vries, Correns, and Tschermak von Seysenegg. Not only did Mendel's work make sense to him in terms of the breeding and crossbreeding experiments that he and his students had been conducting but the law of segregation, based on the recognition of discrete units of inheritance, was in full accord with his ideas on discontinuous variation. Although he had already drafted his lecture, Bateson immediately rewrote it to include a summary of Mendel's experiments and results. It was an association with Mendel's work that would remain for the rest of Bateson's life.

Bateson began his lecture to the Royal Horticultural Society, which was published later that year in the society's journal series, with the following comments:

> An exact determination of the laws of heredity will probably work more change in man's outlook on the world, and in his power over nature, than any other advance in natural knowledge that can be foreseen.
>
> There is no doubt whatever that these laws can be determined. . . . It is rather remarkable that while in other branches of physiology such great progress has of late been made, our knowledge of the phenomena of heredity has increased but little; though that these phenomena constitute the basis of all evolutionary science and the very central problem of natural history is admitted by all. . . .
>
> No one has better opportunities of pursuing such work than horticulturists.

Bateson followed these opening comments, in which he encouraged fellow horticulturists to partake of this research, by citing Francis Galton as the first scientist to try to collect an appreciable amount of data to demonstrate laws of heredity. Whereas others before and after Galton had provided single cases of the transmission of one character or another from parent to offspring, it was only, Bateson pointed out, through the study of a large number of cases that any law of heredity could be demonstrated. It was also only by the application of statistical methods that these laws could be explicated. For, after all, these laws should be predictive: Given such and such characteristics in parents, what combinations of features would be expected in the offspring? Galton's contribution to this field of investigation, which he developed from his studies on human stature and coat coloration in basset hounds was, Bateson reminded his audience, the "law of ancestral heredity." It was through the application of this law that one could calculate how much heritable information was passed on to one generation from each preceding generation—not just from parent to offspring but from each ancestral generation as well. (In later discourses, such as in his book *Mendel's Principles of Heredity,* which would be published nine years later, Bateson was overtly critical of Galton's law of ancestral heredity.)

After giving Galton due historical recognition, Bateson went directly to the heart of the lecture: the work of Gregor Mendel, of which he had become initially aware by reading Hugo de Vries's second rediscovery article of 1900, the one published in the journal series of the German Botanical Society. Clearly, Bateson understood perfectly Mendel's experiments and the results forthcoming from them: the law of segregation and the 3 to 1 ratio in hybrids. But he was not so sure of de Vries's understanding of Mendel's work, for in a footnote to the published version of his lecture to the Royal Horticultural

Society, he questioned de Vries's comprehension of Mendel's law. Before summarizing Mendel's experiments and results to his audience, Bateson made the following comment:

> These experiments of Mendel's were carried out on a large scale, his account of them is excellent and complete, and the principles which he was able to deduce from them will certainly play a conspicuous part in all future discussion of evolutionary problems. It is not a little remarkable that Mendel's work should have escaped notice, and been so long forgotten.

Putting Mendelism into Evolution

Thanks to the efforts of Bateson, no one in the biological sciences would ever again ignore Mendel's work, although it would still be many years before the level of opposition to it subsided. In 1902, in collaboration with Edith Rebecca Saunders, who had been engaged in plant-breeding experiments at Bateson's property, Bateson published a report to the Evolution Committee of the Royal Society in which he and Saunders summarized and explained Mendel's experiments and conclusions and provided their own experimental corroboration of his results. Most important, perhaps, it is in this publication that Mendel's laws were expanded to encompass all the living, sexually reproductive world, in which male and female parents contribute genetic material equally to offspring. Armed with the knowledge that discontinuous characters segregate and the realization that there are dominant and recessive character states, Bateson and Saunders could, for example, discuss polydactyly (the abnormal development of extra digits, such as toes) versus normal development of toes in both chickens and humans using the same language with which they also discussed wrinkled versus round seed coats in common garden peas.

Most of the characteristics that Bateson and Saunders as well as other experimental breeders and anatomists studied could, it seemed, be conceptualized as appearing in one of two different states: either dominant or recessive. As was the case with round versus wrinkled seed coat in garden peas, or fewer versus normal finger-bone number in humans, one character state was dominant over the other. In elegant simplicity, Bateson and Saunders pointed out what became part of the bedrock upon which the field of genetics was grounded: that character states for a particular category of characteristic, such as finger-bone number or seed-coat configuration, sometimes come in pairs of alternative states, and that knowing this was pretty much all that was needed in order to untangle most of the important aspects of inheritance.

In order for biologists to better communicate in a language of inheritance, Bateson and Saunders expanded the lexicon of terms used by embryologists and cytologists. Embryologists and cytologists referred to germ cells as gametes and to the fertilized egg gametes produce as a zygote. Since gametes from each parent contribute "unit-characters existing in antagonistic pairs," they suggested that a pair of such antagonistic characters should be called allelomorphs. If, in the zygote, the allelomorphs of a pair differed—one being dominant and the other recessive—Bateson and Saunders recommended calling the zygote a heterozygote to reflect its possession of different allelomorphs, each allelomorph being responsible for a different morphological character state. If both allelomorphs of a pair were the same—both being either dominant or recessive for the same morphological character state—then the zygote should be referred to as a homozygote. Later, as the language of inheritance became fine-tuned, *allelomorph* would be shortened to *allele* and the nouns *heterozygote* and *homozygote* also made available as adjectives, *heterozygous* and *homozygous.* Bateson and Saunders introduced several other terms to describe other types of allelomorphs, but their use was short-lived. Even Bateson did not employ these terms in his major work on genetics, *Mendel's Principle of Heredity,* which was published only seven years after his joint report with Saunders to the Evolution Committee.

In addition to adopting, modifying, or introducing new terms into the nascent language of an emerging discipline, Bateson and Saunders sought to systematize the representation of successive generations of related individuals. In this way, professional as well as academic plant and animal breeders would be able to communicate using a standard format. Just as Linnaeus had sought to formalize a schematic way of producing a classification, so Bateson and Saunders strove for uniformity in how experimental breeders represented and distinguished parental generations from their offspring, or filial, generations. In characteristic simplicity, Bateson and Saunders suggested that the capital letter *P* should be used to refer to the parental generation, the generation that was the beginning of the experiment or breeding process. Generations of offspring that followed this first-defined parental generation were to be identified by the letter *F,* referring to *filial.* The first filial generation would be F_1, the second F_2, the third F_3, and so on. If generations preceding the first parental generation were discussed, they were identified as P_2, P_3, P_4, and so on to represent the grandparental, great-grandparental, and great-great-grandparental generations. This simple and straightforward approach to representing genetically successive generations, which has been in use ever since, was modestly put forth not in the text but in a footnote to the conclusion of the report to the Evolution Committee.

The breadth and thoughtfulness that Bateson and Saunders brought to bear on the implications of Mendel's laws of inheritance were astonishing and should serve as a lesson to all of us who endeavor to shed light on aspects of the evolutionary process. For example, although it was not a major theme in their report to the Evolution Committee, Bateson and Saunders did raise the question of how Mendel's laws of inheritance might be applied to the question of what, precisely, constitutes a species in nature. It would have been a simple matter to ignore the question altogether. But Bateson had himself been concerned with the problem of defining a species. In fact, after his general opening remarks in *Materials for the Study of Variation,* this was the first topic of substance that he tackled. Of course, Bateson was not the first naturalist to wrestle with the species problem. John Ray, Linnaeus, Buffon, and the father of at least physical anthropology, Johannes Blumenbach, to name but a scant few, had also recognized that this was a slippery subject, even if the concept of the species had become the basic unit of classification. But, in large part because of his advocacy of the importance of discontinuous variation, Bateson felt obliged to attempt to deal with this sticky issue.

As Darwin had cataloged examples of variation, so Bateson struggled to find out and delineate the causes of variation. While Darwin had accepted the reality of species in nature, Bateson sought to better understand how to identify and define species. Although Bateson was never able to fully articulate a comprehensive definition of a species, his constant agonizing over the problems of species definition and identification did help focus attention on the nature of the problem:

> The forms of living things are diverse. They may nevertheless be separated into Specific Groups or Species, the members of each such group being nearly alike, while they are less like the members of any other Specific Group. . . .
>
> The individuals of each Specific Group, though alike, are not identical in form, but exhibit differences, and in these differences they may even more or less nearly approach the form characteristic of another Specific Group. . . . [I]n the case of many Specific Groups which have been separated from each other, intermediate forms are found which form a continuous series of gradation, passing insensibly from the form characteristic of one Species to that characteristic of another. In such cases the distinction between the two groups for purposes of classification is not retained.
>
> . . . but though now a good many such cases are known, it remains none the less true that at a given point of time, the forms of living things may be arranged in Specific Groups, and that between the immense majority of these there are not transitional forms. There are therefore between these Specific Groups differences which are Specific.
>
> No definition of a Specific Difference has been found. . . . But the forms of living things, taken at a given moment, do nevertheless most certainly form a discontinuous series and not a continuous series. This is true of the world as we see it now, and there is no good reason for thinking that it has ever been otherwise. So much is being said of mutability of species that this, which is the central fact of Natural History, is almost lost sight of, but if ever the problem is to be solved this fact must be boldly faced. There is nothing to be gained by shirking or trying to forget it.
>
> The existence, then, of Specific Differences is one of the characteristics of the forms of living things. This is no merely subjective conception, but an objective, tangible fact.

Bateson and the Species Problem

It is obvious that Bateson had begun his struggle with the species problem early in his intellectual career. According to a letter that he wrote to his sister on July 19, 1890, the manuscript for *Materials for the Study of Variation* had been accepted by the editors of the MacMillan Publishing Company that year, four years prior to its publication. Equally obvious is the fact that he had been wrestling with the problems of delineating not only species in nature but also the process or processes that lead to the origin of new species.

In the first case—the delineation of species in nature—Bateson acknowledged that species are distinct entities that are usually distinguishable from one another. In order for a species to be distinct, there must be at least one defining character of that species—something that would be reflected in the form, or the morphology or characteristics, of its members. As to what that feature or those features might be, it was clear to Bateson, as it had also been to Darwin, whom he included as a fellow victim of this quandary, that more often than not these are features to which no significant utility can be ascribed, such as differences in the number of hairs or spines, pattern of scales, details of the shell, or aspects of the male's penis. In other words, features that taxonomists typically end up using to separate species, especially very similar species, from one another are not vital, as best as can be determined, to the existence of the species. Rather, the characteristics taxonomists often use to distinguish even very similar species from one another are seemingly useless and trivial. Echoing the general sentiments of de Vries, Bateson had to conclude that species differences are not those of adaptation.

In the second case—the process or processes leading to the advent of new species—it appears that Bateson was suggesting that new species can arise by the splitting of existing species. This seems evident when he alludes, in various instances, to the existence of intermediates between what, if only each extreme of the distribution were known, present themselves as two distinct species, each distinguished from the other by its own specific characteristics. Since the passage I quoted a few pages earlier on species is at the very beginning of an almost six-hundred-page opus dedicated to demonstrating discontinuous variation, to developing a theory of evolution based on this natural phenomenon, and also to rejecting natural selection and environmental change as the causes of evolutionary change, it is obvious that Bateson's "force" behind the cleaving of a parent species into two distinct daughter species is not the environment, or even the physical separation of different populations of the original species. Whatever the ultimate cause or causes of discontinuous variation, it is the phenomenon of discontinuity in nature that for Bateson holds the key to the mystery of species. As he wrote a few pages after that quote: "On this hypothesis, therefore, Variation, whatever may be its cause, and however it may be limited, is the essential phenomenon of Evolution. Variation, in fact, *is* Evolution." But here variation is discontinuous, not the continuous variation that was fundamental to Darwinism.

By the time Bateson had written the report to the Evolution Committee of the Royal Society with Edith Saunders—which, of course, was undertaken prior to its publication in 1902—not only had he mastered the essentials of Mendel's laws of inheritance but he had also begun to think about evolution in terms of Mendelian inheritance:

> [W]ith the discovery of the Mendelian principle the problem of evolution passes into a new phase. It is scarcely possible to overrate the importance of this discovery. Every conception of biology which involves a knowledge of the physiology of reproduction must feel the influence of the new facts, and, in their light, previous ideas of heredity and variation, the nature of specific differences, all that depends on those ideas must be reconsidered, and in great measure modified.

And Bateson and Saunders did indeed try to reconsider one of the most critical problems facing an evolutionary biologist: figuring out in a biologically meaningful way what a species is and how to recognize one in the flesh, so to speak. They addressed this problem from the perspective of inheritance, which actually has a flip side. For not only can one ask questions about inheritance, as Mendel and others did, investigating the transmission of features from parent to offspring, but one can also consider questions about the nontransmission of features: those instances when potential mates cannot contribute successfully to the production of offspring. In order to try to crack the species nut, Bateson and Saunders took the following approach.

First, they acknowledged that a great number of characteristics can be passed on from parents to their offspring without consequence to the offspring. In the case of Mendel's garden peas, neither the height nor the seed color of either parent had any bearing on the success of a mating. These kinds of features were freely exchangeable between individuals—just as it makes no difference what the eye and hair color or height and body build of potential human mates are. But, Bateson and Saunders asked, in addition to this kind of characteristic and its ultimate genetic basis are there some characteristics and, consequently, their genetic underpinnings, that are not so interchangeable? If so—if there are indeed certain kinds of features and genetic units that do not allow for successful mating or the production of reproductively viable offspring—this would take Mendel's laws of inheritance out of the realm of considering what happens from one generation to the next and place it squarely in the face of the question of determining what, exactly, a species is, at least in terms of defining species by what they can or cannot do reproductively vis-à-vis one another.

Bateson and Saunders were very careful to distinguish between the continuity of inheritance from one generation to the next and the discontinuity of inheritance between groups of individuals that might be called species. With regard to the former, the principles of Mendelian inheritance and the properties of dominant and recessive allelomorphic states are sufficient to describe the process. But, as far as Bateson and Saunders were concerned, species cannot be defined in terms of these aspects of inheritance precisely because they relate to the continuity of a species, not to its origin. Rather than approach the evolutionary question of species identification through the eyes of successful inheritance, Bateson and Saunders proposed that it is through the study of sterility—the inability of individuals to participate in ongoing inheritance—that naturalists will understand what species are. As they envisioned it, there might be something genetic about individuals from different species that would interfere with cross-species production of reproductively viable offspring—even though, at another level, individuals of different species would share the same kinds of easily interchangeable features, such as height and body build or hair, eye, or seed color. Although Bateson and Saunders did not pull out all the stops and suggest that all hybrid offspring of parents of two different species would be sterile, they definitely believed that "some degree of sterility on crossing is . . . one of the divers properties which may be associated with Specific difference."

In this last quote I purposely omitted the word "only": "only one of the divers properties." I did so in order to make two points. The first is to set the record straight about the complexity of Bateson's thoughts on species. Beginning in the 1940s, with the intellectual development of the grand evolutionary synthesis that was to bring together genetics and paleontology through the common language of evolution, various systematists—most notably the evolutionary biologist and bird taxonomist Ernst Mayr—criticized Bateson, along with other evolutionists who had wrestled with the species problem, for putting undue weight on the criterion of sterility as the defining quality of species difference. However, it is obvious from the series of quotes presented in the preceding pages that Bateson was quite aware of the possibility that this was not the be-all and end-all of the species problem. For, after all, and this is the second point, he and Saunders did allow that sterility "is only *one* of the divers properties which *may* be associated with Specific difference" (emphasis mine). Morphology, behavior, and ecology, all of which Bateson discussed and pondered at one time or another in his writings, are some other sources of possible species difference—which continue to rank among twentieth-century systematists and evolutionary biologists, Ernst Mayr included, as high-priority discriminators for attacking the problem of defining species.

As for sterility and its potential role in delineating species, it is true, as Mayr has reviewed, that, at least as early as Buffon's writings, the question of the reality of species in nature had been addressed through consideration of the production of sterile hybrids—the varieties of domesticated dogs that were not being considered different species by some taxonomists because, size and other distinguishing features aside, dogs of different varieties could still produce reproductively viable offspring. But it would be unfair to lump Bateson with Buffon and others, because only the former scholar tried to understand sterility in terms of the fundamental aspects of Mendelian inheritance: whether an individual could produce viable gametes, or sex cells.

More specifically, as Bateson and Saunders discussed at some length in their report, the viability of a gamete would depend on whether the characters, or genetic units, that an individual inherited from its parents could be divided properly when producing its own gametes. In other words, can the pairs of allelomorphs an individual possesses as a result of getting one allelomorph of each pair from its parents be sorted out, or segregated, in the production of that individual's own gametes? This concern is not dissimilar to more recent twentieth-century studies on the genetic incompatibility of individuals with different numbers of chromosomes. Although these individuals might be able to produce offspring, the latter might be unable to produce offspring because they could not sort out the chromosomes properly in their gametes. To their further credit, Bateson and Saunders did not rule out the possibility that individuals from different species can produce offspring that are reproductively viable—offspring that, in their terms, can "divide up the characters among their gametes."

In a nutshell, then, it appears that no matter how difficult his struggle to articulate with precision a definition of a species (which, mind you, is no easier for today's systematists), Bateson, in either his solo writing or in the report to the Evolution Committee that he wrote with Edith Saunders, was quite twentieth-century when the century was still in its infancy. He envisioned species as being discrete entities in nature. But he was also prepared to accept that there would be cases in which intervening populations of intermediates would bridge the morphological or genetic gap between two species. Today, this would be discussed as a hybrid zone between two markedly differing groups, which, if found on their own, would taxonomically be treated as separate species. On a genetic level, at least in terms of his understanding of the subject through Mendelian inheritance, Bateson conceived of species as being in some way genetically incompatible with one another—because, he reckoned, if there was something to species difference, it might not be heritable across species boundaries. On the other hand, he was open to the possibility that differences that might delineate species at one level might not be sufficiently severe at the genetic level to prevent the production of reproductively viable offspring. In these perceptions of nature, Bateson was clearly ahead of his time, at least if one uses the 1940s, the decade of the grand evolutionary synthesis, as the yardstick. Although Ernst Mayr, one of the most visible champions of these ideas during the formative years of this synthesis and thereafter, wrote at great length and detail in his support and explication of them, it is largely to William Bateson that we must turn for their intellectual roots.

Forever Mendelism

It might seem inappropriate to use the term *genetics* in this part of the history of the field of genetics, for the early twentieth century was still a time of discovering and deciphering the most fundamental aspects of inheritance. But I am not too far off the chronological mark because, again, it was Bateson who coined the term "genetics" in a letter that he wrote to the prominent English Cambridge University zoologist Professor Adam Sedgwick, on April 18, 1905. His inspiration for doing so came, so his wife, Beatrice, recounts in her memoir, from his being asked that same year "by the Vice-Chancellor of the [Cambridge] University to draw up a scheme for a 'Quick Institute for the study of Heredity and Variation.' " But he was growing tired of using the phrase "heredity and variation" to describe the field in which he labored and, in apparent response to a similar feeling from Sedgwick, proposed the word *genetics* as a pithier alternative:

> If the Quick Fund were used for the foundation of a Professorship relating to Heredity and Variation the best title would, I think, be "The Quick Professorship of the study of Heredity." No single word in common use quite gives this meaning. Such a word is badly wanted, and if it were desirable to coin one, "GENETICS" might do. Either expression clearly includes Variation and the cognate phenomena.

As with so many of his other research plans, even those that were invited for submission, Bateson's hopes of institutional backing for an established laboratory and research station for the pursuit of studies in genetics went unrewarded. The study of heredity and variation was passed over by the authorities at Cambridge in their establishment of the Quick Professorship of Protozoology in 1906—this in spite of Bateson's many accomplishments and honors, including his being awarded the distinguished Darwin Medal by the Royal Society in 1904. In a letter to a colleague, Bateson wrote:

> Heredity and Variation failed, as proposals generally do, not by reason of opposition, but for want of support. I am responsible in so far as my strength has been insufficient to bring about already the revolution in biological study that sooner or later must certainly come to pass. By this time I ought to have been able to make it obvious that such an endowment should be given to "*Genetics.*" As I said, I have failed my contemporaries; with posterity I hope to be more successful.

The year after Bateson coined the term "genetics," Walter Weldon, a onetime supporter during Bateson's student years at Cambridge but a long-time professional adversary on the subjects of continuous versus discontinuous variation and the role of the environment and natural selection in producing evolutionary change, died unexpectedly. Upon hearing the news, which came to him while he was away from Cambridge, Bateson wrote to Beatrice, recounting his history with Weldon from the good to the bad. He portrayed his suffering at Weldon's hands in the following passage:

> How big a disturbance this will make in our area [of research] I hardly yet know. If any man ever set himself to destroy another man's work, that did he do to me—and now suddenly to have one of the chief preoccupations of one's mind withdrawn, leaves one rather "in irons," as sailors say.

The death of the professor and chair of the department of zoology at Cambridge University in 1907 provided Bateson with another opportunity to apply for an appointment that would elude him. When his friend and colleague Adam Sedgwick got the position, Bateson wrote a warm letter of congratulations. But he would not have to live and work in Cambridge until the burden of yet another disappointment eventually lifted. The previous year, Henry Fairfield Osborn, a prominent American paleontologist and an advocate of evolution, had nominated Bateson for the honored Silliman lectureship at Yale University. Although Bateson was afraid that he would not be able to both work on his current book, *Mendel's Principles of Heredity,* and put together the course lectures that would by contract also be published, he accepted the offer from Yale. From July to November 1907, he was received with great appreciation everywhere he went to lecture in the United States. While he was at Woods Hole Marine Biological Laboratory in Massachusetts, he wrote to his wife that "[t]hese have been wonderful days." Later the same day, he wrote with obvious pleasure that his address to the International Zoological Congress being held in Boston had been a "big success" and that the crowd had been great and "overwhelmed with enthusiasm." And the following day he wrote: "The Americans are rather absurd in their hero-worship and one has continually to remember that they keep a constant procession of heroes on the march. . . . But after the years of snubbing it is rather pleasant to get appreciation even though in an overdose." A politically sensitive individual as well, he also commented: "The tyranny of religion and temperance is constantly making itself felt in the U.S.A."

Having not had the anticlimax at Yale that he had anticipated after his resoundingly successful lecture tour, Bateson was fully prepared for an emotional letdown upon returning to England and to Cambridge University. His expectations were, unfortunately, rewarded when he was offered the belittling position of reader of zoology. Sedgwick convinced him to take the post. In February 1908, Bateson learned of an anonymous donor's contributing to the establishment of a five-year professorship of biology at Cambridge, and, against the seemingly endless trend of disappointment of his career, he was elected to this position in June. Again, in contrast to the history of his career, Bateson resigned from this position two years later to become the first director of the John Innes Horticultural Institute at Merton. Within two years of his leaving Cambridge, the now-permanent position at Cambridge that he had vacated was renamed "professor of genetics." But while Bateson was still at Cambridge as the professor of biology, his long-awaited book, *Mendel's Principles of Heredity,* was finally published.

If Bateson had done nothing more than recount in detail the history of genetics through the rediscovery of Mendel's experiments and publish his English translations of Mendel's two articles, those, in and of themselves, would have been historically significant. But, in addition, he strove, in *Mendel's Principles,* to understand as many of the normal and common, as well as strange and novel, cases of inheritance of characteristics as possible in the context of Mendelian inheritance. By this time in his career, with his enormous experiential as well as literary background in both plant and animal experimental breeding, his appreciation of significant questions in genetics was rather sophisticated for the time, even if, on occasion, his understanding of some of the answers was off base or just plain incorrect.

To begin with, one has to put oneself in the place of those explorers, such as Bateson, who labored at the ground level of the development of a new discipline. In the case of genetics, the kinds of problems Bateson and other experimental breeders labored with fall into the realm of what became known as population genetics. Since so many of the cases presented in *Mendel's Principles* had to do with following the transmission or nontransmission of characters from one generation (the P generation) to others (F_1, F_2, F_3 generations and so on), it is obvious that the questions asked were being directed at the populational level: What happens from one generation to the next of a population or populations within a species. For example, if you are a plant breeder and you want to ensure the production of a variety of plant that has a certain petal color or number, you would want to know which features are dominant or recessive and how they behave when their underlying genetic bases are passed on from one generation to the next. Or if you are an animal breeder and you want to know what to expect when you breed certain cows for particular characteristics, such as meat versus milk, and whether you can create an animal that is abundant in both attributes. Or if you are a medical clinician and you want to predict with some certainty the likelihood that a particular anomaly, such as brachydactyly—the reduction in finger- and toe-bone number—will crop up in future generations of a family. With regard to the latter condition, brachydactyly, Bateson pointed out that this was a definite case of a repeated, or meristic, character being attributable to the simple mechanism of Mendelian inheritance.

What makes the studies of Mendel, Bateson, and others like them so remarkable is that they had to work back through a known genealogy to try to figure out which character states—such as lean versus fat, or tall versus short, or red versus yellow, or more versus fewer toes and fingers—were dominant and which were recessive. As daunting as this task might be, the life of a population geneticist would be a lot simpler if the only thing he or she had to worry about was plain old dominant versus recessive. But Bateson was not content to let the cases that did not fit this simple dyadic model slip through the cracks. His drive as a scientist and researcher was seemingly inexhaustible. As Beatrice Bateson wrote in her memoir:

> Research was one long delicious adventure to him. He was patient, painstaking and ingenious—the drudgery was nought compared to the exhilarating thrill of treasure-trove, which sure enough awaited him. And yet as he worked, in the white heat of excitement, judgment sat within him cool and critical. Emotion could not compel him to unwary haste.

And so, through his own and collaborative labors, as well as his scrutiny of the breeding results or medical genealogical records of others, Bateson pushed past the simple duality of a system of inheritance limited solely to the absolutes of being dominant or recessive to pursue the not so simple aspects of inheritance. For instance, there was the problem of incomplete dominance, in which, as is the case in certain strains of chickens, the heterozygote for a dominant feature cannot be distinguished from the homozygote for the dominant feature. He also understood that there was a situation to be unraveled with regard to characteristics being transmitted by one versus the other sex, as in male baldness or color blindness. Although males may be bald or suffer from the inability to distinguish between red and green, the alleles for these attributes are passed on through the female. Bateson discussed cases of what we now refer to as sex-linked features as "sex-limited" features. He also recognized that there was, in general, something different at the genetic level between males and females: that there was something that caused femaleness and there was something about being female that could repress features that appeared to be dominant in the male.

Unfortunately for Bateson, he was among those in the teething profession of genetics who were not convinced that chromosomes were definitely part of the picture of genetics and inheritance. True, chromosomes, which reside in the nuclei of cells, could at that time be visualized microscopically. And, as was elegantly documented in the first years after the turn of the century by Walter Stanborough Sutton, a graduate student working in the zoology lab of Edmund B. Wilson at Columbia University, their fates could be followed during cell replication and then cell reduction in the process of producing gametes with only half the material of the zygote. But although it was known that chromosomes were passed from one generation to the next by way of their presence in the nuclei of parental gametes, there was still some question as to where, exactly—in the nucleus, in the cytoplasm of the cell surrounding the nucleus, or in the proteins that could be detected on chromosomes—the stuff of inheritance lay.

As for the chromosomes themselves, Bateson and others did not entirely reject a reason or cause for their existence. Among the most potentially exciting discoveries of the early twentieth century was the realization that, as was first discovered in various insects, there were differences in chromosome number between gametes that produced female offspring and those that yielded male offspring. As Bateson summarized the available evidence in *Mendel's Principles,* it was the female-producing gamete that possessed one more chromosome than the male-producing gamete. If a female-producing gamete had n number of chromosomes, the male-producing gamete had $n-1$ chromosomes. When two female-producing gametes united, they formed a zygote with $2n$ chromosomes, which would produce a female. When a female-producing and a male-producing gamete united, they formed a zygote with $2n-1$ chromosomes, which would produce a male. Since the gametes of a female individual will always have n chromosomes, the smoking gun of sex determination, or at least sex association, appeared to be the male-producing gamete with *n–1* chromosomes. But as logical as this seemed, given the available evidence at the time, it would turn out to be the case that, with the exception of certain insects such as ants and bees, in which a viable individual can develop from an unfertilized egg, males and females have the same number of chromosomes, both in their body, or somatic, cells (*2n*) and in their gametes (*n*). Female gametes typically have two large X chromosomes, whereas male sex cells have one large X and one tiny Y chromosome.

Since the role of chromosomes as carriers of the genes encoding for the characteristics that make up an entire, functioning individual was unclear at the time, it was also uncertain whether the supposed extra chromosome of females was actually involved in determining the sex of the individual or was the result of being female. Because the role of chromosomes in heredity was uncertain, Bateson tried to adduce the genetic basis for an individual's sex in the same way that he and others approached any other problem in Mendelian inheritance: by working backward from observations. Based primarily on studies of sex-linked, or sex-limited, characteristics, such as color blindness in humans, horn growth in sheep, and different varieties of the currant moth, Bateson made three inferences: Sex is inherited according to Mendelian principles; femaleness is the dominant character state; and, therefore, females are heterozygous for sex. If femaleness is dominant to the recessive state of maleness, this would also explain why color blindness, for example, is essentially unknown in females: Femaleness masks, or suppresses, color blindness. The same argument could be made for the absence, or at most the development, of reduced horns in female sheep: Femaleness masks horn growth. As for the genetic basis for becoming one sex or the other, Bateson argued that it was the dominant female factor that made an individual a female. A male, lacking any such factor, became male by default.

Given the essentially nonexistent direct knowledge at the time of what, exactly, constituted the carriers of genetic information, Bateson's attempts at deciphering the genetics of sex were certainly courageous if not ingenious. But he, as well as others who thought that femaleness was dominant over maleness, would eventually be proved wrong. As it was later discovered that both males and females have the same number of chromosomes, so it would be demonstrated that the tiny male Y chromosome determines the sex of the individual bearing it, turning embryos that would otherwise develop into females into males.

As with other aspects of his long and productive but, these days especially, often overlooked career, Bateson advocated ideas, formulated by either himself or others, that would remain central to population genetics. In *Mendel's Principles* alone were the concepts, for instance, of incomplete dominance, sex-linked characters, and a genetic basis for maleness versus femaleness. There was also, for the first time, a graphic scheme for presenting the alleles of male and female parents as they would be in the gametes and the possible combinations these alleles would take in the next generation when the gametes united to form zygotes. This schema was the brainchild of Reginald Crundall Punnett, one of Bateson's longtime collaborators on the study of inheritance, especially of characteristics in poultry. This form of presentation became known as the Punnett square.

As Punnett had originally conceived of it, and as it was published for the first time in *Mendel's Principles,* with full credit duly given, the idea was to have a simple way of depicting Mendelian inheritance of alleles from the hybrid F_1 generation to the F_2 generation. First, each allele of the hybrid male F_1 and the hybrid female F_1 parents for a particular heterozygote trait is put in a square. If each heterozygous parent is *Tt,* then a capital *T* for the

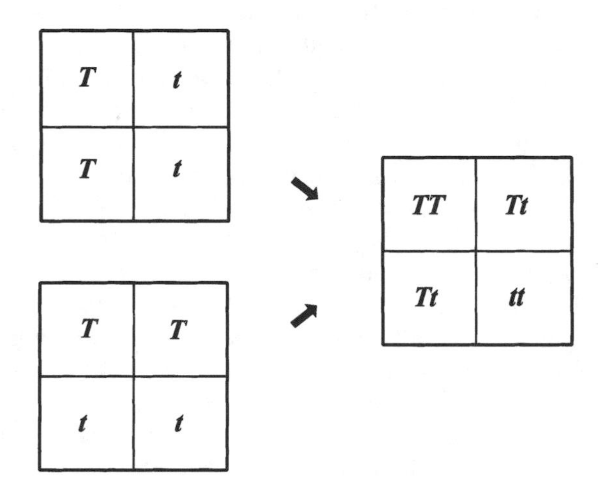

Reginald Punnett's approach (later called the Punnett square) to representing how inheritance works. The two large squares on the left represent two hybrid parents. As hybrids, each parent has in its body cells a pair of dominant (capital letters) and recessive (lowercase letters) alleles, and each is characterized by the trait associated with the dominant allele. Since, during meiosis, a male turns one cell with a full complement of genetic material into four viable sex cells (each of which has half the genetic material), he can pass on any one of two dominant and two recessive alleles to an offspring. Although a female produces only one viable sex cell during meiosis, she also has the potential to pass on a dominant or a recessive allele to her offspring. The four smaller squares within the large squares on the left represent sex cells. Since each parent will contribute one allele to an offspring, the fertilized egg will have pairs of alleles. The pairs of letters in the small squares in the large square on the right represent the possible combinations of dominant and recessive alleles that an offspring could inherit. The chance that an offspring will have at least one dominant allele (meaning that it will express the trait of the dominant allele) versus a pair of recessive alleles (meaning that it will express the trait of the recessive allele) conforms to Mendel's ratio of 3 to 1.

dominant character state would be put in a box and a lowercase *t* for the recessive character state in another. Since, as was known by the early twentieth century, a male produces four gametes at a time through cell division—a cell first doubles and then divides, producing two daughter cells like itself, and then the two daughter cells divide in half, producing four gametes, each with half the original number of alleles—he would be represented by four boxes, two bearing the dominant allele and two having the recessive allele. Since the female hybrid had been bred from the same stock as the male, and one is interested in seeing the possible combinations of this female with the male hybrid, the female hybrid would also be represented by four boxes, two with the dominant character state and two with the recessive character state.

The next step is to overlay one box of four smaller boxes on the other, as if each gamete from one parent had united with a gamete from the other parent. The result is another box made up of four smaller boxes. But each smaller box now contains the letters for two alleles: for instance, *TT, Tt, Tt,* and *tt* (which happens to be, as expected, Mendel's 3 to 1 ratio). The idea of the Punnett square can be applied to more complicated cases, in which, for example, two or three pairs of alleles are being studied.

The Punnett square is presented in a slightly different way these days. The separate alleles of one parent are listed along the top and the individual alleles of the other parent along the side of a large square. This square is then subdivided into as many smaller boxes are there are alleles. Then, proceeding from the upper-left-hand corner of the large square, the symbol for an allele from the top and one from the side are placed in the smaller box that corresponds to their intersection. After doing this for each potential pair of alleles—one from the top row and its potential mate from the side column—all the smaller squares will reveal the possible combinations of all alleles from both parents.

Although Bateson never faltered in his belief in discontinuous variation and the importance of repeated parts, he underplayed these themes while discussing the details of each experimental or natural case, saving his discourses on his particular theoretical views of evolution for other sections of the monograph. As for the matter of discontinuous variation, Bateson was actually rather brief, taking a mere two pages or so to reaffirm the reality and potential evolutionary significance of discontinuous variation (the stuff of evolutionary change) as opposed to the Darwinian notion of continuous variation, which de Vries had come to refer to as fluctuation. Armed now with the insights of Mendelian inheritance, Bateson was more convinced than ever of the significance of discontinuous variation, since, according to Mendelian principles, "[w]e now can see that the discontinuous variations are in the main the outward manifestations of the presence or absence of corresponding Mendelian factors, and we recognize that the unity of those factors is a consequence of the mode in which they are treated by the cell-divisions of gametogenesis [the production of gametes].” Because the units of inheritance are discrete, the features they produce must also be disjunct.

Bateson went from there to address the criticism levied by the Scottish engineer Fleeming Jenkin against Darwin’s proposal that evolutionary change occurs through the gradual accumulation of infinitesimally small differences among individuals. In 1867, Jenkin published a review of *On the Origin* in which he argued that gradual evolution by natural selection could not work because any new trait would be swamped and eliminated through matings of the vast majority of individuals who lacked this small, but slightly different, trait. Clearly, if one adhered to blending inheritance, as did Darwin and virtually everyone else at the time, Jenkin’s objection made a lot of sense. Consequently, it provided a solid threat to Darwin’s conception of gradual transformation.

In a series of letters that Darwin wrote to the English botanist Joseph Dalton Hooker in 1868 and 1869, he bemoaned the many errors and oversights in his earlier editions of the *Origin.* He was blatantly weary from the nasty assaults of some of his contemporaries, such as the elder English statesman of comparative vertebrate anatomy Sir Richard Owen, as well as the many more scholarly objections raised by various scientists, including the most prominent of German botanists, Karl Wilhelm Nägeli. Darwin knew that he would have to deal with them all in his revisions of the *Origin.* But, as he wrote to Hooker, he was particularly struck by Fleeming Jenkin’s comments on the lack of viability of gradual evolution by means of natural selection:

> It is only about two years since the last edition of *Origin,* and I am fairly disgusted to find how much I have to modify, and how much I ought to add; but I have determined not to add much. Fleeming Jenkins *[sic]* has given me much trouble, but has been of more real use to me than any other essay or review.

The American anthropologist Loren Eiseley has suggested that it was Jenkin’s review that, in 1871, drove Darwin to write in *The Descent of Man:* “I now admit . . . that in the earlier editions of my ‘Origin of Species’ I perhaps attributed too much to the action of natural selection or the survival of the fittest.” However, a reading of the entire passage from which this fragment of a quote is taken, which has to do not with the transformation of species but with the reality of features being adaptive, makes it clear that this is not the case. Darwin actually wrote:

> [B]ut I now admit, after reading Nägeli on plants, and the remarks by various authors with respect to animals, more especially those recently made by Professor Broca, that in the earlier editions of my “Origin of Species” I perhaps attributed too much to the action of natural selection or the survival of the fittest. I have altered the fifth edition of the “Origin” so as to confine my remarks to adaptive changes of structure.

The effect of Jenkin's criticism was that Darwin would, and did, increasingly invoke the role of the inheritance of acquired characteristics in the evolutionary transformation of species. But, as Eiseley rightly pointed out, and as Bateson had long before noted, Darwin would probably not have been troubled by Jenkin's objection had he been aware of Mendelian inheritance. As Bateson put it: "The notion that a character once appearing in an individual is in danger of obliteration by the intercrossing of that individual with others lacking that character proves to be unreal; because in so far as the character depends on factors which segregate, no obliteration takes place." As we all learned because of Mendel, the permanence of characters prevents them from being obliterated by being swamped by other characters—which does not mean that characters may not change, but that they would not behave as they would if blending inheritance were the rule.

As for the differentiation of repeated parts, Bateson had commented that brachydactyly, the abnormal reduction of finger and toe bones, was a definite case of Mendelian inheritance of repeated, or meristic, structures. For, after all, the digital rays of fingers and toes represent two versions of repetition: the side-by-side arrangement of fingers and toes along hand and foot, respectively, and the size-decreasing arrangement of bones within each finger and toe. But Bateson now broadened his speculations on how repeated parts may arise to embrace the normal and the abnormal:

> If upon the same individual, parts may as an abnormal occurrence present the same differentiation which is known to be characteristic of dominant and recessive, may not the differentiation *normally* existing between repeated parts of the same individual be a phenomenon of segregation? Why, for instance, may not the differentiation normally existing between petal and leaf, or between the appendages of arthropods, or any other meristically repeated parts, be due to a segregation acting amongst somatic parts as amongst gametes?

Going further than he did in his first book, *Materials for the Study of Variation,* Bateson expanded upon the idea that the differentiation of repeated morphological parts is due to a process of developmental duplication, or division. A normal process of division leads to the expected number of repeated parts. An abnormal process of division leads to a different number of repeated parts. He also added the possibility that developmental division leading to repeated parts is analogous to cell division and the segregation of alleles as seen in the production of gametes. Within pages of the above quote, he developed this theme to equate the process and the result of germ-cell and somatic-cell division: There are "divisions by which similar parts are divided from each other, and differentiating divisions by which parts with distinct characters and properties are separated." Repetition of parts is here reduced to a simple phenomenon that is akin to cell division. Bateson likens the difference between what is played out within a repeated series versus the advent of another meristic structure to the differences between related individuals of successive generations versus differences that constitute discontinuous variation.

As we shall see in the final chapter of this book, regardless of the place in history that Bateson should or should not occupy because of his contributions to the field of genetics, which he formally named, his ideas on inheritance and the formation of repeated parts, although in need of updating, are still germane to questions of genetics, development, and evolution.

8

Rediscovering Darwin

> *The principle of heredity and its necessary implications constitute the only assumption that is necessary for the evolutionist to make in order to go ahead on a sound basis with a presentation of the evidences of evolution. Give him this one point, and he asks no further concessions.*
>
> —Horatio Hackett Newman (1925)

Population Genetics Comes of Age

The wake of the turn of the century brought with it revelations on the constituents of the cell and their behavior during cell duplication and gamete production. These insights would have brought a smile to the lips of the German comparative anatomist and zoologist August Friedrich Leopold Weismann. For it was he who, almost half a century earlier while in the process of dismantling Darwin's theory of pangenesis, had proposed that the stuff of heredity lay in the nucleus of the cell, that this stuff, or "tendencies of heredity," is passed on from each parent through its gametes, and that the individual and its gametes, which in turn also contain "tendencies of heredity," develop from the parentally inherited "tendencies of heredity." Weismann was aware of the existence of structures within the cell nucleus, which fellow cytologists referred to as "loops" (which would later be identified as chromosomes). But the role of these nuclear inhabitants in inheritance was still unknown.

From the time of their discovery, chromosomes posed a vexing problem to cytologists. For a relatively long period—relative to the life cycle of a cell—nothing is easily visible within the nucleus. Then, at a certain point, as if by magic, chromosomes, which were always there, begin to emerge as fine filamentous structures, tangled in a meaningless loose weave. At another point in the cycle of the cell, the wall of the nucleus disappears and the chromosomes thicken as they also assume what for all intents and purposes looks like a purposeful alignment of pairs, often within the central region of the nucleus. Associated with the lined-up chromosomes is a network of thin strands that form a fat-in-the-middle, spindle-shaped structure that lies at a right angle to the series of chromosomes. Then, as the cell begins to divide, some of the chromosomes go off to one pole of the filamentous spindle into one new daughter cell and other chromosomes travel to the other pole of the spindle into the other new daughter cell. It is still debated whether the strands of the spindle pull the chromosomes to opposite poles or the chromosomes sort of "slide" along the strands of the spindle to its poles.

By the turn of the twentieth century, it was known that each parent contributes one set of an equal number of chromosomes to the zygote, thereby providing the full, doubled, complement of chromosomes found in the body, or somatic, cells of the offspring as well as in the cells that will eventually divide to produce gametes. The result of gamete production is sex cells that have half the chromosome number otherwise found in somatic cells. The theory at the time was that, during the phases of cell division, the paternally and maternally derived chromosomes of each pair separated and each ended up in a different gamete.

But the theory—that the union of sperm and ovum brought together paternal and maternal chromosomes that later separated during the offspring's production of gametes—was not fully documented until 1901–1902. And the person who did so was Walter Stanborough Sutton, who was a graduate student in the laboratory of Edmund Beecher Wilson, the Da Costa Professor of Zoology at Columbia University. It was then and there that Sutton was studying cellular division in the great lubber grasshopper as part of his research toward a doctoral degree—a degree that financial circumstances would not allow him to continue to pursue. After leaving graduate study for a few years in business, Sutton returned to Columbia University to study for a doctorate in medicine, believing this field would provide him with greater financial security than would being a university professor or a researcher. His career as a medical doctor and surgeon was brought to a premature close by his death in 1916, at the age of twenty-nine.

In studying the cells of the great lubber grasshopper, one of the things that Sutton noticed was that the eleven pairs of chromosomes typical for this species of insect were of different sizes. What was more curious, though, was that every time the cells divided—which would, of course, separate the pairs of chromosomes—the resultant daughter cells also had their chromosomes arranged in pairs distinguished by size. Sutton was encouraged to speculate beyond his actual observations by the work done by the German cytologist Theodor Boveri on the sea urchin, which seemed to suggest that a proper pairing of maternal and paternal chromosomes is necessary for the normal development of the organism. Indeed, Boveri had produced abnormally formed sea urchins by experimentally introducing two sperm and, consequently, an extra set of chromosomes, into a sea-urchin egg. The resultant zygote was then a victim of improper combinations of chromosomes.

Since the only sea-urchin progeny that grew abnormally were those with the extra chromosomes and the abnormal combinations of chromosomes they produced, Boveri suggested that the chromosomes themselves must somehow be involved in development. With regard to his studies, Sutton speculated that the constant complement of chromosome pairs of differing sizes that he observed was in some way responsible for the physiological maintenance of the great lubber grasshopper.

By following the behavior of the chromosomes in the pre–sex cell, or spermatogonial, stage of gamete production, Sutton was able to document

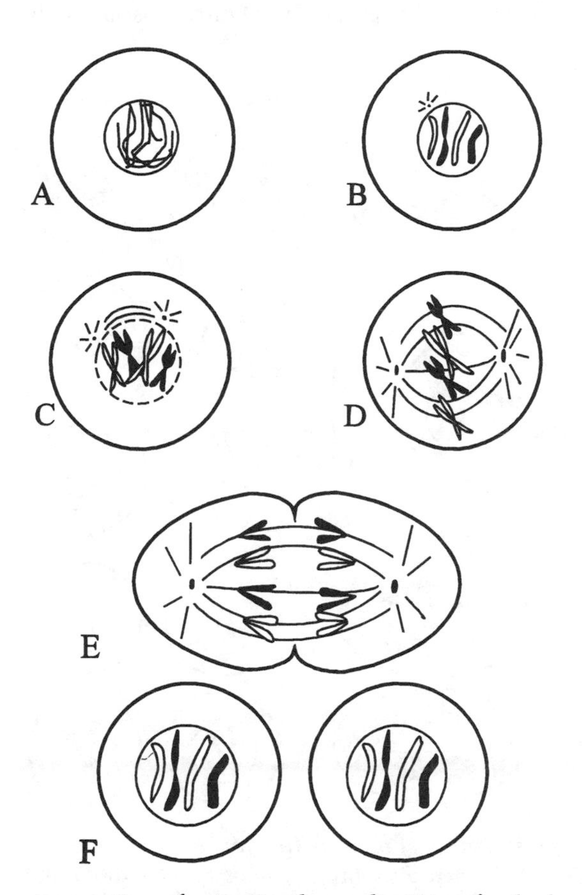

Schematic representation of mitosis: the replication of a body cell and its chromosomes. In the nucleus in the center of the cell, the filamentous chromosomes (A) become thicker (B) and double in number (C). The membrane around the nucleus of the cell disappears (C) and the spindle emerges (B) and enlarges (C) to extend from one end of the cell to the other (D). The pairs of chromosomes line up along the middle of cell (D) and then come apart, with a chromosome from each pair moving toward one end of the spindle as the original cell begins to divide (E). In the end, there are two cells (F), each with the same number of chromosomes as the original cell.

how these cell-nucleus structures came to their final state in the sex cells of the great lubber grasshopper. First, the spermatogonia, which are essentially just like somatic cells in terms of appearance and number of chromosomes, go through a process of duplication, just as somatic cells do. Consequently, the chromosomes become visible and thicken, they double in number, they line up along a midline axis of the cell, and then, as the cell is in the process of splitting into two daughter cells, one pair of the now two pairs of chromosomes goes off along the strands of the spindle into one of the new daughter cells, and the other set of paired chromosomes does likewise into the other daughter cell. If Sutton had been studying chromosome behavior in somatic

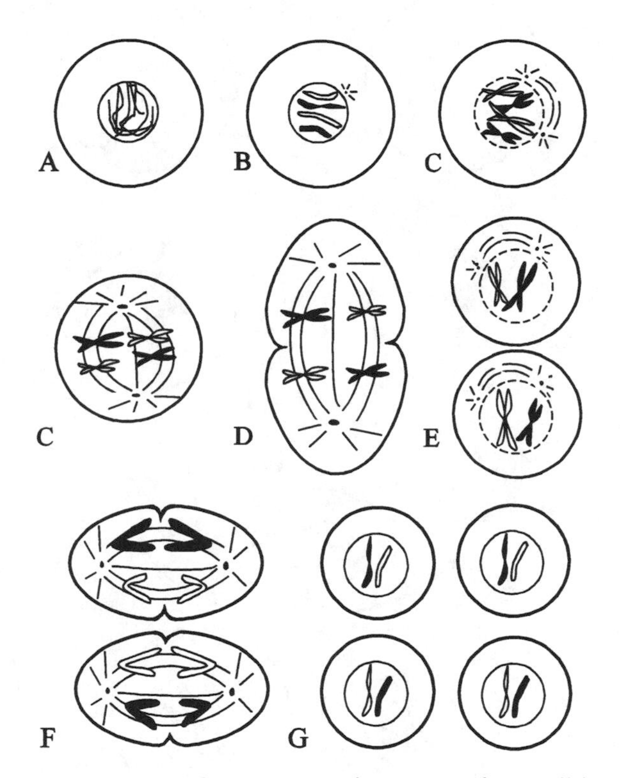

Schematic representation of meiosis (the formation of sex cells) in the male. As in mitosis, the filamentous chromosomes thicken and double as the spindle emerges and the nucleus membrane disappears (A–C). The chromosome pairs line up in the middle of the cell (C) and move toward opposite ends of the spindle as the original cell begins to divide (D) into two daughter cells of equal size (E), each with the same number of chromosomes as the original cell. The pairs of chromosomes of each new daughter cell then line up along the middle of the cell and come apart, with separate chromosomes moving to opposite ends of the spindle as each cell begins to divide into two cells (F). In the end, there are four viable sperm cells of equal size, each with half the chromosomes of the original cell (G).

cells, nothing more would have happened. There would be two new somatic cells, each with the correct number of chromosome pairs. Cytologists referred to the process by which somatic cells are produced as mitosis.

But the formation of gametes—a process the cytologists called meiosis—requires that a sex cell has only half the number of chromosomes of a somatic cell. In order to achieve this reduction in chromosome number, the cell does not simply divide in half. Instead, it goes through first a phase of duplication and then a phase of reduction. Consequently, the initial phase of meiosis looks very much like mitosis: At the end of this phase, there are two daughter cells, each of which has a full complement of chromosome pairs. But it is the second phase of meiosis that produces gametes. The chromosome pairs of the daughter cells line up along a midline axis of each cell and a spindle materializes. Then, as each daughter cell is beginning to split into two new cells, each pair of chromosomes splits, with one chromosome going along the strands of the spindle into one new cell and its now-separated chromosome partner going along the strands of the spindle into the other new cell. In terms of sperm production, one cell with a full complement of chromosomes ultimately gives rise to four gametes, each of which has half the chromosome number of a somatic cell.

The formation of an ovum is procedurally the same as that of a sperm, but the process yields only one viable egg. In the great lubber grasshopper, for example, the pairs of chromosomes of the initial egg-producing cell, or oogonium, become visible, line up along a midline axis, and double. But instead of the cell dividing into two daughter cells of equal size, one set of chromosome pairs becomes captured by a tiny, insignificant cell, in which one pole of the spindle is located, and the other set of chromosome pairs comes to reside in a comparatively huge cell, in which the other end of the spindle is anchored. The pairs of chromosomes of the huge cell as well as those of the tiny cell then line up within each respective cell and the individual chromosomes of each pair move away from one another, following the strands of the second-formed spindle. The tiny cell divides into two equally tiny cells, each with half the full chromosome number. As for the pairs of chromosomes in the huge cell, they split, with the partner of each chromosome pair going off along the strands of the spindle into a different daughter cell, each of which then has half the full number of chromosomes. But the huge cell itself divides unequally into a tiny cell and another huge cell. At the end of the process of meiosis involved in female gamete production, there are four cells, each of which has half the full number of chromosomes. But three of these cells, which cytologists called polar bodies, are tiny, while the fourth is comparatively huge. And it is the huge cell, with its store of cytoplasm surrounding the nucleus, that becomes the viable egg. With the union of a male gamete, a sperm, with a female gamete, an ovum, the full complement of chromosomes is restored and the process of growth and development and expression of characters of the resultant offspring, including the eventual production of its own gametes, proceeds apace.

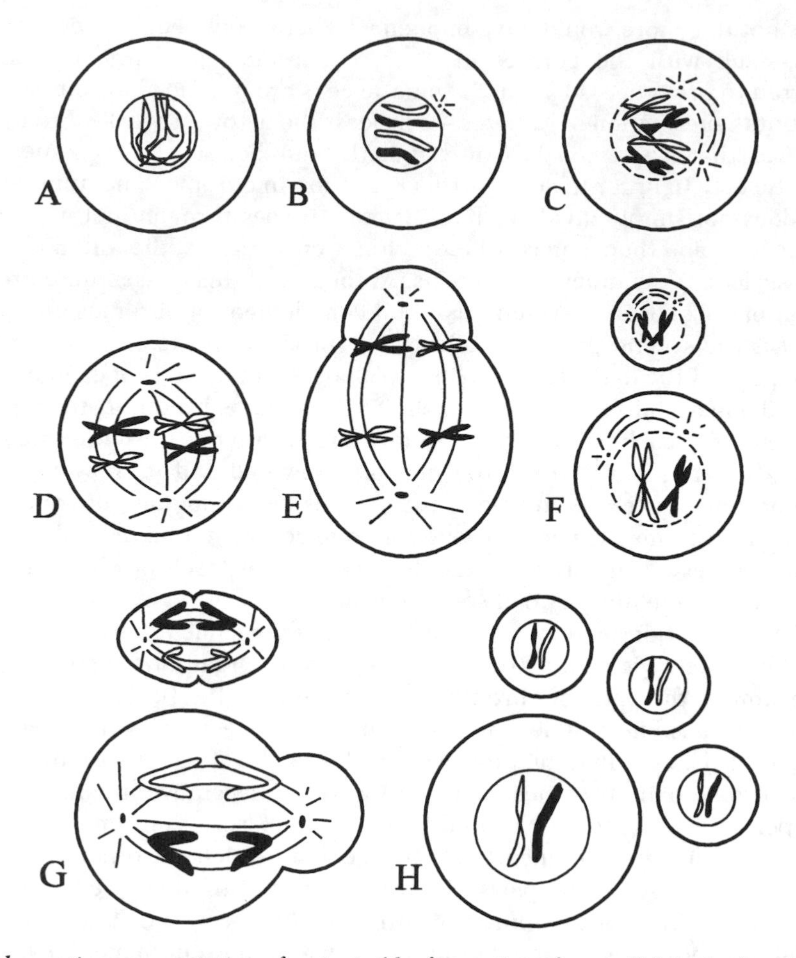

Schematic representation of meiosis (the formation of sex cells) in the female. As in mitosis, the filamentous chromosomes thicken and double as the spindle emerges and the nucleus membrane disappears (A–C). The chromosome pairs line up in the middle of the cell (D) and move toward opposite ends of the spindle as the original cell begins to divide (E) into two daughter cells of unequal size (F), each with the same number of chromosomes as the original cell. The pairs of chromosomes of each new daughter cell then line up along the middle of the cell and come apart, with separate chromosomes moving to opposite ends of the spindle as each cell begins to divide into two cells (G). The smaller cell divides into two cells of small and equal size. The larger cell divides into a large and a small cell. In the end, there is only one large egg cell and three small cells (polar bodies), each with half the number of chromosomes of the original cell (H). Only the large egg cell (ovum) is capable of being fertilized.

From his studies, Sutton became convinced that if each parent contributed its own set of chromosomes to the resultant zygote, then each pair of chromosomes in the somatic cells, as well as in the spermatogonia and oogonia, was composed of one maternal and one paternal chromosome. If this assumption was correct, Sutton further speculated, then, when that individual came to produce its own gametes, maternally and paternally derived chromosomes would eventually separate and end up in different sex cells. And if this was true, Sutton continued to theorize, and there really was some kind of connection between the morphology and the physiology of an organism and its chromosomes, then there was available a mechanism not only for bringing together dominant and recessive allelomorphs in individuals of one generation but also for redistributing these allelomorphs into the offspring of the next generation.

In a personal note contributed to a small volume that Sutton's family published privately in 1917, after the young geneticist-turned-physician's untimely death, Wilson told of his introduction to Sutton's ideas on chromosomes. The professor admitted that when Sutton first tried to explain his observations and speculations to him he had not been fully receptive, in spite of the fact that he himself had been one of the early promoters of the notion that the nucleus somehow carries the stuff of heredity. Nevertheless, Sutton was persistent, and within the year, during a continuation of their collaborative work first in North Carolina and then in Maine, managed to explain successfully to, if not fully convince, his professor of his hypothesis of chromosomes and Mendelian inheritance. Sutton's speculations and those of others who were also studying chromosomes at the time quickly became known as the chromosome theory.

That there was something potentially important about chromosomes was indicated by the apparent discovery, also in 1902, the year of Sutton's first article, that these structures within the nucleus were related to an individual's being of one sex or the other. As with so many "discoveries" or "hypotheses" that have survived into the present, the conclusion may still stand, even though the basis upon which it rested was incorrect. In this case, it was through the then-state-of-the-art microscopic techniques of Clarence Erwin McClung that the extra chromosome that supposedly occurs only in females was found to be present in some sperm. All female gametes McClung studied possessed this extra, or accessory, chromosome. When a sperm containing this accessory chromosome united with an ovum, the resultant offspring was female. When a sperm that lacked this extra chromosome united with an ovum, the offspring was male. If, it seemed to some geneticists, a chromosome was indeed associated with the development of an individual's sex, then the other chromosomes might have a relationship to other aspects of the individual's morphology. In 1905, Edmund Wilson himself and one of his colleagues and collaborators, Dr. Nettie Maria Stevens, would verify the presence of an accessory chromosome in various beetles and weevils as well as in true sucking bugs. They identified this accessory chromosome as the

X chromosome, or sex chromosome. To reflect the apparent (but untrue) lack of a sex chromosome in males, Wilson and Stevens used an *O*. Employing these letters in a Mendelian shorthand, females would be XX and males XO.

But through independent study of other insects, Wilson and Stevens also discovered that the males of these species did not carry an odd number of chromosomes. Indeed, in the cells of these males there was a pair of chromosomes composed of a normal-size X chromosome and a much smaller chromosome, which Wilson called the Y chromosome. In these species, females were expectedly XX and males were XY for the sex-chromosome pair. During meiosis and the production of gametes, these chromosomes behaved just like other pairs of chromosomes despite the great disparity in their size. Because the Y chromosome was either passed from male to male or, purportedly, some males lacked a chromosome (represented by the zero) but the X chromosome could be transmitted by both males and females and inherited by males and females alike, Wilson was able to demonstrate sex-linked inheritance of different kinds of characters. In particular, he pointed to a crisscross pattern of sex-linked inheritance, in which, because of the mode of transmission of the X chromosome, daughters were like the father and sons like the mother. Sex-linked features typically appear to zigzag between the sexes from one generation to the next.

But while there was a growing suspicion on the part of some geneticists that chromosomes were indeed involved in inheritance, and that these nuclear structures behaved in accordance with the Mendelian law of segregation, there was a problem that even Sutton was forced to acknowledge shortly after the publication of his impressive paper. It seemed obvious once one began to think about it that the number of "units of inheritance" that would go into the formation of an organism had to be greater than the number of chromosomes that that individual possessed. According to Wilson's historical account, this problem formed a large part of the reason why various cytologists, geneticists, and embryologists trying to understand inheritance could not embrace the chromosome theory and therefore sought other, smaller elements of the cell as the bearers of heredity.

Another problem that had to be sorted out was the matter of linkage. When Mendel performed his experiments with varieties of the common garden pea, he was able to follow the inheritance of single characteristics, such as seed coat or flower color. Even if it was not Mendel's brainchild, the law of independent assortment—the second law of inheritance attributed to the Austrian monk—was predicated on the notion that characteristics are not linked to one another and that the inheritance of one feature is essentially totally arbitrary relative to the fate of inheritance of another character. But as more and more studies on inheritance in different kinds of plants and then animals were carried out, it became obvious that such simple, single-character-based inheritance was not necessarily the rule. Sometimes when one feature was experimentally selected for, another feature was also inherited, no matter how hard the breeder tried to disentangle one characteristic from the other.

Mendel was fortunate to have chosen the common garden pea as his experimental organism. The common garden pea's characteristics do follow the law of independent assortment. But, as Bateson and Correns pointed out was the case in other plants, and other experimenters would subsequently find to be true of many other plants as well as of animals, many features are linked hereditarily: You cannot have one characteristic (or more) without the other characteristic (or others). For advocates of the chromosome theory, linkage of characteristics was additional proof that the units of heredity lay on the chromosomes themselves. If, they argued, the genetic units responsible for the development of features were strung out along chromosomes, then one would expect to find that some of the units on a single chromosome were linked by virtue of their being contained on the same larger structure—the chromosome—that was passed on from gamete to zygote.

Although it was still a theory in the making, the theory of the chromosome was gaining some popularity, especially among cytologists and geneticists in the United States. Ironically, one of the early detractors of the importance of chromosomes, as well as of Mendelism and Darwinism, was Thomas Hunt Morgan. After leaving the faculty of Bryn Mawr College to join Columbia University, Morgan had his office and lab in the same department of zoology as Edmund Wilson. Eventually, Morgan would turn out to be perhaps the greatest supporter not only of the role of chromosomes in inheritance but also of Mendelism and Darwinism.

Morgan: The Dissident

In order for Mendel and twentieth-century experimental breeders to understand the mode of inheritance of the features they were interested in, they had to figure out the process by working backward from the final results. Experimenting with the common garden pea, Mendel produced thousands of hybrid offspring of a number of successive generations in order to both calculate the percentages of features expressed in any one generation and try to determine which features were dominant and which were recessive relative to one another. Later, Bateson and his team of collaborators did likewise with a host of plants and animals. Indeed, there was no other way to collect these basic data except by testing the predictions experimentally.

The rediscovery of Mendel's results in 1900 was not, however, immediately embraced by everyone. The skeptics raised the point that the hybrid ratio of 3 to 1 was never actually and definitively achieved—that, although the experimental results were pretty close, the 3 to 1 ratio was still an approximation. There was also the problem that all characters were not passed individually and at random from parent to offspring. Instead, some features were somehow inherited together as a package. Consequently, these linked characteristics did not follow what had become known as Mendel's second law, the law of independent assortment. As a result of increasingly detailed knowledge of chromosome behavior during cell duplication and cell reduction, and the possibility that chromosomes both bore the units of heredity and conformed to Mendel's law of segregation, the chromosome theory garnered its share of supporters, particularly in the United States. But in the midst of this rash of discovery and speculation about cells and their parts, modes of inheritance, and evolutionary processes stood the American embryologist turned geneticist Thomas Hunt Morgan. He remained unconvinced that Mendelian inheritance had any significance much less relevance to an understanding of evolution. As late as 1909, Morgan wrote:

> In the modern interpretation of Mendelism, facts are being transformed into factors at a rapid rate. If one factor will not explain the facts, then two are invoked; if two prove insufficient, three will sometimes work out. The superior jugglery sometimes necessary to account for the results may blind us, if taken too naively, to the common-place that the results are often so excellently "explained" because the explanation was invented to explain them. We work backwards from the fact to the factors, and then, presto! explain the facts by the very factors that we invented to account for them. . . . I cannot but fear that we are rapidly developing a sort of Mendelian ritual by which to explain the extraordinary facts of alternative inheritance.

Morgan's critique of Mendel's law of segregation involving dominant and recessive features began with the recognition of the "surprising fact," known to all breeders and geneticists, that a tall pea bred with a short pea will produce a tall pea. Morgan also acknowledged the simplicity of Mendel's explanation for this: that these alternatives character states of dominant and recessive are associated with the different germ cells of the parents. This assumption, Morgan suggested, then led to the modern factor hypothesis, based on which, for example, such a Mendelian could feel justified in referring to separate and distinct "tall-factors" and "dwarf-factors."

Morgan's thoughts of 1909 would hardly have been predicted from his monograph of six years earlier, *Evolution and Adaptation.* There Morgan seemed for all the world to be one of Mendel's most appreciative supporters, commenting at one point that "[t]he theoretical interpretation that Mendel has put upon his results is so extremely simple that there can be little doubt that he has hit on the real explanation" and, at another, that "there can remain little doubt that Mendel has discovered one of the fundamental laws of heredity." But his 1909 rejection of Mendelism as it had become reincarnated in the form of the factor hypothesis was complete:

> We assume that the tall-factor and the dwarf-factor retire into separate cells after having lived together through countless generations of cells without having produced any influence on each other. We have come to look upon them as entities that show a curious antagonism, so that when the occasion presents itself,

> they turn their backs on each other and go their several ways. Here it seems to me is the point where we are in danger of over-looking other possibilities.

For Morgan, the notion of factors of inheritance—in other words, of units of heredity—represented a return to the days of preformationist ideas, which witnessed the debates between the Animalculists and the Ovists. As someone who came to genetics after studying embryology, Morgan favored a hypothesis of epigenesis, which he could discuss at a more fundamental level than the German embryologist Caspar Wolff had been able to do in his theory of epigenesis. Although Wolff was an intellectual pioneer in the early days of embryology, he could study only the development and differentiation of cell layers and organs. But Morgan could delve to deeper levels of consideration.

From genetics, one learned that there were two different character states, dominant and recessive. To Morgan, dominant and recessive represented alternative states of stability without any intermediates between them. He came to think in these terms because of his own breeding experiments. Although hybrids often inheritated alternative character states, according to Mendelian rules, sometimes they did not. Consequently, a hybrid could show characteristics of both parents in equal representation. As an example, Morgan crossed strains of chocolate and black mice and produced hybrid offspring that were chocolate from the waist down, and black from the waist up. As for his theory of epigenesis, as far as he was concerned in 1909, the attainment of any given configuration—whether it be height, color, or an unblended combination of two character states—was determined by the conditions in which the developing individual found itself:

> The egg need not contain the *characters* of the adult, nor need the sperm. Each contains a particular material which in the course of development produces in some unknown way the character of the adult. . . . [W]e are not justified in speaking of the materials in the germ-cells as the same thing as the adult characters until they develop.

As an embryologist and a geneticist, Morgan was trying to deal with the underlying bases for morphology from the perspectives of inheritance as well as of embryonic and subsequent organismal differentiation. Clearly, combining both concerns in this scientific mission was not an easy task, as Bateson and others who had also come into genetics by way of embryology had learned. Although Morgan would eventually shun all attempts to understand inheritance through the study of any aspect of morphology, his thoughts, while attempting to maintain integrity between the two disciplines, led him to reject the Darwinian evolutionist's reliance on selection and adaptation. In apparent contradiction, his rejections of Darwin's proposals derived their strength in part from the invocation of Mendel's ideas.

In his monograph of 1903, *Evolution and Adaptation,* which he wrote while still on the faculty of Bryn Mawr College, Morgan mounted an all-out attack on Darwin's theory of evolution by means of natural selection. Following in the footsteps of others before him, Morgan rejected the utilitarian notion of adaptation that was so crucial to Darwin's argument for the origin of species: that features arise in order to serve some purpose for adapting their possessor to its environment, and that natural selection will choose those individuals whose characteristics make them better adapted, or more fit, than other individuals.

As for natural selection, Morgan also could not accept Darwin's analogy between artificial selection, as practiced by breeders, and anything in the real world of nature. His reason was simple if not also repetitive of others before him: "[N]ew species comparable in all respects to wild ones have not been formed, even in those cases in which the variation has been carried farthest." Because Darwin was well aware that many features appear to the naturalist to lack any adaptive significance for survival, he went to great lengths to formulate the theory of sexual selection, which would explain the evolution of "nonadaptive" features, such as the male peacock's outrageous tail. But, Morgan asked, how can we decide what is and what is not adaptive for any organism? "If, on the other hand," he suggested, "we assume that the *origin* of the responses has nothing to do with their value to the organism, we meet with no difficulty in those cases in which the response is of little or of no use to the organism." As the historian Peter Bowler has pointed out more recently, without sexual selection to pick up the slack where natural selection failed to explain the origin of characteristics that had no visible adaptive value for survival, Darwin's entire theory of evolution fell apart.

Morgan paralleled Bateson when he invoked Mendel's studies to reject Darwin's theory. Following Mendel, characters should be seen as discontinuous and not conforming to Darwin's view of continuous variation. Darwin had rejected discontinuous features, such as those that would create "monsters" and "sports," as presenting to natural selection the variants from among which to choose the better adapted. Because Darwin believed in continuous variation, he thought that individuals within a species could be arranged to form a morphological and behavioral continuum. But this approach required each individual to be evaluated in terms of the sum total of his or her features. An appreciation of discontinuous variation, on the other hand, focused attention on individual characters.

Morgan was also reminiscent of Bateson when he broached the problems of the origin of species and species identification. Like Bateson, and foreshadowing the concerns of population geneticists and evolutionary theorists of only a few decades later, Morgan saw the species problem as one that involved first interfering with genetic continuity and then achieving genetic discontinuity. In his musings on the origin of species, Morgan ruled out the importance of the occasional sport. He did so on the grounds that if a sport and another individual of the same species did mate (which he did not consider impossible), the resultant hybrids would have to self-fertilize or breed among themselves in order to keep the sport present in subsequent generations. And, in turn, this was necessary for a new form to become established. Otherwise, as he believed was more likely, the recurring parental characters in each generation would swamp the sport. One way to effect the origin of a new species would, of course, be possible if more than one sport arose. However, Morgan continued, there would still be the problem of introducing some mechanism of infertility between the individuals with the novelties and those that were still in the guise of the original parent species. But there was yet another way out of the origin of species dilemma:

> If, however, a species begins to give rise to a large number of individuals of the same kind through a process of discontinuous variation, then it may happen that a new form may establish itself, either because it is adapted to live under conditions somewhat different from the parent form, so that the dangers of intercrossing are lessened, or because the new form may absorb the old one. It is also clear, from what has gone before, that the new form can only cease to be fertile with the parent form, or with its sister forms, after it has undergone such a number of changes that it is no longer able to combine the differences in a new individual. This result will depend both on the kinds of the new characters, as well as the amounts of their difference. This brings us to a consideration of the result of De Vries, who has studied the first steps in the formation of new species in the "mutations" of the evening primrose.

Morgan, in embracing a version of Hugo de Vries's mutation theory, also came to reject the notion that evolution occurs gradually, through a process in which natural selection picks and chooses from among myriad small variations—fluctuating variations, as they were called—among individuals. For Morgan, a new feature could arise as a result of a single mutation. And if a novel characteristic could arise by way of a single mutation, so, too, could a new species. Consequently, as de Vries himself had earlier argued, Morgan concluded that if natural selection played any role at all in the overall workings of evolution, it was minor and could be relegated to events subsequent to the introduction of evolutionary novelty by way of mutation.

But for Morgan perhaps the biggest stumbling block in accepting Darwin's theory of evolutionary change was the fact that it was based on a struggle for existence, a battle among individuals, a war of nature. As Darwin had framed it, this struggle begins at the very beginning—with the production of the potential members of the next generation, of which only a certain number will survive—and so on in the elimination of individuals at any and every stage of an individual's life. As Darwin put it:

> It is good thus to try in our imagination to give any form some advantage over another. . . . All that we can do, is to keep steadily in mind that each organic being is striving to increase at a geometrical ratio; that each at some period of its life, during some season of the year, during each generation or at intervals, has to struggle for life, and to suffer great destruction. When we reflect on this strug-

> gle, we may console ourselves with the full belief, that the war of nature is not incessant, that no fear is felt, that death is generally prompt, and that the vigorous, the healthy, and the happy survive and multiply.

Morgan had no time for Darwin's sentimentality, and said so in addressing this passage from Darwin:

> The kindliness of heart that prompted the concluding sentence may arouse our admiration for the humanity of the writer, but need not, therefore, dull our criticism of this theory. For whether no fear is felt, and whether death is prompt or slow, has no bearing on the question at issue—except as it prepares the gentle reader to accept the dreadful calamity of nature, pictured in this battle for existence, and make more contented with their lot "the vigorous, the healthy, and the happy."

Morgan summarized his dissatisfaction with Darwin's arguments on selection, adaptation, and the struggle for existence in many hard-nosed comments strewn liberally throughout *Evolution and Adaptation,* of which the following is only a sample: "The destruction of the unfit, because they can find no place where they can exist, does not explain the origin of the fit." Like Bateson, Morgan rejected the essentials of Darwin's theory of evolution by means of natural selection, his ideas on sexual selection, and his views on adaptation. But, again like his English counterpart, the American embryologist and geneticist still held Darwin in great regard for having focused attention on the fundamental questions of evolution. And Morgan reiterated as much in the closing paragraphs of *Evolution and Adaptation.* But he concluded that monograph with a summary of his main points: that animals and plants do not change in order to become better adapted to their surroundings; that some species are, in fact, not well adapted in some ways for their surroundings; that the struggle for existence, if true, would not allow such instances of less than perfect adaptation; that, on the other hand, some organisms, or features of organisms, seem to be too well or overly adapted for their surroundings; that, consequently, it is improper to evaluate features and adaptation from a utilitarian perspective; and, finally, that the existence of features that appear to be either less or overly perfected can be better appreciated without the constraints of selection pressures and a struggle for existence.

Morgan ended with the following thoughts:

> If we suppose that new mutations and "definitely" inherited variations suddenly appear, some of which will find an environment to which they are more or less well fitted, we can see how evolution may have gone on without assuming new species have been formed through a process of competition. Nature's supreme test is survival. She makes new forms to bring them to this test through mutation, and does not remodel old forms through a process of individual selection.

Morgan: The Champion of Darwinism and Mendelism

In December 1909, in his presidential dinner address at the annual meeting of the American Society of Naturalists in Boston, Morgan began to air questions that presaged the eventual breakaway of genetics from the rest of biology, and especially from comparative anatomy and embryology. He began by warning his fellow naturalists that, for all their concerns about the natural world surrounding them, they were far from understanding not only the causes of evolution but also the origin and evolution of adaptation. Paleontology might be able to document a history of evolution through fossils, but it could not solve these problems. Comparative anatomy might now make most sense in an evolutionary context, but this discipline could not solve these problems. Embryology, the field in which Morgan had originally trained, may have gone from the anti-evolutionary stance of von Baer through the radical evolutionary formulations of Haeckel to providing insights into the relationships of organisms, but it could not resolve these problems. On this occasion, he chose to emphasize questions about evolution and adaptation.

Morgan contrasted what he considered to be the two competing views on organismal evolution. One school of thought, that of the selectionists, believed in the occurrence of chance or accidental variation, upon which natural selection then acted. As such, a new feature was not seen as being adapted a priori to the role it might later play in the survival of the individual. Morgan ventured to state that this was the majority viewpoint. The other school could not accept the possibility that evolutionary events might occur accidentally or randomly, especially in the case of the origin and evolution of *Homo sapiens.* As Morgan characterized these evolutionists: "They *feel* that some more direct and intimate relation must exist between the origin of a new part and the use it comes to subserve."

Morgan then proceeded to engage in a mind game of sorts with the word *chance,* which was often used in an evolutionary context. He pointed out that *chance* can mean "accidental" or "of unknown cause." This is the way in which Darwin, for example, used the word when he discussed chance variation: A feature arose by unknown cause. But afterward the survival of the feature was dependent on natural selection. There was also another meaning of the word *chance,* as in the phrase "I chanced to be there." As Morgan saw it, once a feature came into being, its persistence, or survival, was merely a matter of chance as to whether it found itself in a compatible environment. And this, he told his audience, and not the way Darwin saw it, was the basis of evolution. Consequently, despite the title of his book, *On the Origin of Species by Means of Natural Selection,* Darwin had not made a case for the origin of species at all. Instead, he had merely described the workings of natural selection in the origin of the adaptations of plants and animals. (Had he thought of it, perhaps Morgan would have changed the title of Darwin's book to *On the Origin of Adaptation by Means of Natural Selection.*) For naturalists and systematists studying the origin of species, Darwin's theory of natural selection was of no help whatsoever. In fact, it served only to conflate the very real and important differences between processes of adaptation and of evolution—the clear distinction between which, Morgan firmly believed, must be maintained if any progress in evolutionary science was to be made.

In contrast to his previous, uncompromising dismissal of the notions of adaptation and selection, Morgan now promoted the idea that evolution could result from adaptation, which, in turn, was associated with apparently directed, or purposeful, change. Without overtly invoking mutation, Morgan suggested that the initial variation or novel feature of what would later be seen as an adaptation would have been introduced by chance. As such, the beginning of an evolutionary sequence of adaptational change was devoid of any notion or intent of purpose. Once a feature had been selected for, however, it might then be considered to have a purpose, insofar as one can actually know such things. By chance, then, a "purposeful adaptation"—identified as such in retrospect—happened, or chanced to occur, at the right time and in the right place. But having achieved this first step, "the dice," as Morgan phrased it, "are loaded," "the subsequent events are rendered more probable," and "[e]volution along adaptive lines would be a consequence of the very processes that variation has initiated."

Although Morgan was now moving somewhat toward suggestions that might pretend to accept various Darwinian notions, he continued to maintain his distance from the role of the struggle for existence in provoking evolutionary change. "Is the battle always to the brave—for the brave is sometimes stupid—or the race to the swift, rather than to the more cunning?" Morgan asked. For, he continued, "[a]n individual advantage in one particular need not count much in survival when the life of the individual depends on so many things—advantages in one direction may be accompanied by failures in others, chance cancels chance."

The reason, according to Peter Bowler, that Morgan did not go fully Darwinian, even when he really did come to embrace the essence of gradualism, right down to the level of natural selection acting on minor variation, was due to his stand on human evolution and, in particular, on the origin of human races. A struggle for survival in the Darwinian sense could lead to the conclusion that some races were better adapted or fitter and, consequently, "more evolved" than others. It was an adherence to the notion of fierce competition resulting in a racial hierarchy that gave rise to social Darwinism and a belief in the supposed superiority of some races over others. For Morgan, however, racial "advance" was accomplished "not by [some individuals] supplanting their fellows, for each advantage to be gained, but by combining with them." In essence, it would appear that, in a Morganian context, races would be neither meaningful nor, even if they were, permanently distinct entities. In other words, "we should expect advance in the human races to take place not by every man's hand being raised against his neighbor, nor by the picking out of a few choice individuals in the way the breeder produces new varieties of corn, horses, pigeons and pigs, but we should expect advance to take place in those parts of the world where there is a good stock to start with, and an environment that calls forth in that stock favorable variations in excess of unfavorable ones." For Morgan, survival was not a matter of struggle but a mechanism for ensuring success when and where success was most needed. To wit, Morgan repeated a well-worn phrase: "In evolution nothing succeeds like success."

In closing his presidential address, Morgan reiterated his plea that understanding evolution and adaptation should be the primary concern of naturalists. "The time is past," he warned, "when it will be any longer possible to speculate light-heartedly about the possibilities of evolution, for an army of able and acute investigators is carefully weighing by experimental tests the evidence on which all theories of evolution and adaptation must rest." "To them," Morgan concluded, "belongs the future."

It certainly seems that Morgan, at least, took heed of his own admonition. Beginning in 1910—the very year that his presidential address to the American Society of Naturalists as well as an article expressing his doubts about the role of chromosomes in inheritance were published—Morgan plunged forever into experimental genetic studies. The organism with which Morgan chose to work was the common fruit fly, known by the genus name *Drosophila.* This insect was becoming the organism of choice among geneticists because it was accessible, easy to breed, and about fifteen species had so far been recognized by taxonomists.

To most of us, the fruit fly is a slightly annoying insect that seems to materialize out of thin air just when the bananas' skins are getting a bit mottled. A fruit fly even looks a bit funny, having large eyes visible to our naked eyes and a body that would seem too big and plump for its wings to carry. But to geneticists the fruit fly is the perfect experimental lab animal. For one thing, it takes only about twelve days for members of a new generation to hatch, mature, mate, and lay their own eggs, of which the female sometimes produces a thousand. Such a rapid turnaround between generations of enormous numbers of individuals meant that the experimental population geneticist could, under controlled breeding parameters, produce thousands of consecutive generations of thousands of fruit flies within a relatively short period of time.

Also important for early genetic studies was the fact that more than 400 races, or subspecies, of the most common species of fruit fly, *Drosophila melanogaster,* and more than 125 strains of the relatively common species *D. ampelophila* were known. Each subspecies was recognized by at least one peculiarity, which often had something to do with eye color, wing shape, or number of bristles on the upper body, or thorax. But even though there was a feature or features specific to each subspecies, as members of the same species individuals from one subspecies could interbreed with those from another without problem.

From a different but still practical standpoint, a fruit fly has relatively very large chromosomes but typically only a small number of chromosome pairs—often only four pairs of chromosomes, although species with three, five, and six pairs are known—of which one pair, of course, represents the sex chromosomes. Consequently, a fruit fly's chromosomes are not only relatively easy to study under the microscope—the cells can be picked on a pin from the animal and put right onto a microscope slice—but there are not that many chromosomes to compare among individuals of the same generation and individuals of consecutive generations.

By the time Morgan published *A Critique of the Theory of Evolution* in 1916, he had moved from Bryn Mawr College to Columbia University, where he was installed as professor of experimental zoology. His *Critique* consisted of a series of lectures that he had given at Princeton University in February and March of that year. By then, he had embraced the theory that chromosomes bore the units of heredity and that chromosomes and the units of heredity they carried behaved in a Mendelian fashion. He had also changed his thoughts on evolution. Although he did not, and never would, adopt Darwin's notion of natural selection, he did elaborate on a version of the concept and speculate on its role in evolution. However, the idea that a "struggle for existence" in nature was fundamental to evolutionary change remained anathema to him.

It was in *A Critique* that Morgan promoted genetics as the only gateway to understanding evolution and the processes of evolution. Although he acknowledged the relevance of other areas of the biological sciences to evolutionary studies in general, Morgan found fault with all of them. Paleontology was fraught with the problem of incompleteness, of those awful gaps in the fossil record that prevented demonstration of the expected picture of gradualism and continuity. Comparative anatomy was predicated on trying to arrange organisms from the simple to the complex, as if this represented an evolutionarily real sequence. But the spontaneous appearance of independent variations in strains of fruit flies convinced Morgan of the contrary: that simply arranging fruit flies in a sequence of, for instance, diminishing wing size would mask the unrelated mutational events that had led to wings of different sizes. Morgan's original discipline, embryology, revealed only the evolutionary insight that organisms share certain phases of early development. Although these long-standing fields of biological study provided "circumstantial evidence for organic evolution," in the final analysis they merely set the stage for genetics. For it is only through genetics that one can "observe evolution going on at the present time, i.e.[,] . . . observe the occurrence of variations and their transmission." And with the conviction that "[t]his has actually been done by the geneticist in the study of mutation and Mendelian heredity," Morgan abandoned his embryologist's background forever.

Perhaps the most important element in Morgan's change of heart about the role of chromosomes in heredity was that he actually did alter his intellectual position. This is not a commonplace occurrence in the life of a scholar, or of any of us, for that matter. In science in particular, an individual often spends a lifetime defending a particular brainchild. And in doing so, one often thwarts alternative ideas developed by others. But Morgan was able to make the break with his own past, and not only to embrace the chromosome theory—which would still remain a theory even when he had finished updating it—but also to add considerably to it.

A large part of the impetus for Morgan's conversion to the chromosome theory and to a version of Darwinism came from his own studies on fruit flies. But Morgan's experience, and especially his monographic treatments of genetics and evolution, would surely not have progressed as easily or as quickly were it not for his collaborators—Alfred Henry Sturtevant, Hermann Joseph Muller, and Calvin Blackman Bridges. The results of their labors on fruit-fly genetics, and the inroads they made into the realm of chromosomes and inheritance, were published in 1915, in *The Mechanism of Mendelian Heredity,* which served as the basis for Morgan's later monographs.

By chance, one of the strains of fruit fly that Morgan and his group studied mutated into an eyeless form. As they bred more generations of different strains of fruit fly, Morgan noted other, seemingly spontaneous and chance mutations. Compared to the typical fruit fly found in the wild, which for comparative purposes the experimenters identified as the wild, or normal, type, different mutant strains of fruit fly cropped up from time to time in the lab for no apparent reason. These mutants differed from the wild type in, for example, aspects of body color and size, wing shape and size (and even absence), eye color and size (and even absence), antenna size and shape, number of abdominal segments, and distribution of body bristles on thorax and abdominal segments.

Although the word *gene* had been introduced into the literature of evolutionary biology by the Danish botanist Wilhelm Johannsen in 1909, Morgan did not use it in 1916 when he discussed factors, or units, of heredity. Perhaps this was because, like Bateson, Johannsen had also not been convinced that units of heredity—his "genes"—were associated with chromosomes, or even that such units of heredity were actually correlated with the expression of a specific character. By 1914, however, Morgan's Columbia University colleague Edmund Wilson had not only accepted *gene* as a synonym for "factor" or "unit of heredity" but had also used the word in the context of being a feature on a chromosome that was somehow involved in producing a physical characteristic. Unfortunately, at least in the published version of the distinguished Croonian Lecture, which he gave to the Royal Society in London in 1914, Wilson misspelled *gene* as "gens." By 1922, Morgan would end up using *gene* instead of all the alternatives he had enlisted previously, among them "factor," "unit," "unit of heredity," and "factor-unit." From this point on, except when quoting from someone else's work, I, too, will use the word *gene.*

Having now come to champion the chromosome theory, Morgan pushed further in his efforts to demonstrate that there was more than just a one-to-one relationship between a single chromosome and a single character—the possibility of which had been the stumbling block to the chromosome theory that had troubled Walter Sutton the most. If this were true, then, for example, fruit flies with only four chromosomes should have only four inheritable characters, of which one would be the sex of the individual. Instead, Morgan and his collaborators were able to demonstrate with as much circumstantial evidence as possible that a chromosome could carry many genes. And they did this by suggesting on which chromosome a particular gene might lie.

Fortunately for Morgan and his research team, different strains of fruit fly of the same species had chromosomes of different shapes and sizes and could even differ in having a single chromosome instead of a pair of chromosomes. Consequently, when Morgan and his collaborators pursued their studies of what happened genetically from one generation to the next, they could study the pairing (or lack of pairing) of chromosomes and actually see which chromosomes had come from which strain. In this way, they were not only following the course of inheritance of a particular feature and determining its dominant and recessive character states; they were also studying the inheritance of these character states via the transmission of a particular chromosome from parent to offspring.

One of the first successes Morgan and his colleagues achieved in this new line of research came from studying crosses between a strain of fruit fly called "Diminished-bristles" and the wild, or normal, type of fruit fly, which had normal bristles. It turned out that the mutant strain had only one fourth chromosome instead of a pair. In addition, this single chromosome was small. So when an individual of this strain was crossed with a individual from a strain with the normal complement of four pairs of chromosomes, it was easy to distinguish between offspring that had inherited the fourth chromosome of the mutant strain and those that had not. When an individual of the Diminished-bristles strain was crossed with an individual from the eyeless strain, an offspring with only a single fourth chromosome would be eyeless. When, however, an eyeless individual was crossed with a normal individual, their offspring were normal for eye development because the normal condition was the dominant character state. The logical explanation for the offspring of a Diminished-bristles-eyeless cross was eyeless was that the otherwise recessive character state—"eyeless"—was itself on the fourth chromosome and that, since there was only one fourth chromosome in these individuals, the recessive state could be expressed as if it actually were the dominant character state. By following other cases in a similar fashion, Morgan and his colleagues could begin to draw a map—which they called a chromosome map—of genes on each of the four chromosomes.

But there was another way, Morgan realized, of hypothesizing not only on which chromosome a particular gene might lie but also the relative positions of genes to one another on a particular chromosome. Morgan's insight was inspired by the work of the Belgian cytologist F. A. Janssens, who had contributed perhaps more than anyone else to an understanding of the behavior of chromosomes during cell duplication as well as during gamete production. Janssen's observation, which he published in 1909, was simply this: Sometimes matched-up chromosomes twisted around one another and appeared to fuse at a particular place. When, during mitosis and meiosis, they pulled apart to go off to different poles, the chromosomes would break at the point of fusion. The result was that each chromosome was now a combination of part "original" and part "other" chromosome.

Janssens observed that fusion between associated chromosomes could occur at one of two stages: when the original pair of chromosomes duplicated, as well as prior to the separation of a pair of chromosomes. Sometimes, while twisting around one another, a part of one chromosome (or a part of one chromosome pair) crossed over a portion of its partner. Janssens simply called this mode of chromosomal interaction "crossing over," a terminology that is still used today. Not all instances of crossing over led to the exchange of parts between associated chromosomes. Sometimes even tangled chromosomes would pull apart and remain intact. But when crossing over did lead to the exchange of parts between chromosomes, it obviously meant that the sequence of alleles along each resultant chromosome had been altered. To Morgan, the shuffling around of stretches of genes along a chromosome was a very important source of introducing variation into the population. The fact that chromosomes can cross over and exchange alleles would explain why even in Mendel's simple pea experiments, hybrid crosses always approached, but never equaled, the 3 to 1 ratio.

Schematic representation of chromosome crossing over during meiosis. Proceeding from left to right, a chromosome pair doubles; parts of chromosomes break and reattach to other chromosomes; as the chromosome pairs move away from each other, the chromosomes stay rearranged; the two separate chromosome pairs now have different chromosome segments with potentially different genes, or alleles, along them.

As for figuring out the locations of genes relative to one another along a chromosome, Morgan and his team started with the assumption that crossing over could occur anywhere along the length of a pair of chromosomes. If this is so, then the likelihood of a break occurring between two genes will be greater the farther apart the genes are from one another. The closer together, or more closely linked, two genes are, the greater will be the likelihood that if a break occurs, the two genes will lie on the same piece of chromosome. By crossbreeding strains of fruit fly that were known for certain features, Morgan and his colleagues could calculate how far or close together the genes under study were by seeing which genes cropped up either together or apart in the offspring and with what frequencies they did so. Through an exhaustive series of studies, Morgan and his co-workers discovered that for all four hundred subspecies of *Drosophila melanogaster,* characters are inherited in four groups, which, of course, is exactly the number of chromosomes a fruit fly has. When Morgan and his team looked at other species of fruit fly, they found the same correspondence: There were the same number of linkage groups as chromosomes.

But while, albeit with much painstaking and time-consuming experimental lab work, Morgan and his team were able to put together chromosome maps for the various species of fruit fly they studied, they also discovered that there was not always a one-to-one relationship between a gene and a morphological characteristic. Sometimes it appeared that a physical characteristic was tied to more than one gene. And yet in other instances it was pretty clear that one gene had an effect on more than one characteristic or region of the body, or that a gene was responsible for a significant feature as well as for characteristics that were less vital for survival. Because of these observations, Morgan was very cautious about how the correlation between genes—his "factors"—and their physical manifestations should be described: "It can not too insistently be urged that when we say a character is the product of a particular factor we mean no more than that it is the more conspicuous effect of the factor."

As an interesting aside in their pursuit of character-linkage groups and the relative locations of the genes for these characters on the chromosomes, Morgan and his colleagues stumbled into the species problem. For years, taxonomists had considered *Drosophila melanogaster* to be a species that contained a plethora of races, or subspecies. When Morgan began his research using the fruit fly, he did, too. But after a while it became apparent to experimenters using fruit flies that supposedly represented the species *D. melanogaster* that some matings produced offspring that were sterile. Upon further investigation, it also became apparent that there were small but identifiable physical differences between individuals that, when crossed, produced infertile offspring—this in spite of the fact that the four pairs of chromosomes of all of these individuals, indeed of all fruit-fly races assigned to *D. melanogaster,* were identical in size and shape. Eventually, fruit-fly geneticists used minute morphological differences, as well as the criterion of producing sterile offspring, in isolating a new species, *D. simulans,* from the long-accepted species of *Drosophila, D. melanogaster.*

As was foreshadowed by Bateson, taxonomists and systematists would now no longer be able to rely solely on morphology for identifying species. Within a matter of decades, morphology would, for all intents and purposes, disappear from many of the debates on defining species. This was especially true in the case of neontologists. A neontologist studies living organisms and, for that reason, can pursue genetic and reproductive studies on living organisms. Paleontologists, however, were left in a bind, because fossils are only petrified morphology and at best are often only fragments of the whole organism. As such, a paleontologist who adheres too firmly to a morphological definition of species difference would appear to be hopelessly outdated unless he or she tried to infuse a bit of genetics and talk of reproduction into the discussion.

Although Morgan had moved away from de Vries's mutation theory, he still seemed to think that these mutations, given how suddenly and spontaneously they arose, reflected a form of discontinuous variation. But Morgan now believed that if discontinuous variation was valid, it was only at the level of the character itself. At the genetic level, he saw mutations within strains of the same species of fruit fly as being of minor consequence. In addition, he maintained that such minor differences, or variations, were irrelevant to the distinctions that exist among species in the wild: "I do not think such a group of types differing by one character each, is comparable to most wild groups of species because the difference between wild species is due to a large number of such single differences." Having accepted that the only mutations relevant to evolution were those that produced minor or fluctuating variation, it was inevitable that Morgan should come to embrace the idea that evolutionary change occurred gradually by the accretion of small mutations.

As for the tempo of evolutionary change, Morgan proposed a genetically based model that echoed Darwin's view of evolution: Evolutionary change occurs gradually through the accumulation of small variations, which are produced by minor mutations. Perhaps the most important experiment in guiding Morgan to this position was the breeding of a strain of the species *Drosophila ampelophila,* in which a mutant with truncated wings appeared unexpectedly.

Typically, the end of the wing of the truncated wing mutant was sometimes squared off. Starting with the first individuals of this new mutant strain, in which the wings varied in length from normally being longer than to much shorter than the abdomen, Morgan's collaborator, Muller, worked for three years selectively breeding generations of individuals for shorter wing length. Eventually, Muller produced generations of truncated wing mutants in which some individuals had wings that were significantly shorter than the body. Buoyed by these and other experiments, Morgan felt assured in telescoping the significance of Muller's lab-based selective experiments to envelop evolutionary events as they might occur in nature. If, Morgan suggested, change could be brought about in the lab by the manipulation of generations of mutant strains, it was conceivable that, given enough time, evolutionary change could take place in the wild, with real species and in a similar fashion. He boldly declared: "Evolution of wild species appears to have taken place by modifying and improving bit by bit the structures and habits that the animal or plant already possessed."

Morgan's extrapolation from his laboratory experiences to what he conceived of as being the situation in nature is an application of the concept of uniformitarianism, which had been formulated by eighteenth-century British geologists such as Charles Lyell. Although Lyell was initially one of the most adamant of anti-evolutionists, through Thomas Huxley's urging he eventually came to be one of Darwin's most important supporters. In fact, it was Lyell who helped Darwin when the younger naturalist Alfred Russel Wallace formulated a similar theory of evolution by natural selection while suffering a malaria attack one night in Borneo. Lyell had the abstracts of the two evolutionists' papers read back-to-back at the same scientific meeting in 1858. As for uniformitarianism, the premise is that the processes that one observes acting in the present are the same processes that acted in the same way in the past.

If you were a geologist, like Lyell, and tried to imagine how the Colorado River had produced something as deep as the Grand Canyon, you might conclude that it had been a very long process because the depth of the canyon does not change drastically from day to day, or even from year to year. Lyell's application of uniformitarianism led him to believe that geologic change was gradual. Darwin embraced Lyell's gradualism and applied it to his vision of evolutionary change. If Darwin had been a uniformitarian like de Vries, and had been impressed by the abrupt appearance of change, then he, like the Dutchman, and Huxley as well, would have favored a saltational model over that of gradualism.

Morgan observed that, as spontaneous as they were, the mutations that arose in his fruit flies produced only minor variations. Because he believed that the differences that exist among species should be of greater magnitude than these minor variations, his uniformitarian approach led him to conclude that evolutionary change of the sort that produces species occurs "bit by bit." Population geneticists who followed in Morgan's footsteps also applied a uniformitarian approach to their interpretation of the evolutionary significance of their lab studies. Ergo, evolution, as seen from the perspective of these geneticists, was gradual.

Many evolutionary biologists today who embrace a neo-Darwinian view of evolution derive some justification for their theoretical position from the application of uniformitarianism. Why, they ask, should one invoke other explanations when a perfectly good mechanism of inheritance and change has been demonstrated in the lab? But it should be remembered that uniformitarianism only specifies the extension of present-day observations into the past. It does not specify lab experiments on fruit-fly genetics. Consequently, geneticists like de Vries and Bateson, as well as the early, anti-

Darwinian Morgan, were also being informed by a uniformitarian doctrine when they suggested that variation was discontinuous and that change was abrupt. In these instances, the properties of Mendelian inheritance and the sudden appearance of sports and monstrosities were the observations that these geneticists extrapolated to the past.

Having now, however, taken a gradualistic view of evolution, Morgan had to formulate his ideas on how the potential for change, once introduced via the mechanism of spontaneous and initially nonutilitarian mutation, would be dealt with and developed into adaptive evolutionary change. Darwin had promoted natural selection as the agent—the initiator and developer—of change and adaptation. Morgan thought otherwise.

The problem Morgan had with Darwin's model of natural selection was that it, in concert with a struggle for existence, was supposed to be responsible for evolutionary change. With selection choosing from whatever variation was present among individuals of one generation, individuals of the next generation would certainly be affected. For example, if increased height was the characteristic being selected for, the tallest individuals of one generation would contribute disproportionately to the next generation, which would then, on average, be taller. If the tallest individuals from this generation contributed disproportionately to the next generation, then individuals of the latter would, on average, also be taller. But, as Morgan made clear, this process could not go on indefinitely. Selection would not be able to produce a generation whose members collectively exceeded the greatest height available in the original generation. "Selection," Morgan concluded, "has not produced anything new, but only more of certain kinds of individuals." But, he added, "[e]volution . . . means producing more new things, not more of what already exists." And it is at this point in the process that the concept of mutation entered into Morgan's theory: Mutations, even though they seemed to produce only minor or insignificant variations, nonetheless introduced the novelty upon which a kind of natural selection could act.

But just because mutation produced variation does not mean that the novel feature was necessarily of any benefit or utility to the individual or individuals bearing it. A new feature could, as Morgan pointed out, be "neither advantageous nor disadvantageous, but indifferent." If such an "indifferent" characteristic were to arise by mutation, he speculated, "the chance that it may become established in the race is extremely small, although by good luck such a thing may occur rarely." "If through a mutation a character has an *injurious* effect, however slight this may be," Morgan went on, "it has practically no chance of becoming established." But, he was quick to point out, "[i]f through a mutation a character appears that has a *beneficial* influence on the individual, the chance that the individual will survive is increased, not only for itself, but for all of its descendants that come to inherit this character. . . . It is this increase in the number of individuals possessing a particular character, that might have an influence on the course of evolution."

Although this latter thought may sound essentially Darwinian, Morgan believed that his view of selection differed from Darwin's. He characterized Darwin's stance on struggle and selection as providing as much a means for eliminating the unfit as for promulgating the fittest. In contrast, he saw his view as being neutral. "Does the elimination of the unfit," he asked, "influence the course of evolution, except in the negative sense of leaving more room for the fit?" Instead, as Morgan conceived of evolution and selection:

> New and advantageous characters survive by incorporating themselves into the race, improving it and opening to it new opportunities. In other words, the emphasis may be placed less on the competition between the individuals of a species (because the destruction of the less fit does not *in itself* lead to anything that is new) than on the appearance of new characters and modifications of old characters that become incorporated in the species, for on these depends the evolution of the race.

Further, he went on:

> Evolution has taken place by the incorporation into the race of those mutations that are beneficial to the life and reproduction of the organism. Natural selection as here defined means both the increase in the number of individuals that results after a beneficial mutation has occurred (owing to the ability of living matter to propagate) and also that this preponderance of certain kinds of individuals in a population makes some further results more probable than others. More than this, natural selection can not mean, if factors are fixed and are not changed by selection.

With Morgan, genetics left its disciplinary partners in the study of evolutionary biology—embryology, comparative anatomy, paleontology—at the doorstep and proclaimed itself the language of evolution. And it is this language that would be further developed and refined, leaving the rest of the evolutionary sciences scrambling to figure out how to incorporate it and subsequent genetically based evolutionary hypotheses into their endeavors. Armed with the gene, geneticists seemed to have the key to the study of both the past and the future.

9

Genetics Goes Statistical

The object of statistical science is to discover methods of condensing information concerning large groups of allied facts into brief and compendious expressions suitable for discussion.

—Sir Francis Galton (1907)

The Return of the Biometricians

When Walter Weldon died in 1906, William Bateson lost a vicious and almost lifelong antagonist. The bitterness that arose between these two scholars could be traced to two counts of disagreement. One was whether variation was continuous or discontinuous. Weldon tenaciously held to the former, Darwinian view, whereas Bateson embraced discontinuous variation as a logical extension of Mendelian principles of inheritance. But Weldon also found fault with Bateson's approach to matters evolutionary and hereditary because it was neither steeped in mathematics nor justified by attempts to predict inheritance by statistical methods. The latter criticism came from Weldon's having taken a stand in favor of Galton's notion of ancestral inheritance, which had been demonstrated statistically. Rather than analyze inheritance only between parent and offspring, Galton argued that one must take the entire series of previous generations into account. To be sure, Bateson was well aware of his rather naive understanding of mathematics and statistics. But in spite of possessing what to Weldon and others in his camp was a fundamental flaw, Bateson was still able to seek and gain support from the father of biometricians, Francis Galton. Among other things, Bateson was an early member and then head of the Evolution Committee of the Royal Society, which had been Galton's idea to begin with.

Although Galton's calculations of inheritance through a succession of generations were themselves seriously flawed, they became popular among evolutionists, such as Weldon, as a counterargument to Mendelian inheritance, which focused on inheritance from the parental to the filial generation. Weldon and like-minded scholars also considered Mendelism a threat to the Darwinian account of evolution, which posited change through the accumulation of myriad small, continuous variations. In this regard, the intellectual alliance between Weldon and Galton is indeed a curiosity, because the latter scholar was a firm believer in discontinuous variation, which was the intellectual offspring of Mendelism.

The weight of what came to be known as Galton's law in directing the development of biometry and statistics in evolutionary studies of inheritance did not, however, derive solely from Galton himself. Rather, the impact of Galton's law came from a reworking of the calculations behind this law by the English statistician Karl Pearson. After initially being at least a sympathizer of Bateson's, Pearson, who was born two years before the publication of the first edition of *On the Origin of Species,* came to align himself with Weldon in opposition to Mendelism and discontinuous variation.

In his attempt to offer an alternative explanation for the mechanism of inheritance, Pearson proposed that just as there was slight variation between offspring and parent, so, too, was there minor variation within an individual with regard to its structures, or organs. This idea was not intended to be applied to the level of comparing, say, an individual's heart with its kidneys. Rather, Pearson saw variation in those elements that occurred with some abundance as part of an organism's systems, such as an animal's blood or sex cells or a plant's leaves. While we can identify, for example, any maple leaf as having come from a maple tree, we are also aware of the fact that no two maple leaves are identical. "Like" structures can also be slightly different from one another. Taking sex cells into consideration, Pearson speculated that the more similar, or correlated, in likeness the organs within an individual were, the more similar would be that individual's offspring. But Pearson also gave his theory a strange twist. He promoted the idea that these organs were undifferentiated—which, of course, makes no sense at all. Anyone could see that organs—whether true organs or cells—were differentiated structures whose inheritance conformed to Mendelian principles.

Pearson's homotype theory, as he called it, was to fall the way of many early attempts at developing a working model of inheritance. But the kind of detailed collecting and analysis of data that went into his formulation of the theory would become the cornerstone of his life's research: namely, deriving the statistics from the data-dependent statistical tables for calculating how correlated, or independent of one another, two or more items of the larger set are. In fact, the coefficient of correlation still used today in many statistical tests of correlation is called Pearson's *r,* which a young admirer, Ronald Fisher, named after the senior mathematician. Fisher, with whom Pearson eventually collaborated on the correlation tables, was the one who perfected the latter's calculations of the correlation coefficient itself.

In his efforts to dismiss, if not entirely dismantle, Mendelian inheritance, Pearson argued that at the level of the hybrid or F_1 generation, even random mating would not provide the vehicle necessary for introducing evolutionarily significant variation. In fact, he stated, random mating would actually tend to maintain stability in the population. Curiously, when discussing the F_2 generation, which was produced by crossing hybrids, Pearson inadvertently stumbled onto a point that would become significant to the next wave of statistically minded population geneticists: that it was indeed segregation of characters in the offspring of the hybrid generation that could provide the possibility of variation, or, as he put it, of the "unfamiliar."

As long as Mendelism was associated with discontinuous variation, the biometrical Darwinian evolutionists could not and would not accept it. But through the mathematical argumentation of the English statistician George Udny Yule in 1902, the possibility that continuous variation was compatible with the laws of Mendelian inheritance became theoretically viable. At first, Yule argued that if dominance was assumed, Mendel's hybrid 3 to 1 ratio was actually a special case of Galton's law of ancestral heredity. If, however, according to Yule's calculations, one considered a more epigenetic situation, wherein heritability of characters and the effects of the environment were taken into account, there was then complete harmony between the laws of Mendelian inheritance and Galton's law. But it was when he proposed the "multiple factor hypothesis" that Yule was able to argue mathematically that one actually could entertain the reality of continuous variation. It was simple. One had only to accept that even though variations are discrete and discontinuous, the differences are actually so minute that, for all intents and purposes, variation is effectively continuous. In any case, Yule maintained, variation was nowhere near the order of discreteness and discontinuity to which de Vries and Bateson ascribed species differences. By diminishing the significance of discontinuous variation to Darwin's level of "fluctuating variation," not only could Mendelism and Darwinian evolution be reconciled but they could also be harmonized with the concept of ancestral inheritance.

With the majority of experimental and theoretical geneticists coming to accept the coexistence of Mendelian inheritance and Darwinian natural selection, work began in earnest on trying to characterize mathematically and with statistical predictability the effects over time of selection on the frequencies of alleles in populations. The question was: When, under the conditions of random mating within a population, would two allelic states of the same gene—or, as was becoming accepted jargon, two alleles at the same place, or locus—come into equilibrium in the Mendelian ratio of 3 to 1? The relationship was finally resolved by the English mathematician and statistician Godfrey Harold Hardy. In fact, according to William Provine's historical account, the problem was brought to Hardy early in 1908 by his former cricket buddy Reginald Crundall Punnett, the very same collaborator of Bateson's who had earlier worked out the Punnett square.

In the case of a two-allele system, Hardy calculated that with random mating, equilibrium between the two alternative states would be achieved in the course of a single generation. Having reached equilibrium meant, of course, that the frequencies and expression of these two alleles would remain the same thereafter in subsequent generations. It turns out that Hardy's calculations on a two-allele system under random mating had actually been anticipated by half a year by a German physician, Wilhelm Weinberg. Within another few years, Weinberg also figured out mathematically when equilibrium would be achieved in the case of multiple alleles at a single locus and then at multiple loci. Unfortunately for Weinberg, his publications were overlooked, although not nearly as long as Mendel's, by the primarily English-speaking scientific community, which came to refer to the equilibrium equation as Hardy's law. Only much later would this law, which is still widely taught in evolution courses, be called the Hardy-Weinberg equilibrium principle, or equation.

But the Hardy-Weinberg equation was actually relevant only to questions of maintaining stability in the frequencies of alleles in successive generations of a species or population. What was needed was a demonstration of the effects of selection—the intensity of selection pressure—on *changing* the frequencies of gene expression from one generation to the next within a species. To this end, it was Punnett, again, who was responsible for the next development in the theoretical mathematical consideration of inheritance. It was he who prodded yet another scientist, the Cambridge University, Trinity College mathematician H. T. J. Norton, to see if he could figure out how rapidly or slowly the gene frequency of a population would change given a certain level of selection pressure. Norton rose to the challenge. He produced a statistical table that Punnett featured in his monograph of 1915, *Mimicry in Butterflies.* It was the very subject of this book that had prompted Punnett to ask the question of Norton in the first place. Punnett suspected that the evolution of a mimicking attribute would have to take place rather quickly if it were to be at all advantageous to its possessors.

The upshot of Norton's labors was that, at least on the theoretical mathematical level, it seemed that even low-intensity selection pressures could shift gene frequencies considerably over the span of what he considered to be relatively few generations. For example, if a dominant allele has a 10 percent selective advantage over its recessive counterpart in a population, it would take only seventy generations to reduce the frequency of the recessive allele from about one-half to less than a fortieth. The historian William Provine has argued that Punnett, by way of his solicitation of Norton's mathematical skills, had inadvertently undermined his own theoretical position. Like Bateson, Punnett championed the importance of Mendelism and discontinuously variable traits in affecting evolutionary change. Norton's calculations provided apparent evidence to the contrary. According to Norton, change could be brought about relatively rapidly by the combination of selection and Mendelian inheritance of the genes underlying the seemingly continuously variable physical characteristics. Regardless of the accuracy of Provine's interpretation of the significance of Norton's efforts, it is the case that statistics had become a part of theoretical population genetics.

The elevation of theoretical statistical population genetics to a more sophisticated level took place largely through the work of two Englishmen, Ronald Aylmer Fisher and John Burdon Sanderson Haldane, and one American, Sewall Wright.

R. A. Fisher, Natural Selection, and Theoretical Population Genetics

Ronald Aylmer Fisher came to associate with the biometrician Karl Pearson at a very young age, just a few years after he had discovered and devoured with fascination the elder academic's papers on evolution. After solving the problem of the correlation coefficient in 1914, which he called Pearson's *r,* Fisher joined Pearson and the two worked together for a time during 1915 expanding the utility of statistical correlation tables. The impact of this work was that other scientists could then use Pearson's *r* in testing the relevance of their results. But the relationship between the younger and the older mathematician was not to last more than another three years.

Pearson and Fisher's short-lived association eventually showed signs of intellectual incompatibility at the most basic level of evolutionary understanding, even though each would have characterized himself as a Darwinian. Each scholar accepted natural selection as playing a significant role in affecting evolutionary change. And each accepted the hypothesis that change occurred gradually by way of natural selection's acting on continuous variation within the species. But they were otherwise worlds apart. Pearson, like Darwin, was an adherent of blending inheritance, which, as conceived, was by itself capable of providing new combinations of variants with each new generation for natural selection to choose among. Consequently, Pearson could not embrace Mendelian inheritance. In contrast, Fisher's approach to evolution combined statistical methods, natural selection, and gradual change within the context of the newly informed Mendelian interpretation of continuous variation.

Fisher's first publication in the statistical realm of natural selection and gene frequencies appeared in 1918. Although Fisher was unaware of this at the time, his effort was a rehash of Wilhelm Weinberg's earlier equilibrium work. By 1930, however, when he published the first edition of what would become one of the most influential works on theoretical statistical population genetics, *The Genetical Theory of Natural Selection,* Fisher had outstripped much of his competition. From the very preface to the first edition of this classic work, Fisher held back nothing in the presentation of his case:

> Natural Selection is not Evolution. Yet, ever since the two words have been in common use, the theory of Natural Selection has been employed as a convenient abbreviation for the theory of Evolution by means of Natural Selection, put forward by Darwin and Wallace. This has had the unfortunate consequence that

> the theory of Natural Selection itself has scarcely ever, if ever, received separate consideration.
>
> . . . [A]dvocates of Natural Selection have not failed to point out, what was evidently the chief attraction of the theory to Darwin and Wallace, that it proposed to give an account of the means of modification in the organic world by reference only to "known," or independently demonstrable, causes. The alternative theories of modification rely, avowedly, on hypothetical properties of living matter inferred from the facts of evolution themselves. . . . The present book, with all the limitations of a first attempt, is at least an attempt to consider the theory of Natural Selection on its own merits.

As he proceeded to outline his case for natural selection, Fisher made it inescapably clear that anyone who hoped to understand this process had to embrace Mendelism, which he saw as "the great advance which our generation has seen in the science of genetics." As a statistician and a mathematician, Fisher thought it ironic that, earlier in the century, such a great advance as Mendelism found its biggest detractors among the biometricians working in genetics. The problem now, Fisher stated, was the gulf that existed between biologists and mathematicians in terms of which scientist had the better skills and imagination to undertake evolutionary studies. Clearly, he admitted, it was not a matter of who was more intelligent, a biologist or a mathematician. But there was something about the training of a mathematician, Fisher confessed, that led to a particular process of imagination that was lacking in the biologist:

> The ordinary mathematical procedure in dealing with any actual problem is, after abstracting what are believed to be the essential elements of the problem, to consider it as one of a system of possibilities infinitely wider than the actual, the essential relationships of which may be apprehended by generalized reasoning, and subsumed in general formulae, which may be applied at will to any particular case considered. Even the word possibilities in this statement unduly limits the scope of the practical procedures in which he is trained; for he is early made familiar with the advantages of imaginary solutions. . . . The most serious difficulty to intellectual co-operation would seem to be removed if it were clearly and universally recognized that the essential difference lies, not in intellectual methods, and still less in intellectual ability, but in an enormous and specialized extension of the imaginative faculty, which each has experienced in relation to the needs of his special subject. I can imagine no more beneficial change in scientific education than that which would allow each to appreciate something of the imaginative grandeur of the realms of thought explored by the other.

Just as Darwin before him had employed a type of reasoning sometimes referred to as a "thought experiment," so Fisher was now entering into this realm from the perspective of theoretical populational genetics. Darwin, a naturalist, used his imagination to convince the imaginations of his readers that evolution, and specifically gradual evolution by means of natural selection, was not just a real possibility but a reality. How else could he convey the essence of his argument about, for example, the evolution of the vertebrate eye? Or, for that matter, the evolution of a flatfish, such as a sole or a flounder, in which one eye shifts ontogenetically from one side of the head to the same side as the other eye? Because it was not possible to observe the evolution of the eye, or of flatfish, Darwin had to develop a technique of presentation that talked the reader through a logical argument in support of these seemingly impossible and inexplicable evolutionary events. Thought experiments were particularly important in light of the potentially awkward situation of trying to explain how the different stages in the continuum of gradual evolution of the vertebrate eye from the partial to the complete, or in the shift of an eye from one side of the flatfish's head to the other side, were advantageous at each evolutionary stage—or, at least, not so disadvantageous as to be selected against. In lieu of actual evidence, one had to construct an imaginary situation that seemed reasonable enough to allow the skeptic to embrace the larger concepts of an evolutionary process.

Fisher, a mathematician and a statistician, was going to use his imagination to develop definitions and formulas of how he thought the evolutionary process worked. His tool was Mendelism ("Mendelism supplied the missing parts of the structure first erected by Darwin"), and his operator was natural selection. Although he acknowledged Bateson, the individual, as a pioneer in genetics and his role in championing Mendel's work, Fisher was harshly critical of him for being "unprepared to recognize the mathematical or statistical aspects of biology." What this meant to Fisher was that Bateson was also "incapable of framing an evolutionary theory himself." Being a strict Darwinian gradualist, Fisher was committed to the idea that continuous variation was the foundation upon which natural selection acted. Consequently, he had no tolerance either for Bateson's emphasis on change being in the domain of discontinuous variation or for his theoretical interest in the evolutionary significance of repeated parts.

Scattered liberally throughout most chapters of *The Genetical Theory of Natural Selection* are Fisher's assertions on the gradual nature of evolution, proceeding by the accumulation of small continuous variations, as well as his claims that evolutionary change is progressive. For Fisher, progress was evident not only in the evolution of increasingly more complex organisms but also in an organism's fitness to survive. And it was an organism's fitness that preoccupied much of Fisher's mathematical formulations throughout the monograph and informed his redefining of natural selection. For Fisher, fitness was, quite simply, a matter of reproductive success: How many offspring will an individual of a certain age contribute to the next generation?

Seeking an analogy between economics and population biology—such as finding "the growth of capital invested at compound interest" a reasonable example of population growth—Fisher equated the birth of an individual with the loan to that individual of a life. Consequently, if this formulation was accepted, it must follow that an individual's offspring could be thought of as payment of that debt. But the potential of an individual to contribute offspring to the next generation—to pay off the loan of life, so to speak—comes from the genetic variance of the individual. Fisher imagined genetic variance as being the genetic potential of the individual. And an individual's genetic potential, he thought, could be determined by averaging the positive and negative effects of that individual's genes and then measuring these effects against the average of gene effects in the population. Ultimately, Fisher reasoned, "[a]ny net advantage gained by an organism will be conserved in the form of an increase in population."

By putting reproductive success, or fitness, together with genetic variance, Fisher proposed the following as a logically derivative theorem or principle of natural selection: "The rate of increase in fitness of any organism at any time is equal to its genetic variance in fitness at that time." The greater an individual's genetic potential, seen in terms of more genetic variance, the more fit, or reproductively successful, that individual will be. As Fisher commented in the revised version of *The Genetical Theory,* his concept of natural selection "refers only to the variation among individuals (or co-operative communities), and to the progressive modification of structure or function only in so far as variations in these are of advantage to the individual, in respect to his chance of death or reproduction." As Fisher saw it, thinking about natural selection in this way "affords a rational explanation of structures, reactions, and instincts which can be recognized as profitable to their individual possessors."

Fisher believed that his theorem of natural selection took evolutionary change out of the realm of chance, which some critics of Darwinian evolution claimed to be the fatal flaw of the theory. He was convinced that his theorem was correct because of its emphasis on progress—at least progress as would be reflected by the fitness of a species, which, in turn, is measured by the cumulative fitness of the members of that species. The analogy that Fisher chose to defend his concept of natural selection was that of a casino:

> The income derived from a Casino by its proprietor may, in one sense, be said to depend upon a succession of favourable chances, although the phrase contains a suggestion of improbability more appropriate to the hopes of the patrons of his establishment. It is easy without any profound logical analysis to perceive the difference between a succession of favourable deviations from the laws of chance, and on the other hand, the continuous and cumulative action of these laws. It is on the latter that the principle of Natural Selection relies.

Having presented what he considered to be the best argument for natural selection, Fisher then proceeded to discuss adaptation. From his perspective, adaptations were "the consequences to the organic world of the progressive increase of fitness of each species of organism." But Fisher was willing to allow the process of progressive increase of fitness to proceed only at a certain pace. In keeping with his view of change as being gradual in tempo and enacted by way of small increments—with negative consequences resulting from change that was too large or too rapid—Fisher sought to limit the intensity of any given adaptation. Consequently, just as too large a change would inversely, and negatively, affect progress or chance of improvement for members of a species, too great an intensity of adaptation would have adverse effects.

Fisher was concerned that the concept of adaptation had become lost and muddied by the very use evolutionary biologists had made of the term in their attempts to define it by giving examples of it. For instance, simply stating that a giraffe's long legs and neck constitute an adaptation to feeding high up in a tree actually served to obfuscate the meaning of the word. Adding more features to the picture of an adaptation only made the matter worse and less clear. Fisher's solution to the concept of the adaptation problem was to provide his own definition:

> An organism is regarded as adapted to a particular situation, or to the totality of situations which constitute its environment, only in so far as we can imagine an assemblage of slightly different situations, or environments, to which the animal would on the whole be less well adapted; and equally only in so far as we can imagine an assemblage of slightly different organic forms, which would be less well adapted to that environment.

Although Fisher pursued Darwin's arguments for adaptation and evolution by means of natural selection mathematically, he was clearly no less adept at building his theorems and formulas upon circumstances as he imagined them to be in the natural world.

Upon laying his foundation for natural selection and adaptation, Fisher devoted his energies to discussing mutations and the effects of mutations. For although random mating and independent assortment may permit the shuffling around of alleles and afford a virtually limitless source of variation for natural selection to exploit, new variations came only from mutations. In these chapters, Fisher relied primarily on the experimental work of others, using the results of these studies as the basis for developing his mathematical equations of inheritance and potential evolutionary change.

There were many elements of genetic theory, now backed with the results of some convincing experiments, that were becoming commonplace in the field of population genetics, and Fisher embraced them. Among Fisher's contemporaries who contributed experimentally to this growing body of knowledge was the American Sewall Wright. Although he was considered, along with Fisher and J. B. S. Haldane, to be one of the three most influential theoretical population geneticists of the day, Wright, as we shall see, would end up in intellectual rivalry with the other two, especially with Fisher.

Thanks to experiments in population genetic breeding, one important realization that was available to Fisher was that mutations tended to arise in the recessive allelic state. This fact had been more than suspected by Bateson, Wilson, Morgan, and others. In fact, as Fisher commented, "Numerous cases are now known in which several different mutations have occurred to the same gene, and each of the mutant types can replace each other, and the wild type, in the same locus." In order for a mutation to have played any role in evolution, though, Fisher speculated that the recessive mutant had to have become dominant over the wild gene. He characterized the now-dominated wild gene as the "unsuccessful competitor." To reflect the importance he placed on the dominant state in providing the grist for evolutionary change, Fisher specifically identified it as "the dominance theory."

As for which mutations would remain in the collective gene pool of a species, Fisher repeated what had essentially become a commonplace. Deleterious mutations would either be eliminated or kept rare in the population. A disadvantageous mutation, no matter how rare among the members of a species, may, however, be passed on to successive generations of that species. If this was the case, then a disadvantageous mutant gene could still have an effect on that species over time. On the other hand, from the beginning, an advantageous mutation has a better chance of becoming widespread among the members of a species. But even an advantageous mutation—however defined by the beholder—would not necessarily stay and multiply within the species without help, because, as the experiments seemed to indicate, even an advantageous mutation has a tendency to revert to the wild, or original, state. Help in maintaining and spreading a mutation, according to Fisher, came from natural selection.

When a mutation does occur in a gene, the end result will be a new allele. If a random mutation were to occur in a regular body cell of an individual, this would be of no consequence to that individual's offspring. If, however, a mutation were to occur in the cells that give rise to an individual's sex cells, then it could be passed on to the next generation. Since the pairs of chromosomes that exist in all body cells result from the union of two sex cells and the coming together of their chromosomes, there are two possible ways in which the transmission of a mutant allele can be enacted. Only one parental sex cell might be carrying the mutant allele, in which case the offspring would be heterozygous for the mutation. Or each parental sex cell might be carrying the mutant allele, in which case the offspring would be homozygous for the mutation. Since most mutations arise in the recessive state, they would not necessarily be expressed in the heterozygote. But they would always be expressed in the homozygote.

Given these two possibilities, Fisher argued, the heterozygous condition would always be at a selective advantage relative to the homozygous state. The reason Fisher gave for this statement was that the mating of a heterozygote for the mutant allele with a homozygote for the normal condition would produce some offspring that were homozygous for the normal condition as well as others that were heterozygous for the mutant allele. In contrast, the mating of a homozygote for the mutant allele with a homozygote for the normal condition would produce only heterozygotes for the mutant allele. As Fisher saw it, the former situation was preferable, because it kept more variation—genetic variance—alive in the population. In addition, Fisher suggested, the persistence through a succession of generations of individuals that were heterozygous for the advantageous mutation would afford the time necessary for the mutation to evolve from the recessive to the dominant state. However, Fisher had to admit, natural selection acting on a population with lots of homozygotes for a mutant allele could have quite a considerable impact over a short period of time.

Probably another reason why Fisher preferred a model of inheritance that favored the heterozygous condition for the mutant allele was that it led to a model of gradual rather than rapid evolutionary change. In reviewing the effects of artificial selection imposed by the selective breeding of domesticated animals, Fisher pointed out that breeders had effected quite remarkable change in relatively short periods of time just by selecting among the heterozygotes to continue the breed. But, he cautioned his readership, since this is an artificial situation, "[e]volution under such human selection should . . . take place many thousand times more rapidly than the corresponding evolution of recessiveness in nature."

In considering Fisher's contribution to theoretical population genetics, it is imperative that we understand that he assumed but did not demonstrate that the tempo of evolution on all levels was slow and gradual. From this position, he proceeded to derive his mathematical formulations of, for example, selection, adaptive advantage, and evolutionary change. This is evident throughout *The Genetical Theory* and is perfectly clear in the following quotes:

> It is to be presumed that mutation rates, like the other characteristics of organisms, change only gradually in the course of evolution.

> The slow changes which must always be in progress, altering the genetic constitution and environmental conditions of each species, must also alter the selective advantage of each gene contrast.

> [T]he difficulty of effecting any improvement in an organism depends on the extent or degree to which it is adapted to its natural situation. The difficulties which Natural Selection has to overcome are in this sense of its own creating, for the more powerfully it acts the more minute and intricate will be the alterations upon which further improvements depend. The fact that organisms do not change rapidly might in theory be interpreted as due either to the feebleness of selection or to the intensity of adaptation, including the complexity of the relations between the organism and its surroundings.

Guided by the presumption of gradual change, in conjunction with his modeling of change based on the accumulation of small, Darwinian continuous variations, Fisher went on to contemplate the species problems: the nature of a species and the origin of species. With regard to the former topic, he speculated that one could not define a species on the basis of a collective genetic constitution, or genotype, because the genotypes of individuals can differ significantly from one another. On the other hand, considering that all individuals of a sexually reproductive species are capable of exchanging genetic material, one might think of a species as a group of individuals that are so specifically adapted to a particular niche that "an evolutionary improvement in any one individual threatens the existence of the descendants of all the others." This group, so specifically adapted to a particular niche of nature, would, of course, be the descendants of that one ancestor to whom the original beneficial mutation had conferred greater selective advantage and reproductive success.

Fisher recognized that although individuals of the same species may indeed differ in their individual genotypes, by far the vast majority of their genetic loci must be held in common by them all, by virtue of their common heritage. At one time or another, and in distant ancestors, these commonly shared genetic loci must have been advantageous mutations. Clearly, for Fisher, the significant aspect of a group of individuals being members of the same sexually reproductive species was the fact that they shared genes. And this sharing of genes meant that these individuals maintained a potential if not a real reproductive relation to one another. It was the expressed overall genetic similarity among individuals of the same species, Fisher reckoned, that somehow informed the systematist or taxonomist who was trying to identify that species.

Having satisfied himself that one could think of sexually reproductive species in terms of sharing genes, Fisher went on to consider the origin of new species. But he apparently felt that he could not do so without first taking a poke at Bateson. For Bateson, differences between species were discrete and arose from discontinuous variation. This, of course, was anathema to Fisher, who perceived species change, no matter how minuscule, as a constant, ongoing process. From the latter perspective, because each species continues to change linearly over time, it is always on the way to becoming something else. Also from this point of view, the process of diversification of species is merely an extension of this model, with incipient sister species gradually separating from one another genetically and then each new species continuing in its gradual trajectory of change. If discontinuous variation really was at the base of species diversity, then, in Fisher's evolutionary realm, no species would continue to evolve.

With the engine of continual, gradual evolutionary change being fueled by an uninterrupted exchange of genes among members of sexually reproductive species, Fisher had to propose a mechanism that would create diversity. Without some way of breaking the exchange of genes, or the gene flow, species would continue to evolve without end. In this scenario, the minor variations that naturally occur among individuals would be nothing more significant than, to quote Fisher, "the differences in colour between different threads which have crossed and recrossed each other a thousand times in the weaving [of] a single uniform fabric." But there was at least one solution to the problem of speciation, or the fission of species, as Fisher called it. And it was simple: Introduce a geographic barrier that would prevent some individuals of a species from mating with other individuals of that species. Fisher saw geographic isolation as leading to the rise of new species (a process called allopatric speciation), a frequent impetus for producing species diversity:

> In many cases without doubt the establishment of complete or almost complete geographical isolation has at once settled the line of fission; the two separated moieties thereafter evolving as separate species, in almost complete independence, in somewhat different habitats, until such time as the morphological differences between them entitle them to "specific rank."

But geographic isolation was only one possibility in what Fisher saw as an even more important element in species fission, or speciation, and that was the environment or habitat in which the parental species lived. Consequently, environmental instability, or heterogeneity, as he phrased it, was perhaps the major element in the equation leading toward speciation. Environmental instability could be introduced by some kind of geographic barrier imposing itself between members of a species, but it need not be. To Fisher, environment was not limited to the external world but also included an interactive environment within an organism—one that could be created by an individual's genes and cells. But whatever its cause, environmental instability would demand "special adaptations," and these, in turn, would lead to stress in a species. As Fisher imagined it, adaptation to different aspects of a heterogeneous environment could eventually lead to "circumstances unfavourable to sexual union, of which the most conspicuous is geographical distance, though others, such as earliness or lateness in seasonal reproduction, may in many cases be important."

A species whose members are spread over varied habitats will try to maintain genetic continuity ("an active diffusion of germinal material"). But, at the same time, genetic differences among the members occupying the more extreme regions will arise. Since these differences reflect adaptations to local conditions, they will be advantageous only in those particular circumstances. Because these adaptations are only meaningful locally, they cannot be transferred by gene flow or made to work by the translocation of individuals from one place to another. These adaptations confer reproductive success on only individuals who possess them in the right place and at the right time. A species so differently adapted, or "stretched," to use Fisher's image, will, however, still try to maintain gene flow throughout its gradient of individuals. Because these individuals will still be connected throughout the gradient, they will remain members of the same species. The fissioning of a stretched species would have to involve the "elimination in each extreme region of the genes which diffuse to it from the other . . . [and] . . . incidentally the elimination of those types of individuals which are most apt so to diffuse." Fisher described the completion of the beginning of the process of species fission as follows:

> [S]election . . . must act gradually and progressively to minimize the diffusion of germ plasm between regions requiring different specialized aptitudes. The effect of such a progressive diminution in the tendency to diffusion will be progressively to steepen the gradient of gene frequency at the places where it is highest, until a line of distinction is produced, across which there is a relatively sharp contrast in the genetic composition of the species. Diffusion across this line is now more than ever disadvantageous, and its progressive diminution, while leaving possibly for long a zone of individuals of intermediate type, will allow the two main bodies of the species to evolve almost in complete independence.
>
> In cases in which the cause of genetic isolation is not merely geographical distance, but a diversity among different members of the species in their habitats or life history, in connexion with which different genetic modifications are advantageous, the isolation will . . . be increased . . . by whatever type of hereditary modification will minimize the tendency for germinal elements, appropriate to one form of life, to be diffused among individuals living the other form, and among them consequently eliminated.

Fisher's concept of speciation, then, centered around the elimination of gene flow among individuals of the same species. Although complete geographic isolation—an unambiguous physical separation of some individuals from others of the same species—would certainly lead to cessation of gene flow because it would be an effective and obvious barrier to sexual intercourse, Fisher allowed for other possible sources of speciation as well. Among these would be the development of different adaptations among individuals distributed in a heterogeneous environment, as well as the development of different life histories, which could be as simple as "differential sexual response to differently characterized suitors." All of these kinds of responses would eventually "impair the unity of the species."

A curious consequence of Fisher's trying to figure out ways in which gene flow among individuals could be impeded is that he proposed models that fall on both sides of current debates on the mechanisms of speciation. On the one hand, Fisher's emphasis on the elimination of gene flow by the imposition of a geographic barrier represents the model of allopatric speciation later championed by Ernst Mayr and others in the formulation of the grand evolutionary synthesis. Allopatric speciation remains the most popular model in various quarters of evolutionary biology. In this model, species are defined by limiting factors outside their purview. Geographic separation leading to genetic isolation creates individuals who can mate among themselves but not with others with whom they share a common ancestor.

An alternative model to defining species was developed relatively recently by Hugh Paterson. It is commonly known as the species mate recognition system. In this formulation, a species is defined by its members insofar as individuals recognize each other as potential mates. This involves genetic isolation, but it is the consequence of individuals not recognizing other individuals as potential mates, which could result as much from a slight shift in sexual behavior or reproductive timing as from a more blatant morphological change, such as in coloration. Fisher's ideas on this subject are compatible with Paterson's species mate-recognition system.

Fisher's third example of what happens to species—of a species stretched across a heterogeneous environment—represents what became known as a cline. In a cline, the populations of the species at the extremities of the distribution can look and be genetically very different from one another, but they are considered to be genetically linked by gene flow connecting the myriad intervening populations of individuals. Clines—such as between the living yellow and olive baboons of eastern Africa—continue to be used both as an example of speciation in the process and as an example of the impossibility of defining species.

In all three of these examples, it is obvious that Fisher stressed genetic isolation, whether real or potential. And in doing so he turned around Bateson and Morgan's approach to defining species. Bateson and Morgan regarded species as entities in nature whose members cannot produce reproductively viable offspring when they are crossbred with individuals of other species. Fisher, however, identified a species as a conglomerate of individuals that could reproduce with one another. The Harvard evolutionary biologist Ernst Mayr would later incorporate the latter notion into his biological species definition, which became one of the flagships of the evolutionary synthesis. But it is also the case that this definition would lead to problems in defining species in any way other than a genetic one. Clines and fossils provide two good examples of the dilemma of embracing reproductive potential as the linchpin of one's species definition. In the case of a cline, the extremes of a genetic gradient are morphologically dissimilar to one another, and they often cannot produce reproductively viable offspring. In the case of fossils, only preserved morphology is available for study.

Obviously, Fisher left quite a legacy not only to theoretical population genetics but also to the fathers of the grand evolutionary synthesis. He certainly championed many ideas that became incorporated into the synthetic theory of evolution: gradual evolutionary change, continuous variation, adaptation to the environment, selection of advantageous mutations, and genetic isolation leading to speciation. As for population genetics, he left an abundance of mathematical formulations concerning, for example, gene frequencies in equilibrium, mutation rates, natural and sexual selection, and rates of evolutionary change, all of which flowed as much from experimental data as from his particular theoretical bent.

J. B. S. Haldane: Theoretical Genetics, Yes, but Also Development

According to C. Leon Harris's biographical sketch of John Burdon Sanderson Haldane, JBS, as his friends called him, was larger than life—not just intellectually or in his sometimes audacious and at other times reckless behavior but also in size: Early in his adult life he weighed in excess of two hundred pounds. Born in 1892, the son of an equally self-assured physiologist at Oxford University, John Scott Haldane, Haldane was inspired to practice self-experimentation as an alternative to using lab animals. At the age of sixteen, he and his sister, Naomi, who was four years his junior, embarked on breeding experiments with guinea pigs that would eventually lead to their discovering gene linkage. Unfortunately for the young siblings, Thomas Morgan and his lab-weathered collaborators, who had been experimenting with fruit flies to solve the problem of some traits being inherited as a package, gained priority for the discovery of gene linkage upon publishing their results in 1915, in *The Mechanism of Mendelian Heredity.*

After a stint in the British army during the First World War as a demolition expert, with his final year spent recuperating in India from wounds, Haldane went to Oxford in 1919. There, though totally unschooled in the subject—his only degree was actually in the classics—he taught physiology with his father. It was in the early twenties that Haldane turned his energies toward understanding evolution from a theoretical mathematical perspective. True to form, he doggedly pursued his chosen subject, publishing a string of papers on theoretical genetics and natural selection between 1922 and 1932 that would be integrated into a series of lectures he gave in 1931 entitled "A Reexamination of Darwinism." In turn, these lectures would serve as the basis for his influential monograph *The Causes of Evolution.* Although he gave his lecture series only one year after Fisher's publication of *The Genetical Theory of Evolution* and closely on the heels of some of Sewall Wright's most important articles on genetics and speciation, Haldane had already made up his mind as to which aspects of his colleagues' theories he agreed with and which he could not support.

Like Fisher and Wright, Haldane was instrumental in imbuing theories of evolutionary change with Mendelian principles and also in putting mathematics into evolutionary theory and into genetics. Indeed, it was these three scientists who single-handedly neo-Darwinized the theory of natural selection by redefining it in mathematical terms steeped in theoretical genetics and repositioning the significance of its input within an equally theoretical framework of evolutionary change. As Haldane said about himself: "I can write of natural selection with authority because I am one of the three people who know most about its mathematical theory."

Darwin had invoked the importance of natural selection from the very beginning of the evolutionary process, as the instigator as well as the arbiter of change. Because of his belief in a limitless amount of continuous or fluctuating variation within and between generations, and within the context of blending inheritance, Darwin imagined that novelty was produced by natural selection, which would then have its pleasures with the newly emergent features. Fisher, Wright, and Haldane, however, argued that random mutation was the sole cause of novelty. Only after mutation created something new would natural selection act, weighing each new feature in terms of its relative degree of advantage or disadvantage to both the individual and the individual's species. As would be expected of these three egos, each scholar conceived of and modeled selection differently mathematically.

Perhaps more so than his colleagues, and even as late as 1931–1932, Haldane felt compelled to defend Darwinism against those detractors who believed in creation as well as those supposed scientists who continued to embrace Lamarckism, or neo-Lamarckism, as the ideas of use and disuse had by then become known as a result of being updated with the language of genetics. It was a relatively simple matter for Haldane to dismiss the claims of some geneticists that an individual's life of adjustments to its circumstances could be incorporated into its genes and then passed on to the next generation. Each genetic study that had been proposed as proof of the inheritance of acquired characteristics fell by the wayside under Haldane's scrutiny. As for the creationists, the classically schooled Haldane had this to offer:

> I have given my reasons for thinking that we can probably explain evolution in terms of the capacity for variation of individual organisms, and the selection exercised on them by their environment. . . .
>
> The most obvious alternative to this view is to hold that evolution has throughout been guided by divine power. There are two objections to this hypothesis. Most lines of descent end in extinction, and commonly the end is reached by a number of different lines evolving in parallel. This does not suggest the work of an intelligent designer, still less of an almighty one. But the moral objection is perhaps more serious. A very large number of originally free-living Crustacea, worms, and so on, have evolved into parasites. In so doing they have lost, to a greater or less extent, their legs, eyes, and brains, and have become in many cases the course of considerable and prolonged pain to other animals and to man. If we are going to take an ethical point of view at all (and we must do so when discussing theological questions), we are, I think, bound to place this loss of faculties coupled with increased infliction of suffering in the same class as moral breakdown in a human being, which can often be traced to genetical causes. To put the matter in a more concrete way, Blake expressed some doubt as to whether God had made the tiger. But the tiger is in many ways an admirable animal. We have now to ask whether God made the tapeworm. And it is questionable whether an affirmative answer fits in either with what we know about the process of evolution or what many of us believe about the moral perfection of God.

As for one of his counter-creation examples, Haldane believed that the study of parasites provided one of the best arguments for accepting the theory of evolution: "If two animals have a common ancestor, their parasites are likely to be descended from those of the ancestor." In fact, the study of parallel evolutionary histories of hosts and their parasites remains an active research area in evolutionary biology today.

In definite contrast to Fisher, Haldane was willing to entertain the possibility that while evolutionary changes could at times be gradual, they were at other times quite rapid (or, at least, he perceived the different available evidence as leading to these alternative interpretations). Whereas Fisher, in good uniformitarian fashion, never seemed to go beyond the experiences of the present in formulating the theory that was to inform him of the past, Haldane accepted the possibility that the fossil record might provide some insight into evolutionary processes. Consequently, Haldane listened to invertebrate paleontologists' claims of having documented slow and continuous change in the evolution of various mollusk groups, particularly in the fossil oyster of the genus *Gryphaea*.

True, Haldane admitted, at any one time in the evolutionary history of *Gryphaea* one could find specimens of the common type. But one could also still come upon specimens that, though rare compared with the common form, are like the forms that had been the common type at some other time in the evolution of the group. Indeed, Haldane had to conclude from this paleontological evidence, "[e]volution in such cases has clearly been a very slow and almost (if not quite) continuous process, exactly as Darwin had predicted." But, he went on to caution his readers, "the organisms studied in this way are far from representative" because "[t]hey are in general the most successful members of animal associations living in very constant marine or lacustrine environments." As far as Haldane was concerned, "dominant species in a uniform environment are the least likely to undergo sudden change to a new type."

In this regard, as we shall see, Haldane saw himself as being somewhat near Sewall Wright's camp. Almost by default, however, he found himself on the opposite side of Ronald Fisher's fence, in terms of which tempo of evolutionary change—rapid and seemingly discontinuous versus slow and seemingly continuous—was the more frequent, at least among sexually reproductive organisms. Fisher, of course, was wedded to gradual change. Consequently, his theoretical formulations reflected this bias. Haldane and, under certain circumstances, Wright favored rapid change, which they could also effectively model mathematically. From a more descriptive perspective, Haldane likened those examples from paleontology that supposedly documented slow, continuous change to the "large random-mating populations" that Fisher had hypothesized in his modeling of gradual evolution. But Haldane did not consider "large random-mating populations"—which are often "recorded by palaeontology"—to be "representative of evolution in general." Rather, he favored the "less numerous species," those of more modest population size, as being both the more common natural cases in nature and the grist of rapid evolutionary change. Consequently, while the more complete and chronologically continuously preserved fossil records of marine invertebrates may provide better evidence of evolutionary change than their terrestrial counterparts, and some invertebrate groups may document gradual evolutionary change, as far as Haldane was concerned these very cases of slow and continuous evolution would be the exceptions to the rule.

Haldane led his audience toward his understanding of evolutionary change: "Even in the record of the dominant marine forms there are breaks which suggest that some more sudden process was at work." Using ammonites to make his point—ammonites being an extinct Mesozoic era (the "age of dinosaurs") group of often coiled shelled cephalopods, of which octopuses, cuttlefish, and squid are living soft examples—Haldane proposed that even the fossil record provides examples of rapid evolutionary change:

> The palaeontologist can always postulate a slow evolution in some area hitherto unexplored geologically, followed by migration into known areas. But until a continuous series is discovered sceptics may well ask whether the gap, which is not a very vast one, was not bridged by a discontinuous process.

Thirty years later, the British geologist and invertebrate paleontologist A. Anthony Hallam would demonstrate that the evolution of *Gryphaea* was not slow and continuous. Rather, the first representatives of the genus appeared quite abruptly in the fossil record and, once there, the lineage changed little during its long history of existence. Although Haldane accepted the earlier and incorrect paleontological interpretation of the fossil record of *Gryphaea,* in his overall belief that the pace of evolution was rapid and seemingly discontinuous, rather than slow and continuous, he anticipated, at least in part and by three decades, the model of punctuated equilibria, which was proposed by the American invertebrate paleontologists Niles Eldredge and Stephen Jay Gould.

As for how species might arise suddenly, Haldane proposed some very intriguing possibilities. One possible mechanism for producing rapid evolutionary change and speciation was "the isolation of a small unrepresentative group of the population." This situation, involving the isolation of a small, skewed sampling of the parental population, conforms to what Ernst Mayr would ten years later define as a particular type of allopatric speciation, allopatric speciation via peripheral isolates. The notion that the source of speciation is a biased sampling of a small portion of the parental species also became known, thanks again to Ernst Mayr's influence, as the founder effect. Although Mayr acknowledged that speciation by such a mode was possible, he was not convinced that it was the most frequent mode of speciation in sexually reproductive organisms. Rather, he favored a more general kind of allopatric speciation that was compatible with Fisher's notion of large populations of individuals gradually separating from one another. In contrast, the role of allopatric speciation via peripheral isolates was central to Eldredge and Gould's presentation of the model of punctuated equilibria.

But a model of speciation via peripheral isolates relies on small numbers of individuals, and this raises the potentially sticky issue of inbreeding and its apparent deleterious effects on offspring. This was not, however, a problem for Haldane. In fact, he did not even discuss the topic of inbreeding in his considerations of small populational isolates. But he was certainly aware of the pros and cons of inbreeding. The subject of inbreeding and how far one could push it before things went awry with offspring was well known to animal breeders. If nothing else, they would need to know what to expect so that their economically motivated efforts would not be compromised.

Haldane did discuss inbreeding in *The Causes of Evolution,* as when he detailed the specific case of albinism in human populations or outlined a theoretical case for it. Albinism, which exists because of a certain amount of inbreeding, also provided Haldane with an example of how "the proportion of the population bearing a recessive character [and albinism is recessive] shows no tendency to diminish further after the first generation of random mating, unless the character is disadvantageous." Although albinism can hardly be thought of as an advantageous character, Haldane's explanation for the persistence of a recessive trait over a succession of generations was certainly plausible. But, he added in an appendix to the book, a new recessive gene that is not completely recessive has even a better chance of spreading through the population.

As for the theoretical case of inbreeding, Haldane considered that,

> [i]f the only available genes produce rather large changes, disadvantageous one at a time, then it seems to me probable that evolution will not occur in a random mating population. In a self-fertilized or highly inbred species it may do so if several mutations useful in conjunction, but separately harmful, occur simultaneously.

Although he believed this latter case to be rare, Haldane did admit that in the cultivation of wheat, at least, it was the rule rather than the exception. But, as he stated elsewhere, inbreeding would be the only probable mechanism for keeping a single mutation within a population.

There could also, Haldane continued, be a situation in which traits that would otherwise be eliminated under severe competition might be maintained, and even spread throughout a population. If, he suggested, the effect of natural selection were to ease up, not only would this allow new forms to arise and persist but it would also give "many ultimately hardy combinations . . . a chance of arising." Citing a case in a particular species of butterfly, where a new form arose suddenly in the apparent absence of natural selection, Haldane proposed this avenue for introducing evolutionary novelty rapidly and on a much grander scale:

> This seems to have happened on several occasions when a successful evolutionary step rendered a new type of organism possible, and the pressure of natural selection was temporarily slackened. Thus the distinction between the principal mammalian orders seems to have arisen during an orgy of variation in the early Eocene [the Eocene is a geological epoch that is now considered to have lasted from 55 to 35 million years ago] which followed the doom of the great reptiles,

> and the establishment of the mammals as the dominant terrestrial group. Since that date mammalian evolution has been a slower affair, largely a progressive improvement of the types originally laid down in the Eocene.

It is interesting to note that Haldane could entertain both a rapid origin for the major groups of mammals and a slower pace of differentiation within each mammalian group. Consequently, he envisioned the appearance of the parent species of a new group, and the features that distinguished it, as rapid, followed by less profound and more gradual change in the descendants of the founder species. Clearly, rather than investing in a single mode to explain all things, Haldane stands out in contrast to his contemporaries in being open to the possibility of a hierarchy of evolutionary processes as well as to the results of these processes.

As for the rapid pace of evolution, which Haldane obviously felt was the significant element of evolutionary change, he had even more suggestions. "Another possible mode of making rapid evolutionary jumps is by hybridisation," he wrote. Just as breeders had known that some plants arrest meiosis, with the result that offspring will have double the number of chromosomes and, consequently, be reproductively isolated from the parents, so, too, they were aware that hybridization can produce new forms of plants within the space of a generation. But when Haldane discussed hybridization he was not referring to crossing between members of the same species. He was thinking in terms of matings between females and males of different plant or animal species and the possibility of their producing viable as well as fertile offspring. In this situation, he believed that "hybridisation (where the hybrids are fertile) usually causes an epidemic of variation in the second generation which may include new and valuable types which could not have arisen within a species by slower evolution."

In the process of discussing ammonite evolution—his paleontological example of rapid evolutionary change—Haldane also broached the subject of heterochrony, the effects of differential growth rates on the final appearance of the adult, which Gavin de Beer had already begun to emphasize as essential to one's understanding of evolutionary change. In some ammonite species, it appeared that the shape of the shell during the juvenile growth phase resembled the shape of the shell of the adults of earlier forms. This, Haldane explained (most likely following de Beer), was an example of recapitulation, in which development appeared to parallel the sequence of previous ancestors, with each successive "ancestral stage" being tacked onto the end of the last one. The terminal addition of ancestral stages forced what had been adult stages to appear earlier and earlier in ontogeny. But in other ammonite species, shell shape as represented in the juveniles of early forms had apparently been retained in the adults of later forms. Haldane defined this latter phenomenon as neoteny, a process that he described in this particular way: "The preservation, in the adult stage, of what were embryonic characters in the ancestor is called neoteny."

Why did Haldane stray from theoretical mathematical genetics into the realm of paleontology and then into embryology and development? Perhaps it was his lack of schooled training in a particular scientific area that gave him the intellectual freedom to seek insight into evolution through different disciplines—disciplines that had been part of evolutionary biology until Thomas Morgan began the movement to relegate them to the position of merely interesting disciplines and to install genetics as the key to understanding evolutionary processes. Regardless, it is clear throughout his writing that Haldane viewed developmental processes, and their interactions, as basic to evolutionary change, both as phenomena in their own right that affect the morphological outcome of an individual and as phenomena that mediate between the individual at the genetic level and the survival of the individual in the environment in which it finds itself.

To emphasize the point that developmental processes must not be overlooked in considering evolutionary change, Haldane went from the discussion of ammonites to the topic of the introduction of novel features during embryonic development—a process that embryologists called cenogenesis. Haldane found an example of cenogenesis in the mammalian placenta. The homologue of the mammalian placenta is the respiratory organ in the eggs of birds and reptiles, but it has no counterpart in fish. As for examples of other types of heterochrony, Haldane cited Louis Bolk's argument that humans are "fetalized" (that is, neotenic) in certain features, such as skull shape, late-closing cranial sutures, short snout, lack of brow ridges, angle of skull to vertebral column, and lack of a full coat of hair. After reviewing the fossil hominids then known—such as Neanderthal, *Pithecanthropus* from Java, and *Sinanthropus* from China—Haldane reflected on the course of human evolution:

> If human evolution is to continue along the same lines as in the past, it will probably involve a still greater prolongation of childhood and retardation of maturity. Some of the characters distinguishing adult man will be lost. It was not an embryologist or palaeontologist who said, "Except ye . . . become as little children, ye shall not enter into the kingdom of heaven."

Haldane's appreciation of the role of developmental interplay, with, of course, genetics at the base, is clear in his discussion of the evolution of a complex organ, such as the mammalian eye. Not accepting Darwin's thought experiment on how such a complex organ could evolve gradually, Haldane pointed out that "the number of changes must take place simultaneously," possibly through "the simultaneous change in many genes"—or, he might just as easily have said, through the simultaneous change in the interactions of many genes. Certainly, Haldane pointed out, "matters would have been easier if heritable variations really formed a continuum, as Darwin apparently thought." But, as Bateson (whom Haldane did not mention) had recognized, on the morphological level, meristic, or repeated, characters—such as neck vertebrae, petals, and even chromosomes—provide evidence to the contrary. In fact, in apparent sympathy with Batesonian Mendelism, Haldane admitted that "the atomic nature of Mendelian inheritance suggests very strongly that even where variation is apparently continuous this appearance is deceptive."

Although Haldane was equal to Fisher in mathematical sophistication, and in harmony with him at least at the level of promoting a role of significance for natural selection after novelty is introduced by mutation, he clearly differed in his considerations of virtually all other aspects of evolution—especially the rate of evolutionary change, the model of speciation, and the interplay between, as well as the interface from, the gene to the individual in its environment.

Sewall Wright: Isolation, Inbreeding, and the Shifting Balance Theory

Sewall Wright was perhaps best known among his colleagues as a theoretical mathematician extraordinaire. The ease with which he did calculus and path analysis allowed him, for more than half a century, to generate dozens of articles on genetics, natural selection, and evolution. Wright came by his mathematical prowess honestly, if not also genetically. His father was a professor of economics and mathematics (as well as of astronomy and English) at Lombard College, in Illinois. Wright undertook his studies of calculus and analytic geometry with his father on his way to obtaining his bachelor of science degree at Lombard College in 1911. By the end of the following year, he had completed his master of science on the microanatomy of the flatworm at the University of Illinois, and, three years later, he took his doctorate of science in genetics at Harvard University, having studied under William Ernest Castle.

During the early decades of the twentieth century, Castle was the leading experimental geneticist and an early champion of Mendelism in the United States. He had been among the first geneticists in this country not only to investigate a possible connection between Mendelism and evolution but also to test experimentally the effects of selection on Mendelian characters. Through painstaking breeding experiments with hooded rats, Castle was able to demonstrate that without continually introducing new stock into his breeding groups, he could produce offspring with different color patterns just by selecting from within the variability that had originally existed within the species. "From the evidence at hand," Castle concluded, "Darwin was right in assigning great importance to selection in evolution; that progress results not merely from sorting out particular combinations of large and striking unit-characters, but also from the selection of slight differences in the potentiality of gametes representing the same unit-character combinations."

Because his selection experiments involved mating relatives in successive generations, Castle was particularly interested in the consequences of inbreeding. Of course, animal breeders had long been aware of the potential ill effects of constant inbreeding in producing loss of reproductive vigor as well as introducing disadvantageous characters. But Castle had the following to say about how much one could accurately generalize about the situation:

> In the production of pure breeds of sheep, cattle, hogs, and horses inbreeding has frequently been practiced extensively, and where in such cases selection has been made of the more vigorous offspring as parents, it is doubtful whether any diminution in size, vigor, or fertility has resulted. Nevertheless it very frequently happens that when two pure breeds are *crossed,* the offspring surpass either pure race in size and vigor. This is the reason for much cross-breeding in economic practice, the object of which is . . . the production for the market of an animal maturing quickly or of superior size and vigor. The inbreeding practiced in forming a pure breed has not of necessity *diminished* vigor, but a cross does temporarily (that is in the F_1 generation) increase vigor above the normal. . . . We know that inbreeding tends to the production of homozygous conditions, whereas cross-breeding tends to produce heterozygous conditions. . . . Cross-breeding . . . brings together differentiated gametes, which, reacting on each other, produce greater metabolic activity.
>
> Inbreeding, also, by its tendency to secure homozygous combinations, tends to bring to the surface latent or hidden recessive characters. If these are in nature defects or weaknesses of the organism, such as albinism and feeble-mindedness in man, then inbreeding is distinctly bad. . . . [C]ontinual crossing only tends to *hide* inherent defects, not to exterminate them; and inbreeding only tends to bring them to the surface, not to *create* them. We may not, therefore, lightly ascribe to inbreeding or intermarriage the *creation* of bad racial traits, but only their manifestation. Further, any racial stock which maintains a high standard of excellence under inbreeding is certainly one of great vigor, and free from inherent defects.

Castle's interest in the impact of varying levels of inbreeding on the viability of successive generations was a topic that Wright mined when he took his first post, senior animal husbandman at the U.S. Department of Agriculture. It would also figure centrally in Wright's later discussions of processes leading to evolution within a species as well as to the origin of new species.

From 1915 to 1926, while he was based in Washington, D.C., Wright experimentally studied and mathematically modeled the effects of inbreeding on livestock, especially cattle. Inclined, as he was, to pursue topics as thoroughly as possible, he also undertook a historical study of how shorthorn cattle had been bred without genetically impoverishing the line through intensive inbreeding. Wright took the reconstructed genealogy of the breed—which had been produced in England during the early 1800s by Thomas Bates, who started the line from a cow named Duchess—and demonstrated mathematically how Bates could not have succeeded without having inbred and crossbred exactly as he had, choosing just the right cousin or other relative to keep the character of choice present without ruining the breed genetically. Bates, of course, knew nothing of the mechanics of inheritance or of the mathematics behind what he had accomplished purely by feel and gut instinct.

After his stint at the Department of Agriculture, Wright took a professorship at the University of Chicago, which he left in 1955 to become L. J. Cole Professor of Genetics at the University of Wisconsin at Madison. He became professor emeritus at the latter institution in 1960. Although Wright produced the vast majority of his evolutionary papers after leaving Washington, D.C., while he was there he began to formulate the shifting balance theory of evolution, which would become the central dogma of his theoretical stance. As recounted by Wright, the three main influences on his development of the shifting balance theory of evolution were, first, Castle's experiments with hooded rats; second, his doctoral dissertation on the effects of multiple alleles on coat color and hair patterning in guinea pigs; and, third, his studies, while in D.C., on the effects of inbreeding in guinea pigs. It was the latter work that led to his investigation of inbreeding in livestock and to his historical study of the development of the shorthorn breed. His efforts in inbreeding livestock led, in turn, to his delving into the theoretical mathematical consequences of the reduction of genetic heterogeneity—called heterozygosis—of a population over time.

The lesson Wright learned from Castle's experiments was that mass selection—the choosing of a character in large numbers of breeding individuals—could, as had been suggested by Darwin, cause change from within the variability already present in the population. But because there are a great number of heterozygous genetic loci, and because the heterozygous alleles may differ only slightly in their effect, the antagonistic forces of weak selection, recurrent mutation, and diffusion within the population would, theoretically, keep the population more or less in a state of near-equilibrium. Interference with this equilibrium, Wright speculated, would actually have deleterious effects on the population.

Wright's own studies on gene combination taught him a career-long lesson in just how interactive genes truly are: Different gene combinations can provoke different gene interactions and, consequently, manifest themselves differently in the organism. Gene substitutions and allelic mutations would add to the available pool of potential interactive gene combinations. And because of the apparent lack of a direct link between a single gene and a physical characteristic, different combinations of the same genes could have different effects on aspects of an individual as well as on the entire organism.

From Wright's theoretical perspective of the interaction of alleles from several loci, it seemed to follow logically that the seemingly infinite number of combinations of interactive systems that could be postulated between different combinations of genes could be visualized as a field composed of high and low peaks. The high peaks would correspond to combinations of interactive systems of genes that had higher selective values, and the low peaks

Wright's depiction of his shifting balance theory. Like a topographic map, there are higher points or peaks (+), representing combinations of interactive genes with greater selective values, and lower peaks (–), representing combinations of interactive genes with lower selective values. The valleys between higher and lower peaks were the saddles. As the environment changed, so, too, would the selective values of different gene combinations: An interactive gene combination that had a high selective value in one set of circumstances might have a lower selective value in others. (Adapted from Wright)

to those with lower selective values. Wright identified the topography between peaks as the saddle. He suggested that "[t]he elementary evolutionary process becomes passage from control by a relatively low selective peak to control by a higher one across a saddle, at first against the pressure of weak selection but under pressure of selection toward the new peak after the saddle has been passed." From this argument, he then concluded that natural selection, rather than acting on individual characters, as Darwin had suggested, and as others, such as Fisher, continued to maintain, "must somehow operate on combinations of interacting genes as wholes to be most effective." Later in his writings, Wright would refer to his image of selective peaks and intervening saddles as the "selective topography."

The lesson Wright took away from his experimental inbreeding of guinea pigs was that while it was true that intensive inbreeding within small populations could lead to reduction of reproductive vigor and other deleterious changes, less severe cases of inbreeding could fix novel combinations of genes. Once fixed, these new gene combinations might be selected for, without reducing reproductive vigor in the species. Wright believed that the latter situation—moderately intense inbreeding within groups of intermediate size—occurred frequently enough that it was a real source of potential evolutionary change, both within a species and in the origin of species. As he imagined it, the best conditions for evolutionary change—which could mean either shifting the balance within a population from a higher to a lower peak or vice versa, or shifting from a saddle to a peak—came from the fragmentation of a species into many small groups, or subpopulations. As relatively smaller groups, each would be subject to some amount of inbreeding. Each could also, theoretically, differ from the others in that its members would not represent the average genotype of the species as a whole. As subpopulations, however, these small groups were not entirely isolated from one another, which meant that there could still be gene flow between some of them.

The restriction of individuals to small groups—which, as a consequence, would experience a certain amount of inbreeding—could lead to the fixation of different gene combinations and, consequently, to change within each subpopulation. In terms of its own internal equilibrium, each small group, which might represent a slightly different sampling of the species, could theoretically drift farther away from the others genetically and, ultimately, morphologically. This part of Wright's shifting balance theory of evolution became known as genetic drift. If the process of genetic drift went far enough, a new species could conceivably arise. This process would involve first the founder effect, in which the initial small group represented a skewed sampling of the original species. The founder effect would be followed by genetic drift, which would push the group genetically farther away from the original population. Wright also believed that the process of change within the partially isolated subpopulation would occur relatively rapidly. Ultimately, reproductive isolation would be achieved, leading to the advent of a new species. Although he considered that avenues other than geographic separation could lead to reproductive isolation, Wright's articles on the question of "What is a species?" recognized the significance of "the processes by which populations split into non-interbreeding groups."

But until a partially isolated population drifted off to become a distinct species in its own right, its members were still potentially available to crossbreed with other individuals, of other subpopulations, of the species. Because of this possibility, the changes that had become fixed in the smaller populations—because of the inbreeding that partial isolation provoked—could spread to other subpopulations and, ultimately, feed back into the species at large. It was in this manner, Wright speculated, that evolutionary change within a species could then occur over time.

As proposed by Wright, partial isolation could lead to evolutionary change of a species over time, which would not produce species diversity, as well as to the origin of a new species, which would. Wright summarized the entire evolutionary process as he saw it in his article "The roles of mutation, inbreeding, crossbreeding and selection in evolution":

> The course of evolution through the general field is not controlled by direction of mutation and not directly by selection, except as conditions change, but by a trial and error mechanism consisting of a largely nonadaptive differentiation of local races (due to inbreeding balanced by occasional crossbreeding) and a determination of [a] long time trend by intergroup selection. The splitting of species depends on the effects of more complete isolation, often made permanent by the accumulation of chromosome aberrations, usually of the balanced type. Studies of natural species indicate that the conditions for such an evolutionary process are often present.

Before leaving this thought, one should stop for a moment and recognize that Wright had introduced yet another concept into his model of change via intermediate-size subpopulations. Not only did he envision an effect resulting from their partial isolation—the potential to fix novelty and to be able to pass it back into the mainstream of the species—but he also speculated that part of the mechanism of fixing novelty was by way of intergroup selection. This notion contrasted vividly with the general theme of individual selection that Darwin had promoted (although there are times in *On the Origin* as well as in *The Descent* when Darwin ventures into intergroup-selection arguments). But to dyed-in-the-wool neo-Darwinians, such as Fisher and those who embraced his evolutionary speculations, anything other than individual selection was not only anathema, it was not worthy of serious consideration in discussions of evolutionary change.

It is also important to point out that Wright never did embrace Fisher's theoretical mathematical formulation of Darwin's mode of evolution: change gradually accruing over time by selection acting on continuous variation produced by random mating within a large population. From his first critical review of Fisher's articles proposing a theory of evolution based on the role of dominant alleles to his first book review of *The Genetical Theory of Natural Selection* and on through his last publications in the 1970s, Wright was adamant about the insignificance of these theoretical considerations in the course of evolution.

Like Fisher, Wright did believe that mutation was necessary for change, but he did not put all of his eggs into this basket. He believed that chromosomal rearrangements—many versions of which had been recognized shortly after the Belgian cytologist F. A. Janssens discovered that parts of chromosomes can become rearranged by crossing over—as well as hybridization were also important factors that could potentially provoke change.

As for Fisher's notion that the effects of mutation would be realized through the dominant allele, Wright thought that this was probably not a viable suggestion, primarily because a dominant allele would immediately be subject to selection and, especially if it was in any way disadvantageous, eliminated. Wright preferred the idea that since mutations present themselves most frequently in the recessive rather than the dominant state, it is an accumulation of recessive alleles within a population over time that would ultimately provide the fodder for change. As recessive character states, these alleles would not be subject to elimination by selection prior to gaining a foothold in the population. Clearly, given the centrality to evolution that Wright placed on a certain amount of inbreeding in intermediate-size populations, this presumably commonplace situation of mutation arising and being spread in the recessive state would certainly provide the opportunity necessary for these new alleles to multiply.

With regard to Fisher's mathematical modeling of change gradually accruing in a large population, Wright considered this situation to be unlikely from the beginning. First, one of Darwin's, and subsequently Fisher's, assumptions—that all individuals would potentially contribute to the available variability of the species—was not, as Wright saw it, viable. In genetic terms, the Darwinian-Fisherian model would entail random mating. But, Wright argued, in reality there could not be random mating throughout the full extent of a large population. Mating would still be localized within subsets of the population. In fact, Wright pointed out, the larger the species, the more likely it is to spread out over space and to fragment. This would create discrete subpopulations. As for the genetic and selection properties of a large population, Wright had this to say:

> Each gene is held in equilibrium at a certain frequency determined by selection, mutation, etc. Although the variability of the population may be great, the genes of different series combining in different ways probably in every individual, the average condition remains the same as long as conditions are constant, subject to the possibility . . . that wholly novel favorable mutations may disturb the situation. A change in conditions, such as more severe selection, may rapidly change the average of the population, but the change is at the expense of the store of variability of the species and compromises evolutionary advance for a long time following, since there is no escape from fixation except through the slow process of mutation. As previously noted, such change is of an essentially reversible sort and thus not really of an evolutionary character.
>
> Thus it seems that neither a small [very inbred] nor a large freely interbreeding population offers an adequate basis for a continuing evolutionary process.

Certainly, Wright preferred his model of inbreeding within subpopulations of intermediate size as the centerpiece of evolutionary change over those of his competitors. However, he did admit that another element could be applied to his picture of a selective topography of interactive gene systems. Following in the footsteps of Darwin, and in rare agreement with Fisher, Wright invoked environmental change as the impetus for organismal change.

In this evolutionary scenario, the organism tracks the shifting environment. Since, Wright argued, the environment is apparently always in a state of flux, the high and low peaks of selective values would also be unstable. The peaks change because their selective value is defined by the advantageousness of the interactive systems at a particular time and in a specific situation. In this regard, Wright could not imagine how a species in a small selective topographic field could survive under very variable environmental conditions. Eventually, he thought, such a species would become extinct because it would not have the breadth of possible gene combinations to cope with the ongoing and unpredictable environmental change around it. If, however, selection was moderate and a species could occupy a wide selective topographic field, it, like the shifting environment, would be continually on the move. Although this movement would not necessarily produce a new species by transformation—nor would it lead to diversity by way of species splitting—Wright did have this to say about the influence of the environment on species change:

> Here we undoubtedly have an important evolutionary process and one which has been generally recognized. It consists largely of change without advance in adaptation. The mechanism is, however, one which shuffles the species about in the general field. Since the species will be shuffled out of low peaks more easily than high ones, it should gradually find its way to the higher general regions of the field as a whole.

Although Wright's speculations differed as much from Fisher's as did Haldane's, they were the least easy to grasp quickly. As the historian William Provine has pointed out, most of Wright's articles were accessible to an extremely limited audience because he typically published only the sophisticated mathematical modeling of his views on the evolutionary process. Consequently, most biologists who were not trained in mathematics were limited to reading the introduction and conclusion of the majority of Wright's articles. But the real problem that stood in Wright's way in terms of reaching a larger, more sympathetic audience may also have been the fact that he tried to cover too much ground in his theoretical musings. Fisher's presentation, though also mathematically dense, was single-minded in its strictly Darwinian perspective. Haldane was able to write an entire book elaborating upon his theoretical considerations without resorting to math at all. For those interested in the mathematical formulations of his theoretical stance on evolution, he provided them in a series of appendices to his monograph. But no matter how many times Wright presented and re-presented the ideas that contributed to his shifting balance theory of evolution, and printed and reprinted the same diagram of the selective topography, he did little to clarify his original thoughts. Perhaps, in the long run, he was his own worst enemy.

Waiting to Be Asked to Dance: The Choices Facing the Evolutionary Synthesis

Although Wright, Haldane, and Fisher might be thought of as birds of a feather because they all arrived at a theoretical mathematical speculation on evolutionary change by way of experimental breeding programs, in many important ways they were worlds apart. Perhaps the only arena in which they were in some agreement was on the role of mutation in providing the potential for change. They did not, however, agree on how that change would be enacted, or on the exclusivity of mutation as the proximate cause of change. But even while agreeing that mutation played a role in evolutionary change, they differed in their stance on how large or small that mutation and, consequently, its effect, could or would be. They did not even agree on how a mutant, which might introduce the potential for change, would initially be expressed: Would it be in the dominant or the recessive state? Fisher took the former position, expecting the recessive to be converted quickly into the dominant state so that it could be expressed in the heterozygote. Wright and Haldane favored the latter condition, allowing the mutation to spread in the recessive state until it could be expressed in homozygotes.

They also did not see eye to eye on the importance of selection or its target, or even on how rapidly or slowly evolutionary change would be enacted regardless of the role of selection. Fisher was the champion of gradual change. Haldane favored rapid change. And Wright at times invoked one, and at other times the other, tempo of change.

Although all three scientists allowed some degree of interactivity of genes, Fisher was the most neo-Darwinian in that he adhered to the notion of a one-to-one correspondence between a gene and a character, and believed that the character was the unit of selection. Wright was the extreme opposite of Fisher, in that he saw everything in terms of interactive gene systems. Haldane was certainly a bit closer to Wright in his acceptance of gene linkage.

Of the three men, Fisher was most wedded to the notion that a large number of individuals would be involved in the process of change, providing lots of opportunities for random mating to supply a limitless amount of variation for selection to feed on. Wright focused on subpopulations of a species, and invoked inbreeding as the mechanism behind fixing novelty, both in a continually evolving species and in a potentially newly emerging species. Haldane maintained that complete isolation of a small subset of a population was necessary at the very beginning of the process of speciation.

By the mid-1930s, theoretical population genetics had been established. It would continue as a field in which theoretical speculations abounded on

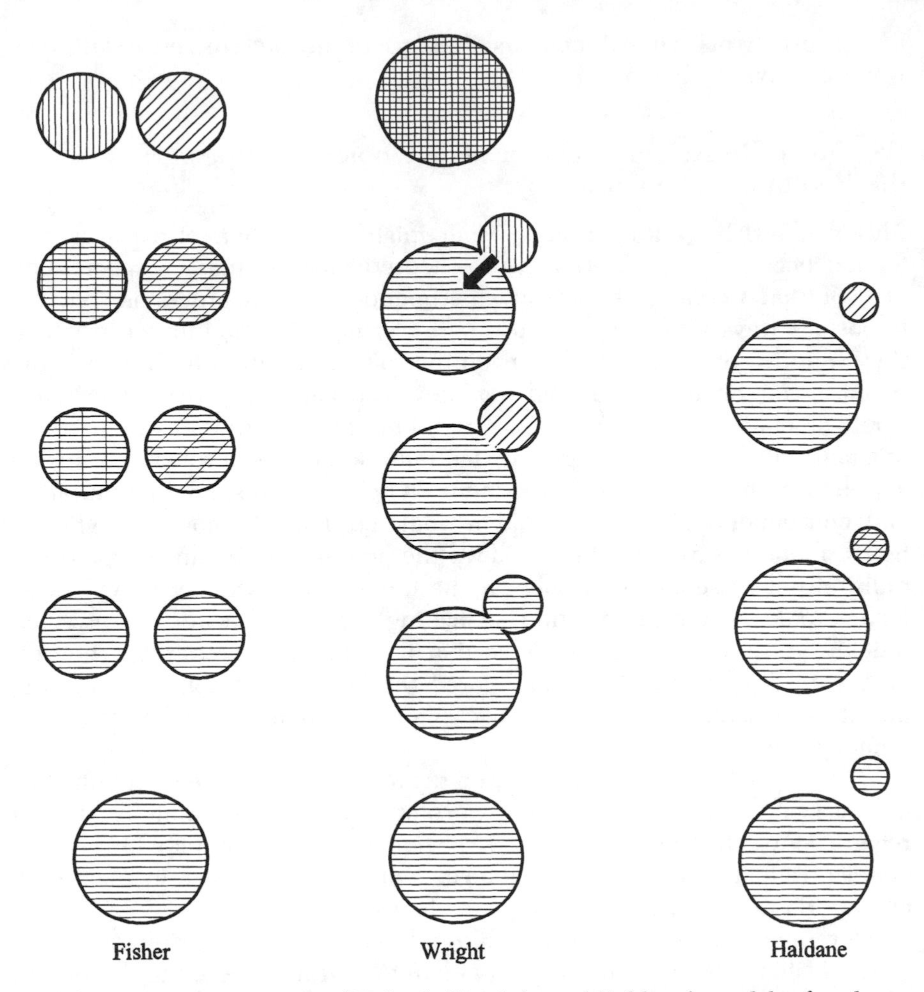

Schematic representation of Fisher's, Wright's, and Haldane's models of evolutionary change and speciation. Fisher maintained that the original species (bottom) would be divided into two large populations in which genetic and, consequently, morphological differences slowly accumulated over time. Wright envisioned a small peripheral part of the parent population accumulating genetic change more rapidly than the larger parent population and then infusing (arrow) the parent population with its genetic novelties. Haldane favored the complete separation of a small peripheral population from the original species, with the rapid accumulation of difference in the smaller population.

such issues as how selection pressures or other hypothesized constraints affected gene frequencies. When they expanded their speculations to larger evolutionary questions, Fisher, Haldane, and Wright went their different ways. Fisher stayed at the level of gene frequencies. Wright envisioned interactive gene systems. Haldane tried to deal both with these different levels and with the role of genes in the development of the organism. All three tried their hands at species definitions and identification. But they did so, again, in somewhat different ways. The intellectual element that they all shared was that of speculation.

The founders of the evolutionary synthesis—the individuals who formed the core of the Committee on Common Problems of Genetics, Paleontology, and Systematics, which was established on February 6, 1943, to address the problems so clearly emphasized in the committee's title—were obviously faced with three, often wildly different theoretical formulations of the evolutionary process and its perceived elements. As Darwin, for example, chose to incorporate a model of gradual evolutionary change into his theory of evolution by means of natural selection, so, too, did the members of this committee make choices among the available hypothetical alternatives. As a result, the alternative theories, which had kept the possibility of intellectual novelty alive through the debates they provoked, were submerged. The synthesis that emerged was, by stark contrast, largely intolerant of criticism and resistant to change.

10

The Origin of Species Revisited

> *Orthodoxy does not more firmly rule in the vestry than in the laboratory, in actual fact. Yet every one will agree that a real and permanent step in advance in science is only taken when some one looks at an old problem and at common facts from a naïve point of view, and applies to them unorthodox methods of research.*
>
> —Raymond Pearl (1915)

Whither Embryology?

With population genetics overshadowing the fields that had been responsible for a general acceptance of evolution—comparative anatomy, paleontology, and embryology—it is fortunate that some exponents of the latter disciplines were not cowed into complete silence. Among those who maintained a presence was the English embryologist and comparative vertebrate morphologist Sir Gavin Rylands de Beer. During the very years that Fisher, Haldane, and Wright were composing their most forceful theoretical mathematical arguments in support of their individual approaches to evolutionary change, de Beer was formulating some of the most interesting ideas on the relation between ontogeny, morphological change, and evolution. Specifically, de Beer was concerned with heterochrony—that is, the developmental interplay between the differential rates of attaining sexual versus physical maturity. Humans, for example, appear to be neotenic. Neoteny occurs when the onset of adult physical growth rates is retarded but the rate of reproductive maturation remains normal. The result is a sexually mature individual that is physically childlike.

In his small but to the point volume entitled *Embryology and Evolution,* which was published in 1930, de Beer mentioned Darwin only once. It was in the context of a chapter that dealt in part with adult variation. For de Beer, such variation constituted differences "between different individuals, varieties, races, and, probably, species." The thrust of de Beer's argument in this chapter was that the variations one sees among adults of a species are typically the features that appear later in an individual's ontogeny. As he put it, "[T]he later a character appears in ontogeny, the smaller as a rule is the change which it produces by its presence." The earlier a character appears during ontogeny, the greater the change—both morphologically and potentially evolutionarily—it will produce. De Beer commented in passing that Darwin also "drew attention to the fact that 'slight variations generally appear at a not very early period of life.' " As for the topic of "slight differences between animals" in general, de Beer was convinced that "genetic research tends increasingly to show that these differences are controlled by Mendelian factors or genes, and that they may concern any and every feature of an animal." But, de Beer continued, "[t]he substitution of one character for another in the adult does not usually involve heterochrony, and produces only small phylogenetic effects."

De Beer's brief mention of genetics did not reflect an ignorance of the basics, or even of more sophisticated levels of understanding, of what was then known. In the introductory comments on ontogeny, heterochrony, and evolution, he presented a more than adequate summary of genes and inheritance. He did so in the context of trying to understand ontogeny. As such, and paralleling in part the earlier suggestions of the embryologist Louis Bolk, de Beer was interested in the "internal factors" of an organism, which "ensure that if the external factors are normal and do evoke any response in development and produce an animal at all, that animal will develop along the same lines as its parent." As such, de Beer's larger picture of ontogeny "required . . . the transmission by internal factors of the capacity to react . . . to new stimuli which evoke one response after another."

This happens in amphibians: Some species of salamander become arrested in the larval gill stage of development because harsh environmental conditions trigger an acceleration in their rates of sexual maturation. Under typical, more quiescent environmental conditions, the internal factors of these amphibians behave normally, with the result that an adult individual—adult both morphologically and reproductively—develops. Another example de Beer cited was the case of the parasitic worm *Polystomum integerrimum.* If the larva of *Polystomum* lodges in the bladder of an adult frog, the parasite takes three years to mature. But if the larva ends up in the gills of a tadpole, it will reach sexual maturity in only five weeks. For de Beer, genes alone do not make the final organisms. He envisioned all aspects of the development of an organism, from its beginning to its final state, as the result of interactions between internal and external factors.

De Beer summarized his thoughts on the relationship between ontogeny and evolution in the following manner: Ontogeny results from external factors' engendering an internal response from an individual. If the external factors are held constant, an individual's ontogeny is determined by the rates at which internal factors express themselves. In turn, changes in internal factors will alter the course of development, resulting in heterochrony.

Evolution does not determine ontogeny. But evolution derives from the introduction of novelty, which results when developmental rates are modified by internal factors that have themselves been altered.

If, de Beer mused, we assume constancy in external factors, "we have now to inquire into the way in which the internal factors produce their effects." He continued:

> These internal factors have recently been subjected to very intense study, but only as regards the mechanism of their transmission from parent to offspring. They are now called Mendelian factors or genes, and are regarded as discrete units situated in or on those universal constituents of the nuclei of cells—the chromosomes. A change induced in one of these genes is called a mutation, and the gene is then fixed in this changed condition until it mutates again. It is now usual to regard these mutations as responsible for the appearance of novelties in evolution.
>
> The way in which the genes are sorted out and distributed between parent and offspring is well known as a result of Morgan's continuation of Mendel's lead, and forms the subject-matter of the science of genetics. In the process of their transmission the genes are carried in the germ-cells, the egg, and the sperm. Germ-cells, like all cells, only arise by the division of pre-existing cells, and the production of ripe germ-cells occupies two cell-generations. It is during these two cell-generations (which culminate in fertilization) that the distribution of genes takes place, and so the geneticist who studies this distribution in hereditary transmission is really following the genes through two cell-generations only. But it takes fifty-six generations of cells to reproduce a body like that of a man out of a fertilized egg (itself a single cell), and during these fifty-six generations the genes are playing their part in company with the external factors in moulding the animal through the successive stages of ontogeny. We now want to know how this part is played.

As far as de Beer was concerned, geneticists knocked on the door but did not cross the threshold to understanding the fundamental processes of evolutionary change. At this point in his argument, de Beer cited first and foremost the work of the German geneticist and embryologist Richard Benedikt Goldschmidt. Most beginning students of evolution today have probably never heard of Goldschmidt. And those who have heard of him have probably heard the worst: that he was the author of the ridiculous idea that evolutionary change arises from mutation of such magnitude that its bearers could be referred to only as "hopeful monsters"—hopefully waiting for their novel selves to become the species of the future. But although Goldschmidt did propose a model of evolutionary change that was based on large-scale mutational effects, and he did call the bearers of these mutations "hopeful monsters," he was also one of the leading geneticists of his day, if not, according to Stephen Jay Gould, of all time.

Goldschmidt was born in 1878, in Frankfurt, Germany, the son of a wealthy merchant. He studied with some of his country's most prominent scientists in the fields of development and comparative morphology. From 1903 to 1913, after receiving his formal education, he remained in Munich to work with one of his professors. Thereafter, in large part because of the efforts of the famous experimental breeder and early geneticist Karl Erich Correns (who had been one of the three rediscoverers of Mendel's work), Goldschmidt became the director of the newly established Kaiser Wilhelm Institute for Biology in Berlin. He then went to the United States on research leave. But, after the outbreak of the First World War, he was placed in an internment camp and then sent back to Germany. Upon his return home, he was able to resume his position as director of the Kaiser Wilhelm Institute. Being Jewish, he was forced out of this post in 1935 when Hitler began his ethnic-cleansing crusade. He was able to flee to the United States, however, and in 1936 he obtained a professorship at the University of California at Berkeley.

After completing the research for his doctoral degree on the development of a particular species of parasitic flatworm, Goldschmidt studied development in various species of moth, particularly the nun and the gypsy moths. He studied everything he could, from the genetics of their sexual development to the formation of the patterns of the veins in their wings to the effects of inducing abnormal development by manipulating external factors, such as heat. With regard to sexual development, Goldschmidt discovered that there were varieties of gypsy moths that differed in their intensity of being female or of being male. He referred to these varieties, which had a tendency to produce intersexual, or part male/part female, offspring when crossbred, as "sex races." In comparing these sex races, Goldschmidt found that in some cases the genes for femaleness were weak compared with the genes for producing males. In other instances, it was the other way around. For example, a female of a weak female race when crossed with a male of a strong male race would produce daughters that were intersexual or even morphologically male in their genitalia. These "sons" had become male as a result of a reversal of the process of developing into a female; that is, in the course of developing along female lines, "maleness" took over and the course of development altered the partly female genitalia into a male's. In other experimental crosses, sons could be produced that were intersexual or even female in their genitalia because of a reversal in sexual development. In the latter cases, the larval moths had been following a course of maleness but had then veered off this developmental trajectory to become female genitally.

In order to make sense of these phenomena, Goldschmidt suggested that there was more to sexual development—and, by extension, to the development of an organism—than the basic Mendelian genetics of inheritance. True, there was in the case of sex determination the X chromosome. But beyond that, there was the matter of differential development, which could produce either a morphological male or a morphological female from the same embryonic primordium of undifferentiated cells. The differences in expression of femaleness or maleness were a consequence of informing, or of influencing, the uncommitted primordium to become the genitalia of one or the other sex. In the case of moths, females were genitally female because the primordium developed into an ovipositor, and males were male because the same tissue developed into a clasping hook. But this general phenomenon of sexual development was not restricted to moths, or even to insects in general. It was also observed in mammals. Female mammals became genitally female in part because the undifferentiated genital primordium developed into the labia majora and the clitoris, whereas in males the tissue of this region became the scrotum and the penis. In hyenas, females cannot be distinguished from males merely by visual inspection because the clitoris is as large and erectile as a penis and the labia majora are fused together into a scrotumlike structure.

The implications of Goldschmidt's studies on the gypsy moth were not lost on de Beer. For him, Goldschmidt's study of intersexuality in gypsy moths was a great boost for heterochrony and its role in evolution, where developmental timing was everything. In those experiments in which he manipulated the surrounding temperature of the moth larvae, Goldschmidt altered the developmental timing and, consequently, the final structure of an organ or organs. Perhaps because Goldschmidt's first study was published in 1929, after the manuscript for *Embryology and Evolution* had been completed, de Beer failed to cite it. However, de Beer did go from his discussion of Goldschmidt's intersexuality studies to a summary of Edmund Briscoe Ford and Julian Huxley's heat-shock experiments on *Gammarus chevreuxi*, a species of brackish-water shrimp.

What Ford and Thomas Henry Huxley's grandson, Julian, had done was alter the eye color of shrimp by changing the water temperature in which the embryos matured. Genes obviously controlled the deposition of black eye coloration. But the rate at which the genes deposited coloration was affected by the temperature in which the developing shrimp found themselves. Noting that this phenomenon was not restricted to animals, de Beer cited as further support of the importance of developmental timing the fact that a primrose exposed to a temperature of 20 degrees centigrade develops red flowers, while a primrose grown in a temperature of 30 degrees centigrade produces white flowers.

That timing is everything in development was well demonstrated by Julian Huxley's solo study on the rates at which amphibians—in this case, frogs and toads—metamorphosed from the juvenile into the adult state. Huxley's study was based on the fact, well known to embryologists, that in vertebrates, hormones, which are chemical substances produced in specific glands, in some way control or influence development. In particular, the thyroid hormone, or thyroxine, had been found to be essential for growth of the limbs as well as of the tongue and the lungs. In frogs, for example, these particular anatomical regions are transformed during the metamorphosis of a tadpole into an adult. By studying the rate at which the thyroid gland matured in various species of frog, as well as in a species of toad, Huxley was able to demonstrate that there was a definite correlation between thyroid-gland maturation and the time of onset of metamorphosis. The faster the rate of thyroid-gland maturation, the smaller the animal's size at the time of metamorphosis, and vice versa. In Huxley's study, the toad metamorphosed during the early summer, before all the frogs, and the large bullfrog was the last to metamorphose, not changing until its third summer.

Never one to miss an opportunity to include humans in his comparisons, de Beer went on to point out that the thyroid hormone is important in the proper development of this vertebrate species as well. Cretinism, for instance, is the result of a deficiency in thyroxine during fetal development. Among the diagnostic features of cretinism are dwarfism and mental deficiency. If the fetal thyroid gland is missing altogether, the expression of dwarfism is dramatic. If the mother has a low level of thyroid hormone, the degree of dwarfing will be less; in fact, the severity of abnormal development can be correlated with the degree of hypothyroidism of the mother. Clearly, humans, whose entire growth process is extremely slow compared with other animals, are just as susceptible to interference of the timing of developmental processes as other organisms.

In his attempt to put together a picture of development that incorporated an awareness of genes, de Beer offered the following explanation:

> Some animals go on developing throughout their life, but most cease altering their shape at a certain stage in ontogeny which is called the adult, and which is characterized by the fact that the reproductive organs are then ripe and ready to propagate the race. Now, the time at which the adult stage is reached is also governed by the rate of action of the genes, either directly as in the gypsy moth, or indirectly by means of hormones as in the case of frogs and newts. It is possible, therefore, for there to be a competition between the genes which control any particular character and those which determine the assumption of the adult stage, and unless the former work fast enough and get in time, the character will not be able to show itself.

In an elaboration of this thought, de Beer proposed a theory that he called clandestine evolution. He envisioned clandestine evolution as a two-part process. First, a novelty might arise early in ontogeny—a phenomenon called cenogenesis. But, by inserting itself so early in the developmental sequence of an individual, this character may not be expressed in the adult. It would remain hidden, or clandestine. Since evolutionists generally accepted modification of the adult as the yardstick of evolutionary change, a cenogenetic event would not be recorded as an evolutionary event. It would be as if nothing had happened. But, then, de Beer asked, what if neoteny were to impose itself on the already cenogenetic organism? In neoteny, the organism reaches sexual maturity while remaining in the juvenile stage of physical development. So if neoteny was imposed on cenogenesis, and the resultant individuals were then preserved in the fossil record, it would look as if a new kind of animal had appeared out of nowhere.

De Beer was quick to point out that this is precisely what one sees in the case of various major groups—such as in the origin of vertebrates, where there are no fossil links connecting this group to their obvious relatives, the invertebrates. Walter Garstang's theory of vertebrate origins was based on the neotenizing of the tunicate. Most groups within Vertebrata or Invertebrata, for that matter, seemed to appear suddenly, without a trail of transitional forms.

While conceiving of clandestine evolution, de Beer was apparently not concerned that he should deal with the problem of only one individual's being neotenized. Why should he, since the very essence of heterochrony made changes available to an entire population of embryos that have been exposed to the same external factors, such as temperature? Perhaps it is because he did not specifically point this out that de Beer's argument that there is a relationship between ontogeny and phylogeny received so little recognition among other evolutionists, particularly those on the committee charged with formulating the evolutionary synthesis.

But de Beer's, as well as Goldschmidt's, early writings were not lost on J. B. S. Haldane. In *The Causes of Evolution,* which was published in 1932, Haldane incorporated the essence of *Embryology and Evolution* as well as the first gypsy-moth studies. In Goldschmidt's work, Haldane found evidence of the interplay between genes and development, especially of sexuality. From de Beer's writings, he saw the overall importance of heterochrony in evolution. He even invoked heterochrony to explain differences among species of fossil ammonites. Haldane also spent considerable time, given that it was such a short book, discussing Bolk's theory of fetalization in human evolution and de Beer's updated version of it. Although Haldane was a practicing experimental geneticist and a mathematical modeler of de Beer's versions of selection and evolutionary change, he was a kindred spirit of de Beer's and Goldschmidt's in that he espoused the importance of the interface between genes, development, and developmental timing.

It is curious that Sewall Wright did not combine the implications of Goldschmidt's and de Beer's work with his. Wright was, after all, a promoter of interactive gene combinations. He also believed that the environment played a role in the selection of available gene combinations. That is, under certain environmental conditions, some gene combinations would be more advantageous, or adaptive, than others and represent a high selective peak on the selective topographic landscape. But if environmental conditions were to change, these gene combinations might end up being of less adaptive value and slip to a lower selective peak. In this scenario, still other gene combinations would emerge as being of higher selective value. In this way, Wright envisioned the environment as the instigator in shifting the balance of the selective topographic field. As such, it would seem to have been an easy connection from here to development as mediator between gene combinations and the outside world. But, it turns out, Wright felt that "[l]ittle importance can be attributed to these [direct influences from the external world] in controlling the course of normal embryonic development, apart from the establishment of a primary simple gradient pattern by trivial differential stimuli, the maintenance of differential conditions at the surface, and of course the supply of the necessary conditions of life."

Wright never did try to think in terms of heterochrony. He did, however, engage in some speculation about what he called "the factors of development" and "the chain of reactions between gene and character." But the real reason he did so was that he disapproved of Ronald Fisher's theory of dominance. Fisher had assumed that modifiers altered the recessive states of heterozygotes to become dominant, as the wild, or normal, alleles already were. Wright felt that the process of dominance could not be so restricted. He accepted Haldane's objection to Fisher's dominance theory: that such modifiers would also modify the already-dominant wild alleles. But he speculated further that such modifiers would also have the potential to affect the homozygous condition of the rare alleles. In other words, a modifier was capable of modifying any allele in its nuclear environment, not just the alleles that Fisher thought should switch over to being dominant.

Wright proposed an alternative to Fisher's dominance theory, which he call the physiological theory. He argued that such a theory was necessary because "whatever the evolutionary mechanism by which a particular gene complex has been reached, the state of dominance of all the genes in the complex must always have a completely physiological explanation."

According to Wright, all development could be traced to only a small number of gene activities from which there was a cascading and broadening of cell and cell-product interactions. His physiological theory began with a gene producing an enzyme that would affect cell metabolism, which, in turn, would influence the constitution of the cell. This alteration would be reflected in cell activity, such as growth, division, and migration, the secretion and absorption of hormones, the stimulation and inhibition of other cells, and the absorption and release of energy. These actions would then be translated into the development of structures, which, in turn, would have the potential to respond to stimuli. And this potential response would ultimately impact the lifestyle and behavior of the individual. Goldschmidt's contribution to Wright's physiological theory came from two aspects of his gypsy-moth research: first, that genetic differences can be translated into different rates of reaction, specifically enzyme reactions; and second, that one might think of genes as catalysts in the reaction process. Wright even praised Goldschmidt for having "presented a number of beautiful examples in Lepidoptera [the insect group that includes moths]" that supported his theory.

Wright also acknowledged Goldschmidt's contributions to a topic that was of broader evolutionary concern. Through his gypsy-moth studies, Goldschmidt came to believe that differences among species must arise abruptly rather than gradually. As Goldschmidt saw it, a Darwinian gradual accumulation of small mutations was responsible only for the differentiation of the various races of the species of gypsy moth he studied. However, the processes that go on at levels within a species are totally separate from the evolutionary processes that produce species. As such, he could not subscribe to the Darwinian notion that subspecies represent species in the making. Goldschmidt referred to the processes that occur within a species as microevolution, and those that lead to the origin of species as macroevolution. He later expanded these thoughts in his monograph *The Material Basis of Evolution*—a volume that would incite attacks from some of the most powerful neo-Darwinians and lead to his expulsion from the arena of evolutionary debate. But while Goldschmidt's ideas were still in the formative stages, Wright, at least, was able to discuss them without condemnation and as interesting alternatives to Darwinian dogma.

The Beginning of a Synthesis?

In 1927, well after the heyday of the basic discoveries in Mendelian genetics, Theodosius Grigorievich Dobzhansky came to New York City from Russia to work on chromosomes with Thomas Hunt Morgan at Columbia University. Upon completing his studies in Morgan's laboratory, Dobzhansky headed off to the West, where he assumed a position at the California Institute of Technology in Pasadena. His research continued to focus on chromosomes in general, and, in particular, on the sterility or fecundity of hybrids that resulted from the behavior of their chromosomes. In part because of these research pursuits, Dobzhansky became deeply interested in evolutionary questions concerning the formation of geographic races, or variants, as well as the origin of species.

In 1935, Dobzhansky took a break from his usual articles—on such topics as the role of temperature and other external factors in the reproduction of fruit flies, or the behavior of the Y chromosome and the process of spermatogenesis in pure and hybrid strains of fruit fly—and turned to the problem of how one defines a species. He began with an epigram from William Bateson: "[T]hough we cannot strictly define species, they yet have properties which varieties have not, and . . . the distinction is not merely a matter of degree." Echoing Bateson's call to arms, Dobzhansky repeatedly made the comment that continuous variation is the exception rather than the rule among living forms. He believed that the "manifest destiny of life [is] toward formation of discrete arrays." He was equally Batesonian in chiding his fellow evolutionists for spending more energy on trying to fill in the gaps between discontinuous forms than in attempting to understand the nature of the gaps themselves. In one sense, a discontinuous world was a boon for taxonomists, because in a classification one can identify discrete groups as species. As Dobzhansky put it, "Discontinuous variability constitutes a foundation of the biological classification."

But while acknowledging that living species appear to present themselves to the biologist as discrete entities, Dobzhansky still had to attack the problem of the origin of species. He started with the premise that "[t]he discontinuity of the living world is constantly emerging from a continuity." In parallel with Fisher, Dobzhansky assumed that species divergence begins in a parent species that is continuously variable, freely interbreeding, and widespread, or panmictic, in its geographic distribution. Since, he argued, "[e]very discrete group of individuals represents a definite constellation of genes," it followed that "a stage must exist in the process of evolutionary divergence, at which an originally panmictic population becomes split into two or more populations that interbreed with each other no longer."

Dobzhansky was aware that there was a difference between the potential to interbreed and the act of interbreeding. Individuals in close proximity may not interbreed even if they could mate successfully in the artificial conditions of a laboratory or a zoo. As such, he proposed that, rather than thinking solely in terms of geographic separation as providing the isolating mechanism between subpopulations, it might be more useful to think in terms of physiological mechanisms that might interfere with interbreeding. He identified two categories of potential physiological isolating mechanisms: those that would make potential parents reproductively incapable of producing hybrid offspring; and those that would make hybrids sterile and prevent them from producing offspring.

With regard to the former category of isolating mechanisms, Dobzhansky offered the following possibilities: avoidance between males and females; interference between the sexes caused by different breeding seasons or their inhabiting different ecological "stations" (which Goldschmidt, who was not cited, had discovered did distinguish some of his races of gypsy moth); inability of sperm and ova to be attracted to one another; incompatibility of sperm and ova, leading to death of or abnormalities in the developing embryo; and in plants, inability of the female's pollen to germinate on the male's stigma.

As for the attainment of hybrid sterility, Dobzhansky suggested the following sources: different chromosome numbers in each parent, which would cause interference with meiosis, or sex-cell production, in their hybrid offspring; interference in sex-cell production in hybrids caused by different gene arrangements in parents with similar numbers of chromosomes; and, citing Goldschmidt, the development of intersexuality in hybrid offspring. Dobzhansky explained that in the second case different gene rearrangements were caused by rearrangements of chromosomal segments, such as by the swapping of segments between chromosome partners (crossing-over), double crossing-over, duplication, flipping of a segment (inversion), or by the unilateral movement of a segment from one chromosome to its partner (translocation).

In the light of these considerations, Dobzhansky proposed that a species could be defined as "a group of individuals fully fertile inter se [among themselves], but barred from interbreeding with other similar groups by its physiological properties (producing either incompatibility of parents, or sterility of the hybrids, or both)." From a geneticist's viewpoint, this definition of a species, which was predicated on reproductive isolating mechanisms, might be testable. It also made sense in terms of explaining how different discrete groups could inhabit the same geographic area. But how, Dobzhansky then asked, does a geneticist's definition of a species fit with a taxonomist's? In spite of the fact that taxonomists did not generally know whether groups they defined as species could interbreed, Dobzhansky felt confident that, in most cases anyway, a geneticist's and a taxonomist's identifications would intersect. Since, he stated, "[i]t is a matter of observation that closely similar individuals habitually interbreed, and those less similar do not," there would be "gaps" between less similar groups. And taxonomists tended to identify the discrete groups they classified on the basis of the gaps, or differences, between them. Only when very similar groups did not actually interbreed would taxonomic and genetic species identification not coincide.

The culmination of Dobzhansky's breeding experiments, his studies on chromosome rearrangements, and his thoughts on the nature of species was the publication, in 1937, of the first edition of his monograph, *Genetics and the Origin of Species.* There he elaborated on the facts, issues, and concepts he had introduced in his species article of 1935.

Although Bateson would not have embraced all of Dobzhansky's assumptions, the latter paved his entrée into the realm of evolutionary considerations of species by first quoting Bateson's comment of a quarter century earlier: "that particular and essential bit of the theory of evolution which is concerned with the origin and nature of species remains utterly mysterious." In the pursuit of his stated goal, Dobzhansky commented that while morphologists had indeed contributed something to evolutionary theory, they had not addressed the species question because of their focus on filling in the gaps between groups and trying to sort out the evolutionary relationships of these groups. He, however, was interested in "elucidating the nature of the discontinuities themselves" and "the mechanisms through which they originated." Given his particular background in population genetics through the study of fruit flies, it is not surprising that Dobzhansky approached the question of the origin of species from the following perspective:

> Since evolution is a change in the genetic composition of populations, the mechanisms of evolution constitute problems of population genetics. . . . Experience seems to show . . . that there is no way toward an understanding of the mechanisms of macro-evolutionary changes, which require time on a geological scale, other than through a full comprehension of the micro-evolutionary processes observable within the span of a human lifetime and often controlled by man's will. For this reason we are compelled at the present level of equality between the mechanisms of macro- and micro-evolution, and, proceeding on this assumption, to push our investigations as far ahead as this working hypothesis will permit.

Goldschmidt had defined microevolution as change occurring within a species, such as would be involved in the formation of varieties, or races. Macroevolution for Goldschmidt was the process of change at the level of the origin of species. Although Dobzhansky did not define these terms in his monograph, he implied throughout that microevolution would encompass the origin of species, whereas macroevolution would apply to the evolutionary radiations that produce the larger groups of organisms into which taxonomists organize species. Whether this is the case is irrelevant, because one of Dobzhansky's major themes is that the processes of change that one can study in the laboratory are representative of those that occur in the wild in tempo and mode, in the origin both of species and of higher taxonomic groups.

Although Dobzhansky sought support in Bateson's writings, his overall vision of evolution was not really Batesonian:

> The theory of evolution asserts that the beings now living have descended from different beings which have lived in the past; that the discontinuous variation observed at our time-level, the gaps now existing between clusters of forms, have arisen gradually, so that if we could assemble all the individuals which have ever inhabited the earth, a fairly continuous array of forms would emerge; that all these changes have taken place due to causes which now continue to be in operation and which therefore can be studied experimentally.

Bateson would never have invoked gradualism as the tempo of evolutionary change. But while Dobzhansky did, he was no Ronald Fisher, either. Like Bateson, Dobzhansky interpreted the particulate nature of inheritance as evidence of genetic discontinuity. But Dobzhansky also embraced the notion that, with the exception of the special cases of plants that multiply their chromosome number between generations, the mutations and their effects that did normally accumulate were often small. Because of this, Dobzhansky believed that discontinuity must by default be minute and the achievement of significant change a typically long-term process. As he stated almost halfway through his book, "The extreme slowness with which the new favorable mutations that might arise in the species from time to time can acquire a hold on the species population has been pointed out already."

Dobzhansky differed from Fisher, and was more reminiscent of Wright, in allowing that "the mode of evolution need not be the same in all organisms, not even related ones." He even went so far as to suggest that selection might not always play a role in change, especially in the differentiation of races, or varieties. Genetics, he reiterated throughout, not selection, was the source of variation. As for selection and the origin of species, Dobzhansky was, again, more like Wright in allowing for the possibility of different combinations of selection and mutation rates to have relevance in different situations, particularly in the face of an ever-changing environment. Genetic drift, for example, might lead to change in some situations without any input from selection at all.

Dobzhansky, like others, saw the environment as playing a potentially significant role in evolutionary change. Because, he remarked, the environment is not static, an organism should never reach genetic equilibrium. To the contrary, an organism, and by extension its own species, will always be in a state of genetic flux. This, Dobzhansky firmly believed, was at the very base of evolution. Otherwise, he stated, "[a] species that would remain long quiescent in the evolutionary sense is likely to be doomed to extinction."

In his review of the genetic levels that could contribute to change, Dobzhansky spent some time discussing mutations, often along the lines taken by Fisher. But although Dobzhansky recognized that mutations at a specific locus had to be included in any formulation of evolutionary change, he was more committed to the idea that gene rearrangements, caused by the rearrangements of chromosome segments during phases of meiosis, played an even more significant role. This, of course, is not surprising, given the fact that his research had largely been concerned with the behavior of chromosomes as they paired from individuals of different races of the same species as well as from individuals of different species. Dobzhansky even identified races of a species of fruit fly, *Drosophila pseudoobscura,* as well as a new species of fruit fly, *D. miranda,* on the basis of chromosome differences. In fact, since *D. miranda* was virtually identical morphologically to *D. pseudoobscura,* Dobzhansky could identify it only on the basis of its chromosomes.

One of Dobzhansky's significant discoveries about chromosomes was the fact that even if they were of different lengths and shapes, they would pair upon fertilization if they shared any loci at all. That is, when individuals from two different species were bred experimentally, a very long chromosome from one parent would pair up with a short chromosome from the other parent at the loci they had in common, even though there might be significant stretches of both chromosomes that had no matches whatsoever. These unpaired chromosome segments would float as loops between the few points that were matches. According to Dobzhansky, the reason these disparate chromosomes came together in the first place was that the few loci they shared acted as attractants. In even more extraordinary cases, such as a cross between the two species *D. pseudoobscura* and *D. miranda,* a chromosome from one parent could end up pairing up with more than one chromosome from the other parent. The reason for this was that the matches for the alleles on one parent's chromosome were distributed among different chromosomes of the other parent.

Using a sausage as an analogy, Dobzhansky described his view of the relationship between genes, gene activity, and chromosomes in the following manner:

> Classical genetics conceive the hereditary materials, the germ plasm, as absolutely discontinuous. The germ plasm is a sum total of discrete particles, the genes. The genes are pictured as independent of each other, both in inheritance

> and in evolution. Each of them can undergo mutational changes, or can separate from its neighbors by crossing over and by chromosome breakage, without affecting the adjacent genes. A chromosome is, then, a sort of sausage stuffed with a definite number of layers of genes, arranged in a definite but fortuitous linear order. A mere rearrangement of the layers within the sausage need not have any effects on the properties either of the layers or of their container.
>
> The discontinuous or particulate nature of the germ plasm is of course beyond doubt. It is a corollary to Mendel's laws, and the existence of the elementary hereditary particles, genes, is as well established as that of molecules or atoms. A rapidly growing amount of evidence indicates, however, that the genes are not quite so impregnable and impervious to the influence of their neighbors as has been thought. The effect of a gene on development is a function of its own structure as well as of its position in the chromosome. A change of the linear order of the genes in a chromosome may then leave the quantity of the gene unaffected, and yet the functioning of the genes may be changed.

Dobzhansky's emphasis on the major role of chromosomal rearrangement in introducing variation and potential long-term change contrasted with Fisher's strictly Darwinian, single-allelic-mutation position. Dobzhansky's view of chromosomes could, however, be incorporated more into Wright's theory of interactive gene combinations. Nevertheless, it was certainly a more unique speculation than any held by the theoretical population geneticists.

Dobzhansky accepted that chromosome rearrangements, and to some extent allelic mutations, were the potential causes of variation within a species. They were not, however, in and of themselves the sources of new species. This is where his physiological isolating mechanisms came in to play. In *Genetics and the Origin of Species,* his chapter on this topic was an expanded discussion, with examples, of the basic outline of isolating mechanisms that he had presented in his article of 1935. Physiological isolating mechanisms provided a break in genetic exchange between groups of individuals of the same original species. This break, Dobzhansky postulated, was necessary if each group was slowly to achieve its own genetic and perhaps morphological identity and assume its new status as a species. Dobzhansky often hinted at species developing from races, or varieties, that had become distinct within a species. But, in contrast to the central position that Ernst Mayr would later place on the evolutionary transformation of races into species, Dobzhansky was clearly more neutral on this point.

It was in his chapter on isolating mechanisms that Dobzhansky mentioned Goldschmidt: "It is a fair presumption that the pessimistic attitude of some biologists . . . [specifically, Goldschmidt] . . . who believe that genetics has learned a good deal about the origin of variations within a species, but next to nothing about that of the species themselves, is due to the dearth of information on the genetics of isolating mechanisms." Clearly, Dobzhansky could not embrace Goldschmidt's contention that species formation resulted from a different genetic source than variation within a species. But neither could he present proof that the genetic basis for variation within a species also led to the advent of a new species—that the same genetics was relevant in micro- as well as macroevolutionary events. Consequently, Dobzhansky had to admit that "[s]o long as the genetics of the isolating mechanisms remains almost a terra incognita, an adequate understanding, not to say possible control, of the process of species formation is unattainable." As such, he was left only to "try to assemble some facts and to outline some suggestions that may throw light on these problems."

In the end, as Darwin had not been able to describe the origin of species, so, too, Dobzhansky could not provide a clear genetic model for the origin of species. Like others before and after him, Dobzhansky was forced to assume the position that if he gave enough examples of allelic mutations, chromosomal rearrangements, hybrid sterility, intersexuality, and geographic variation and distribution of races, his audience would get the feeling that his evolutionary speculations on the origin of species were feasible. But there is a difference between demonstrating that something has happened, or that genes or differential timing of development may affect the final morphology of an organism, and demonstrating how a novel feature arises and is transmitted in such a way as to produce a new species.

Goldschmidt's Systemic Mutations: A Threat to the Synthesis

Richard Goldschmidt would not be denied an attempt at an explanation for the origin of species. In his monograph *The Material Basis of Evolution,* which was published in 1940, three years after Dobzhansky's book, he expounded upon his earlier suspicions of the separateness of processes operating within a species and those involved in the origin of species. In this two-part volume, Goldschmidt dealt first with what he called microevolution and then with macroevolution. As far as he was concerned, all the studies he had done on the morphology, development, genetics, variation, and geographic and climatic distribution of races of gypsy moth were relevant only at the level of understanding what was happening within that one species of gypsy moth, *Lymantria dispar.* In short, allelic mutations and their apparent morphological and developmental consequences could, at best, only help to explain the origin of races and geographic variation.

As for the influence of climate or environment on organismal change, Goldschmidt was equally impressed that his studies served only to explain the origin of regional variation, not the origin of species. When he looked critically at other, similar studies, whether on plant or animal, he was convinced that they, too, documented only the changes that occur within a species over time and in a less than uniform landscape and environment. The examples of the effects of industrial melanism on changing the frequency of wing color in a species of moth, which are so familiar to population geneticists, served only to make his point.

Goldschmidt's contemporaries thought that the often-observed change in frequency of moth wing color from a more widespread lighter pigment to a more common dark color, which seemed to be coincident with the spread of factories in an area, was evidence not only of natural selection but also of the way in which evolution worked. In this case, the factory soot darkened the trees on which the moths sought haven, making the light color a disadvantageous trait. Over time, predation by birds on the more visible moths skewed the frequency toward a typically dark-winged population. If one extrapolated from this example, one might get a sense of how the accumulation of minor changes could eventually produce a new species. For Goldschmidt, however, this was merely an example of how members of a species survive within the limits of the genetic and morphological variability already present. Small, allelic mutations only provided more of the same, low-level, within-species kinds of interindividual variation.

If the accumulation and selection of small mutations were truly the bases for major evolutionary change, Goldschmidt wondered, how could one explain the evolution of "hair in mammals, feathers in birds, segmentation of arthropods and vertebrates, the transformation of the gill arches in phylogeny including the aortic arches, muscles, nerves, etc.; further, teeth, shells of mollusks, ectoskeletons, compound eyes, blood circulation, alternation of generations . . . [many details of various invertebrates] . . . poison apparatus of snakes, whalebone, and, finally, primary chemical differences like [iron-based] hemoglobin vs. hemocyanin [a copper- rather than iron-based respiratory pigment in the blood of various invertebrates], etc."? He stated that examples of plants could also be given. But he did not do so.

In contrast to the effects of small mutations affecting subgroups within species, Goldschmidt suggested that major systemic mutations are responsible for creating the kinds of differences—whether morphological, developmental, or genetic—that distinguish species, and groups of species, from one another:

> The change from species to species is not a change involving more and more additional atomistic changes, but a complete change of the primary pattern or reaction system into a new one, which afterwards may again produce intraspecific variation by micromutation. One might call this different type of genetic change a *systemic mutation,* though this does not have to occur in one step.

As to how systemic mutations come about, Goldschmidt turned to chromosome rearrangements and especially to Dobzhansky's research, particularly that involving chromosomally defined races of fruit fly as well as the newly described species of fruit fly, *Drosophila miranda:*

> A systemic mutation (or a series of such), then, consists of a change of intrachromosomal pattern. This is what is actually found taxonomically (the bridgeless gap) and cytologically. Whatever genes or gene mutation might be, they do

> not enter this picture at all. Only the arrangement of the serial chemical constituents of the chromosomes into a new, spatially different order; i.e., a new chromosomal pattern, is involved. This new pattern seems to emerge slowly in a series of consecutive steps. . . . These steps may be without any visible effect until the repatterning of the chromosome (repatterning without any change of the material constituents) leads to a new stable pattern, that is, a new chemical system. This may have attained a threshold of action beyond which the physiological reaction system of development, controlled by the new genetic pattern, is so basically changed that a new phenotype emerges, the new species, separated from the old one by a bridgeless gap and an incompatible intrachromosomal pattern. "Emergent evolution" but without mysticism! I emphasize again that this viewpoint, cogent as it is and, in my opinion, necessary to an understanding of evolution, is to be understood only after the fetters of the atomistic gene theory have been thrown off, a step which is unavoidable but which requires a certain elasticity of mind.

Systemic mutation became the theme for the second half of *The Material Basis of Evolution,* and Theodosius Dobzhansky became the hero behind its formulation. Virtually every major example of intrachromosomal repatterning that Goldschmidt invoked in support of his argument came from Dobzhansky's solo or collaborative work. Along with promoting systemic mutation as the key to the origin of species, Goldschmidt left no opportunity unfilled to express his belief that "[t]he classical atomistic theory of the gene . . . blocks progress in evolutionary thought." But Goldschmidt also promoted Dobzhansky as a potential sympathizer of his objection, claiming that the latter scientist had reached a similar "impasse" in his investigations of the origin of species. Clearly, Dobzhansky had neither reached this point nor rejected a neo-Darwinian explanation for evolutionary change.

For Goldschmidt, the importance of thinking in terms of systemic mutation lay in recognizing the overall effects of major genetic reorganization on the developing organism. Since the genetic basis for the development of major features would be altered, so, too, would the course of development. More specifically, altering the course of development meant that the timing and relative rates of development would be changed from what would have been the normal state in the previous or ancestral species. As an example, Goldschmidt described at great length mutants of fruit flies he had discovered in which the antennae developed into legs. In normal development, the antennae would, of course, have become antennae and the legs, legs. This suggested to him the existence, early in development, of clusters of cells that could be induced to become, in this case, either antennae or legs. In normal development, these groups of undifferentiated cells are affected by inductive stimuli that produce antennae and legs. In the mutant fruit flies, however, the timing of induction was shifted so that the stimuli affected not only the cell masses that would grow into legs but also those that would otherwise have become antennae. Clearly, although the resultant mutants were "monstrous" in comparison with a normal fruit fly, they arose by a relatively simple alteration of induction and developmental timing.

In addition to providing a number of examples in which systemic mutation produced the development of a structure in an unexpected location, Goldschmidt also spent considerable time discussing the reduction or diminution of structures, such as in the truncation of wing size and the patterning of veins in various insects. Although rudimentary structures were formed, Goldschmidt viewed them as additional evidence of the effects of alteration of timing mechanisms in early phases of development. In these instances, alteration of timing of critical phases of development led to the absence of features. In addition to the development of truncated wings and vein patterning in insects, truncation, or rudimentation, as Goldschmidt called it, could explain the reduction of toes in the evolution of various vertebrate groups, such as single-toed horses and the hypothesized last common ancestor of the two-toed group of mammals that includes sheep and deer.

Goldschmidt referred to the general regulation of growth processes as "homeosis," which was a term Bateson had coined for the development of repeated structures. This is what other embryologists had identified as "heteromorphosis," which literally means the development of different shapes. As for the changes that result from differential rates of regulation and the timing of development, Goldschmidt called them "homeotic" changes. As de Beer had also argued, Goldschmidt saw regulation of developmental timing as providing a simple explanatory mechanism for differences among organisms in their "repeated parts." Homeosis could account for differences in number of vertebrae and teeth, for example. But for Goldschmidt the bottom line where any homeotic change was concerned was that the genetic alteration, which, in turn, affected the timing of induction and development, did not have to be large. Since a genetic change early in ontogeny would have a snowball effect on the developing organism, the change itself did not have to be enormous or have resulted from a large-scale genetic upheaval. A simple chromosomal rearrangement, or a simple mutation affecting a suite of genes, would be sufficient to set off a course of development that would take its bearer to a different end point.

Goldschmidt was unreserved in arguing that a Darwinian approach to within-species phenomena was inapplicable to the study of the origin of species and to groups above the species level. Most neo-Darwinians, and certainly Ronald Fisher, were wedded to the idea that subspecies should be regarded as incipient, potential species. Goldschmidt rejected this out of hand. As he wrote: "Isolation or no isolation, the subspecies are diversifications within the species, but there is no reason to regard them as incipient species." He was convinced that the reason neo-Darwinians had not looked beyond the notion of species arising by a simple transformation of subspecies was both "because a strictly Darwinian view requires such an interpretation, and because it is taken for granted that no other possibility exists."

He blamed the neo-Darwinian's myopic view of evolution on the theoretical population geneticists' mathematical modeling of selection and evolution work. He pointed his finger at Sewall Wright in particular. Goldschmidt saw Wright as the primary culprit because of his mathematical argument "showing that small isolated groups have the greatest chance of accumulating mutants, even without favorable selection." Not that Goldschmidt was against mathematical applications. As he told his readers, he, too, had tried mathematically to work out a case of selection in the nun moth. Unfortunately, he could not do so until Haldane came up with the correct formulas. Goldschmidt's objection to the impact of the theoretical population geneticists on evolutionary thought was predicated on his belief that "biology must be studied *with* mathematics but not *as* mathematics." As he stated further:

> I am of the opinion that this criticism applies also to the mathematical study of evolution. This study takes it for granted that evolution proceeds by slow accumulation of micromutations through selection, and that the rate of mutation of evolutionary importance is comparable to that of laboratory mutations, which latter are certainly a motley mixture of different processes of dubious evolutionary significance. If, however, evolution does not proceed according to the neo-Darwinian scheme, its mathematical study turns out to be based on wrong premises.

In his pursuit of an explanation for the evolution of those kinds of differences that distinguish species, and groups of species, from one another, Goldschmidt was certainly on a collision course with the very scientists he needed to convince. If he could not persuade them, he most definitely should have sought their intellectual support. But his harangues and undiplomatic dismissals of his colleagues' work and their neo-Darwinian predilections led only to a barrage of calumnious attacks. And when Goldschmidt theorized that the bearers of his proposed systemic mutations might be thought of as "hopeful monsters," his fate could not have been sealed better by his worst enemy.

Although Goldschmidt had introduced the notion of "hopeful monsters" in an article seven years prior to the publication of *The Material Basis of Evolution,* it was in the latter publication that he elaborated upon the idea "that mutants producing monstrosities may have played a considerable role in macroevolution." Quite simply, a "monstrosity appearing in a single genetic step might permit the occupation of a new environmental niche and thus produce a new type in one step." Goldschmidt used the tailless Manx cat as an example of a monstrosity in which the lower vertebrae did not fully develop. He suggested that a sole or a flounder was another kind of monster in which a mutation caused the skull to become distorted and one eye to end up on the same side as the other eye; a systemic mutation was also responsible for severely flattening their bodies. Dachshunds were monstrosities in which a particular kind of dwarfism—normal-size torso and head but reduced limbs—caused by a systemic mutation, had been selected for by animal breeders. And *Archaeopteryx,* the fossil that was part bird and blatantly part dinosaur, had been a monster in which feathers had appeared.

Goldschmidt also referred to commonly occurring abnormalities as being representative of the ease and frequency with which monstrosities appear naturally. For instance, he cited the reduction of digits in humans and other mammals; Bateson had earlier called attention to brachydactyly, or the reduction in finger and toe bones in humans. Goldschmidt also referred to hairlessness or reduced eyes in various animals, and, of course, to rudimentation of wings and patterns of venation in insects as examples of the kinds of morphologies that would arise as monstrosities.

For Goldschmidt, these examples provided evidence "of evolution in single large steps on the basis of shifts in embryonic processes produced by one mutation." As far as he was concerned, "there is no reason to assume for such taxonomic traits [such as provided by the examples above] an origin by slow selection of micromutations instead of origin in one large step." An appraisal of the "hopeful monster" concept, Goldschmidt suggested, was now "furnished by the existence of mutants producing monstrosities of the required type and the knowledge of embryonic determination, which permits a small rate of change in early embryonic processes to produce a large effect embodying considerable parts of the organism." The hopeful part of "hopeful monster" was that the new feature or organismal reorganization was actually a preadaptation to a different econiche or life circumstance than the one in which the pre-monster species had been. If the mutation and its developmental product were appropriate for a different situation, then the monster would persist.

By advancing the notion of "one step" changes, Goldschmidt seemed to have forgotten his earlier and more fully argued suggestion that systemic mutation could be achieved by a series of steps. But his predication of "hopeful monsters" on single mutations and single-step, major reorganizational changes would become the focus of his detractors.

Taking his criticism of Darwin too far at this point, perhaps, Goldschmidt pointed out that the former scholar had not embraced the importance of monsters in understanding evolution, while others had done so. One of those whom Goldschmidt boldly cited in support of his thesis was the German paleontologist Otto Heinrich Schindewolf. Schindewolf had published a short monograph in 1936, in which he outlined an anti-Darwinian theory of evolution. Although he later expanded this presentation in a larger monograph entitled *Basic Questions in Paleontology,* which was eventually published in 1950, the essence of his argument was already in place at the earlier date.

As a paleontologist, Schindewolf discussed the gaps in the fossil record as recording not the frustrations of an incompletely preserved fossil record but the realities of the process of evolution. For Schindewolf, the gaps in the fossil record demonstrated clearly the incorrectness of Darwinian gradualism as an explanation of evolutionary change. As he wrote in 1950: "Evolution does not proceed by constructing the major types synthetically, cumulatively, out of individual building blocks collected over a long period of unfolding . . . but rather of profound transformations in the characters of basic organization." For Schindewolf, the ancestor of a new group—whether it would come to have many species, as in the case of birds, or few, as is true of horses and zebras—would be the bearer of a profound reorganization of characteristics. Once this novelty, or suite of novelties, had been established, the descendants of that evolutionarily innovative ancestor would evolve within the constraints of that reorganization. For instance, the feet of the five-toed ancestor of all horses were reorganized to distribute weight along the axis that extends into the middle toe. All subsequent evolution involving toe reduction and loss within the horse group maintained the organization around the central toe. The leaps between stages of horse evolution may not have been as profound as the leap that established the ancestor of the horse group, but this was certainly not a neo-Darwinian view of evolutionary change. Even less neo-Darwinian—actually, it was anti-Darwinian—was Schindewolf's dismissal of selection and adaptation as orchestrating major evolutionary change.

While, as a developmental geneticist, Goldschmidt came to seek support of his systemic mutations from the fossil record, Schindewolf went from his strength in paleontology to genetics and development for his backing. Curiously, the two scholars never met. In fact, in a footnote in his 1950 monograph, Schindewolf commented that he first learned not only about Goldschmidt's work but of the latter's acceptance and praise of his 1936 monograph while reading another colleague's paper. Although his comments were confined to a footnote, Schindewolf could not have been more enthusiastic about Goldschmidt's thesis, and concluding that "Goldschmidt's inferences completely meet the challenge that fossil material appears to me to pose, and that he, a leading geneticist, has presented a complete interpretation that does justice to the tangible, historical phylogenetic data." In turn, Goldschmidt saw in Schindewolf's research on fossils the supportive evidence that his theory needed:

> I need only quote Schindewolf (1936), the most progressive investigator known to me, who showed that the material presented by paleontology leads to exactly the same conclusions as derived in my writings, to which he refers. He elaborates the thesis that macroevolution on a higher level takes place in an explosive way within a short geological time, followed by a slower series of orthogenetic [evolution within a lineage] perfections. . . . He realizes that the conception of preadaptation accounts completely for this type of evolution. He shows by examples from fossil material that the major evolutionary advances must have taken place in single large steps, which affected early embryonic stages with the automatic consequence of reconstruction of all the later phases of development.

> He shows that the many missing links in the paleontological record are sought for in vain because they have never existed: "The first bird hatched from a reptilian egg." . . . Thus we see that the results of paleontology . . . vindicate the thesis which we developed here. It is gratifying that all the disciplines which furnish material for the understanding of evolution—taxonomy and morphology, descriptive and experimental embryology, static and dynamic (physiological) genetics, comparative anatomy and paleontology—supply ample and parallel evidence for a theory of evolution which is more plausible than the neo-Darwinian theory.

Goldschmidt's exuberance would, unfortunately, be short-lived.

The Neo-Darwinians Strike Back

One of the people who reviewed Goldschmidt's *The Material Basis of Evolution* was Sewall Wright. Although he could be caustic in his book reviews, Wright was exceedingly generous in his treatment of *The Material Basis,* especially given the lack of restraint Goldschmidt often showed when criticizing his colleagues.

After summarizing clearly and correctly Goldschmidt's premises and conclusions, Wright admitted that the disagreement between Goldschmidt and neo-Darwinians was not solely a matter of different interpretations of the relative roles of chromosomal rearrangement versus micromutation in enacting evolutionary change. Neo-Darwinians were interested in more than just the kinds of mutation that lay behind evolutionary change, all possibilities of which, he declared, were acceptable. Neo-Darwinians were also more generally interested in the vicissitudes of the evolutionary process. As for the discontinuity between species that prompted Goldschmidt to invoke systemic mutation, this, Wright stated, was a relative matter to neo-Darwinians. Gradualism could equally account for what appear to be "bridgeless gaps" between species. A major problem for neo-Darwinians, however, was the multiplication of chromosomes in various kinds of plants that quickly produced new species. In parrying Goldschmidt's criticism of the neo-Darwinian reliance on gradualism, Wright reiterated his theoretical position on evolutionary rates, which, again, left the door open for any number of possibilities. Because he favored the idea that isolation of subspecies followed by genetic drift and accumulated change ultimately led to intersterility or hybrid sterility between newly emerging species, Wright could not accept Goldschmidt's suggestion that the development of intersexuality was the first step in achieving genetic isolation and then new species.

Wright was most critical of Goldschmidt's reliance on chromosomal rearrangement as the basis for systemic mutation. In fact, he accused Goldschmidt of reinventing the role of genes, which happen to reside on chromosomes, in order to sustain his theory of systemic mutation. As Wright summarized, neo-Darwinians accepted that the action of a gene was independent of its position on a chromosome. Consequently, the effect of a mutation was not tied to the position of the mutant gene on a chromosome. Obviously, chromosomal rearrangement, with the profound effects Goldschmidt hypothesized, was at odds with a neo-Darwinian interpretation. But, Wright's criticisms aside, his review was in general quite evenhanded, and was punctuated with a recognition of Goldschmidt's contributions to the disciplines of developmental genetics and evolutionary theory. For instance, in the middle of the review, Wright stated: "Goldschmidt has played a leading role in bringing home to geneticists the necessity for a physiological interpretation." He concluded his review with the following statement:

> While the reviewer radically disagrees with the author's central thesis, he wishes to testify to the importance of the book. A great store of well-selected data have been assembled from diverse sources, fairly presented and discussed from viewpoints which must be carefully considered by any one interested in the problem of evolution.

Wright's was the last positive evaluation that Goldschmidt would receive from those credited with the formulation of the evolutionary synthesis.

In the same year that Wright's review of *The Material Basis of Evolution* came out, Dobzhansky, who was now back at Columbia University as a professor of zoology, published a revised edition of *Genetics and the Origin of Species.* One of the differences between the first and the second edition lay in Dobzhansky's expansion of the number of examples he had mustered to discuss the topics of gene mutation, chromosomal rearrangement, variation and selection, isolating mechanisms, hybrid sterility, patterns of evolution, and species. But the most notable difference was his blatant hostility to virtually everything Goldschmidt had written. Dobzhansky could not dismiss Goldschmidt's classic studies on intersexuality in gypsy moths, but he did his utmost to attack Goldschmidt on all other grounds, especially in the realm nearest to his own research, chromosomal rearrangement. Dobzhansky's almost uncontainable rage at Goldschmidt's propositions, which derived from the former geneticist's very own work, is evident in the following passage:

> According to Goldschmidt, all that evolution by the usual mutations—dubbed "micromutations"—can accomplish is to bring about "diversification strictly within a species, usually, if not exclusively, for the sake of adaptation of the species to specific conditions within the area which it is able to occupy." New species, genera, and higher groups arise at once, by cataclysmic saltations—termed macromutations or systemic mutations—which bring about in one step a basic reconstruction of the whole organism. The role of natural selection in this process becomes "reduced to the simple alternative: immediate acceptance

> or rejection." A new form of life having been thus catapulted into being, the details of its structures and functions are subsequently adjusted by micromutation and selection. It is unnecessary to stress here that this theory virtually rejects evolution as this term is usually understood (to evolve means to unfold or to develop gradually), and that the systemic mutations it postulates have never been observed. It is possible to imagine a mutation so drastic that its product becomes a monster hurling itself beyond the confines of species, genus, family, or class. But in what Goldschmidt has called the "hopeful monster" the harmonious system, which any organism must necessarily possess, must be transformed at once into a radically different, but still sufficiently coherent, system to enable the monster to survive. The assumption that such a prodigy may, however rarely, walk the earth overtaxes one's credulity, even though it may be right that the existence of life in the cosmos is in itself an extremely improbable event.

Perhaps provoked by Goldschmidt's thesis of systemic mutations, by 1941 Dobzhansky had become a much more hard-core gradualist, who perceived evolutionary change as resulting from the accumulation of myriad small mutations over a long period of time. His appeal to a definition of evolution as meaning "to unfold or to develop gradually"—which harks back to the old embryological definition of the term—certainly attests to an entrenched gradualistic position. Selection now took on a greater significance in Dobzhansky's formulation of a process of speciation that also invoked subspecies, or races, as incipient species. His raising of this latter speculation to the level of fact is coupled with another rejection of Goldschmidt: "The present writer thoroughly agrees with Goldschmidt in that if races are incipient species their differences must be commensurable; in the writer's opinion they are, in Goldschmidt's they are not."

When invoking isolating mechanisms that would lead to the origin of new species, Dobzhansky, again, could not refrain from attacking Goldschmidt. If Goldschmidt was correct, a single mutation could result in reproductive isolation. But since Dobzhansky was now inextricably wedded to micromutation and gradual change, from incipient species formation to the origin of full-fledged species, Goldschmidt could not be correct. In short, between 1937 and 1941, Dobzhansky went from being able to allow for the possibility of evolutionary mechanisms other than those he favored to a position in which everything that did not fit his definition of evolution was rejected.

In the midst of his outpouring of anger at and dismissal of Goldschmidt, Dobzhansky neglected to consider the fact that while Goldschmidt's systemic mutations may not have been observed, neither had the mechanisms of speciation that he, or anyone else, for the matter, had proposed. Rather, Dobzhansky, as others did and would do, took for granted that, with enough time, the kinds of small mutations and changes that were observed in laboratory experiments on fruit-fly population genetics were also capable of producing the degrees of differences that seem to characterize species in the wild. To be sure, there was a certain logic in the belief that it was unnecessary to postulate another mechanism for evolutionary change when one already appeared to exist. This logic also seemed to benefit from the assertion that not only had no other mechanism been observed but that no other mechanism had yet produced species. Nevertheless, it was and still is the case that, with the exception of Dobzhansky's claim about a new species of fruit fly, the formation of a new species, by any mechanism, has never been observed.

The fact that the full process of speciation had not knowingly been observed and studied was not, however, a deterrent to the bird systematist and evolutionary theorist Ernst Mayr. Mayr had left Germany in 1931 to assume a position at the American Museum of Natural History in New York City, where he would organize the museum's extensive bird collections taxonomically, particularly those from New Guinea and the Solomon Islands. In his influential monograph of 1942, *Systematics and the Origin of Species,* Mayr was convinced not only that he could identify species but that he could also identify the particular stage of speciation an emergent species was in at any given time. He also believed that he knew the evolutionary pace of speciation:

> That speciation is not an abrupt, but a gradual and continuous process is proven by the fact that we find in nature every imaginable level of speciation, ranging from an almost uniform species at one extreme to one in which isolated populations have diverged to such a degree that they can be considered equally well as separate, good species at the other extreme. I have tried in a recent paper . . . to analyze this continuous process and to demonstrate its different phases by subdividing it into various stages. I am well aware that these divisions are somewhat artificial and that a poltypic species [one of many stable subspecies] may be in different stages in different parts of its range at the same time. . . . A widespread species is more likely to represent the first stage of speciation than one with a narrowly restricted range.

Mayr, of course, was speaking as a naturalist, not as a geneticist, and this is what distinguishes his book on the origin of species from Dobzhansky's. But Mayr believed that there was congruence between the suggestions of geneticists and his picture of the process of species formation. Like various geneticists—and he specifically cited Sewall Wright in this instance—Mayr accepted "that the accumulation of small genetic changes in isolated populations can lead in the course of time to a new integrated genetic system, of such a difference that it thereby acquires all the characters of a new species, including reproductive isolation." Not surprisingly, Mayr was also a firm believer in subspecies being incipient species. His faith in this scenario grew from his own research on the geographic variation and distribution of varieties of species of birds. Bird species can often be widespread and cluster into easily identifiable races, or subspecies, especially when the distribution is over a series of islands or other geographically disjunct locales.

In his quest to synthesize evolutionary theory from the perspectives of naturalist and geneticist—indeed, to resurrect natural history and systematics into evolutionary theory—Mayr made the following assumptions:

> First, there is available in nature an almost unlimited supply of various kinds of mutations. Second, the variability within the smallest taxonomic units has the same genetic basis as the differences between the subspecies, species, and higher categories. And third, selection, random gene loss, and similar factors, together with isolation, make it possible to explain species formation on the basis of mutability, without any recourse to Lamarckian forces.

Clearly, the assumption of "an almost unlimited supply of various kinds of mutations" was a geneticized version of Darwin's concept of continuous variation with regard both to its source and its expression. The second assumption derived from the application of uniformitarianism: Processes acting today are the same today as they were in the past. The third assumption reiterated the relatively common approach to discussing evolutionary change and the origin of species; namely, if enough examples of possible factors that might lead to shifts in the relative expression of characteristics were put forth, a sense of how the otherwise undemonstrable process of speciation works might be achieved. In spite of his apparent need to reject the Lamarckian notion of use and disuse engendering change, Mayr seemed at times to rely on this kind of explanation in conjunction with a Darwinian emphasis on utility. For example, he wrote elsewhere in this book: "Characters are usually lost rather quickly as soon as they are no longer needed, as demonstrated by the loss of eyes in cave animals, and so forth. . . . More difficult to explain is the survival of conservative characters long after they have lost their apparent usefulness."

In his ruminations on species identification, Mayr did break away from the geneticists, even from Dobzhansky, who had written the preface to *Systematics and the Origin of Species.* Rather than emphasize the attainment of some form of sterility or reproductive failure in the emergence of a species, which focused on the interaction among groups of individuals, Mayr proposed a simpler interruption of interbreeding by geographic separation. He defined what he called "the biological species definition" as follows:

> A species consists of a group of populations which replace each other geographically or ecologically and of which the neighboring ones intergrade or interbreed wherever they are in contact or which are potentially capable of doing so (with one or more of the populations) in those cases where contact is prevented by geographical or ecological barriers.
>
> Or shorter: Species are groups of actually or potentially interbreeding natural populations, which are reproductively isolated from other such groups.

The difference between Mayr's species definition and, for instance, Dobzhansky's was that Mayr placed more emphasis on geographic separation than on physiological separation as leading to reproductive isolation. This would, perhaps, seem too restrictive. Dobzhansky's species concept, as did Fisher's, left open the possibility that barriers other than the most obvious—geographic separation—might lead to reproductive isolation.

The most important element in Mayr's view of speciation was the imposition of a geographic barrier between a subspecies—what he called incipient species—and the parent species. This is evident in the following comment: "A new species develops if a population which has become geographically isolated from its parental species acquires during this period of isolation characters which promote or guarantee reproductive isolation when the external barriers break down." It is in the context of this discussion that Mayr also came to admit that his "definition contains a number of postulates," including "the concept of the 'incipient species.' " In fact, he wrote, "[g]eographical speciation is thinkable only, if subspecies are incipient species." Clearly, this is tautological. At best, under these preconditions, the process of speciation is a self-fulfilling prophesy.

It appears that a major motivation in Mayr's perception and definition of species, and, indeed, in his depiction of the origin of species, was the discrediting of Richard Goldschmidt. As he stated in *Systematics and the Origin of Species:* "The fact that an eminent contemporary geneticist (Goldschmidt) can come to conclusions which are diametrically opposed to those of most other geneticists is striking evidence of the extent of our ignorance." From there on, Mayr went after Goldschmidt.

In rejecting Goldschmidt's systemic mutations in favor of geographic isolation in the origin of species, Mayr enlisted as support on his side of the argument "the taxonomist of birds, mammals, butterflies, and other well-known groups"—in short, virtually all naturalists and systematists. As for proof of the almost singular role of geographic isolation in speciation, Mayr asked, "What is proof?" "Is it not," he continued, "sufficient to point out, as we have done, that the majority of well-isolated subspecies have all the characters of good species and are indeed considered to be such by the more conservative systematists?" Clearly, the implication is that the weight of received wisdom—derived from examples provided by "the more conservative systematists" that subspecies should be regarded as incipient species—was deemed sufficient to negate the suggestions of the few less conservative evolutionary biologists.

In addition to dismissing Goldschmidt's theories and research results by an appeal to a differing majority, Mayr sought to discredit him in other ways. For example, he accused Goldschmidt of purposefully manipulating differences between subspecies of gypsy moth in order "to deceive a specialist of moths." In attempting to undermine totally Goldschmidt's research on a particular species of gypsy moth, *Lymantria dispar*—research that had been accepted by other geneticists, including Dobzhansky and Wright, even if they did not embrace his notions of systemic mutations—Mayr went so far as to state that one should not trust his "sweeping generalizations" because "the taxonomy of the genus is in a state of complete disorder, since the family to which it belongs has at the present time no specialist and the treatments in the leading catalogues . . . are nothing better than uncritical compilations." True, Goldschmidt did use the currently available taxonomy of the larger grouping of species of *Lymantria.* But whether two other species attributed to the genus *Lymantria* are closely related to the species *L. dispar* is irrelevant to Goldschmidt's studies on development and variation within the species *dispar* itself.

The ad hominem attack on Goldschmidt that permeates *Systematics and the Origin of Species* served as the backdrop for Mayr's assertions about what evolutionary biologists really do know. For instance, after discussing de Vries's mutation theory—which was an early formulation of major change deriving from mutation—Mayr commented that, since then, "the species concept has been clarified by the taxonomist, and we know now that species differ by so many genes that a simple mutation would, except for some cases in plants, never lead to the establishment of a new species." From these comments, Mayr went on to discuss Goldschmidt's theory of species originating from systemic mutations:

> To him [Goldschmidt] a species is like a Roman mosaic, consisting of thousands of bits of marble. A systemic mutation would be like the simultaneous throwing out of all the many thousands of pieces of marble on a flat surface so that they would form a completely new and intelligible picture. To believe that this could actually happen is, as Dobzhansky has said in review of Goldschmidt's work, equivalent to "a belief in miracles." It seems to me not only that Goldschmidt did not prove his novel ideas, but also that the existing facts fit orthodox ideas on species formation so adequately that no reason exists for giving them up.

Goldschmidt was fully aware of the consequences of his suggesting major organismal reorganization through systemic mutation. He knew that this was a hit-or-miss proposition in terms of the bearer of the mutation—his "hopeful monster"—being able to survive in the environmental circumstances in which it found itself. But he was also aware that there were certain constraints on reorganization imposed by the preexisting genetic material. He was, after all, committed to the interplay between genetics and development in determining the final outcome of the individual. Mayr's Roman-mosaic analogy took no account of Goldschmidt's deliberations. Nor did Mayr's appeal to the correctness of orthodoxy.

In light of Mayr's assault on Goldschmidt for suggesting that species might originate abruptly rather than gradually, it is curious that his discussion of Schindewolf was by comparison so benign. Mayr even suggested that Schindewolf and Goldschmidt only appeared to be similar in their concept of how species originate. In reality, he said, they were not. Schindewolf, according to Mayr, did not subscribe to the notion that species arise abruptly or, as the latter put it, in step fashion. It is also surprising that it is in the section on Schindewolf that Mayr made the generalization that "[p]aleontologists have too many examples of perfect evolutionary series, leading without obvious breaks from species to species in subsequent horizons, to believe in the instantaneous creation of species." It was, of course, the lack of clear-cut cases of smoothly transitional, gradual evolutionary continua in the fossil record that provoked Schindewolf to view the gaps as conveying real information about evolutionary processes: that "[m]acromutations are the determining factors of evolution," meaning "that phylogenetic advance does not consist of increased differentiation and superficial accumulation of adaptive characters, but rather of profound transformations in the characters of basic organization." Schindewolf may have believed that profound macromutations gave rise to the ancestors of groups of novel kinds of organisms. But the ancestor of any new group, whether large (such as birds) or small (such as horses), would be recognized as a species.

Convinced, as Dobzhansky also was, that the study of processes of population dynamics within a species is sufficient to understand evolution at all levels, Mayr concluded *Systematics and the Origin of Species* with the following:

> [W]e may say that all the available evidence indicates that the origin of the higher categories is a process which is nothing but an extrapolation of speciation. All the processes and phenomena of macroevolution and the origin of the higher categories can be traced back to intraspecific variation, even though the first steps of such processes are usually very minute.

With these closing remarks, Mayr set the tone for what would be called the Evolutionary Synthesis.

The Grand Evolutionary Synthesis

The idea for the Synthesis was proposed by the geologist and invertebrate paleontologist Walter Herman Bucher, in 1941, at the annual meeting of the Geological Society of America. As reported in 1949 by Glenn Lowell Jepsen (Jep to his colleagues), the Sinclair Professor of Vertebrate Paleontology at Princeton University, Bucher "suggested that some of the riddles of evolution, the most significant and basic fact of biology and paleontology, might be solved by a synthesis of the two subjects." According to Jepsen, Bucher was concerned that geneticists "have little time for, and less interest in, the body of solid fact which . . . morphology, taxonomy, and paleontology are accumulating" and that, in turn, paleontologists have a "distrust in the long range significance" of genetics. Embryologists and developmental biologists would not be represented in the Synthesis.

Discussions among geneticists, paleontologists, and systematists that would lead to the Synthesis began in earnest on October 17, 1942, with a meeting in the library of the Department of Zoology at Columbia University. As Jepsen summarized the sentiment among the group: They agreed that paleontologists should learn more about the techniques and the essential results of genetics, that geneticists should become better acquainted with the degree of reliability of stratigraphic and paleontologic methods and the validity of generalizations proposed by paleontologists, and that specialists in the several evolutionary disciplines should learn each other's languages." The plan was that different regional sections of what was called the Committee on Common Problems of Genetics, Paleontology, and Systematics would first meet and then there would be a national meeting of all representatives. Unfortunately, the Second World War intervened between the first regional meetings of 1943 and the larger meeting of 1947, which actually became international. The papers that were presented at the latter meeting were published in 1949, in a volume entitled *Genetics, Paleontology, and Evolution,* edited by Jepsen (chair of the section on paleontology), George Gaylord Simpson (chair of the entire committee), and Ernst Mayr (chair of the section on systematics).

But while the world war may have interrupted the more formal gatherings of those elected to this austere committee, it did not prevent the vertebrate paleontologist George Simpson from trying to do his part in bringing the language of genetics and systematics into paleontology. In 1944, Simpson, who held positions at the American Museum of Natural History and at Columbia University, published *Tempo and Mode in Evolution.*

Experience as a paleontologist, particularly as a vertebrate paleontologist, gave Simpson a slant on evolutionary problems that was different from that of a population geneticist or a systematist working on living organisms. There was, of course, the ever-present question of whether gaps in the fossil record were real, in which case one could end up thinking like Schindewolf, or whether they were artifacts of preservation and merely a nuisance to the business of figuring out the past. Writing six years before Schindewolf published his major opus, Simpson was convinced of the latter scenario:

> When breaks or apparent saltations do occur within lines that are true or structural phyla [broadly, lineages], frequently they can be shown to be due to one of the two causes now exemplified: to hiatuses in the time record caused by nondeposition of middle strata or fossils and to sampling of migrants instead of main lines. Continued discovery and collecting have the constant tendency to fill in gaps. The known series are steadily becoming more, never less, continuous. It cannot be shown that discontinuity between, let us say, genera has never occurred, but the only rational conclusion from these facts is that no discontinuity is usually found and that there is no paleontological evidence that really tends to prove that there is any.

In his attempt to reconcile the discipline of paleontology with the dominant field of genetics, Simpson felt confident in proclaiming that "[t]he paleontological record is consistent with the usual genetical opinion that mutations important for evolution, of whatever eventual taxonomic grade, usually arise singly and are small, measured in terms of structural change." In discussing this topic more fully, Simpson took the position that, though it is impossible to rule out saltation, "[t]here is . . . abundant and incontrovertible paleontological proof that saltation does not always occur . . . that continuity . . . or gradual intergradation commonly occurs between what are certainly good species and genera." In a footnote to this latter statement, Simpson took issue—one of many—with Goldschmidt and Schindewolf. Given his belief in the general picture of evolution as one of gradual change, Simpson rejected Goldschmidt's denial of what he, Simpson, and paleontologists in general knew to be "fact." As for Goldschmidt's citation of Schindewolf as providing paleontological support of the former's thesis, Simpson declared that this was not the case. In fact, Simpson stated, his impression of Schindewolf's work (of 1936) was that it contradicted Goldschmidt's claims. But when, according to Simpson, Schindewolf's conclusions did mirror Goldschmidt's, they were "at wide variance with the consensus of paleontologists and even with some of its own author's other works." A case in point here is the evolution of the horse. Schindewolf thought the fossil record provided evidence of saltatory evolution. Simpson argued that a smooth curve could describe the same data.

As was becoming increasingly articulated among those who represented the disciplines of the Synthesis, Simpson mirrored Dobzhansky and Mayr in believing that the same processes involved in microevolution were at work in evolution at higher taxonomic levels. Simpson's outline of speciation and the origin of groups of new kinds of organisms was also compatible with general models proposed by geneticists and systematists, but with the added twist that it could also be rapid more frequently than a gradual picture promised:

> The typical pattern involved is probably this: A large population is fragmented into numerous small isolated lines of descent. Within these, inadaptive differentiation and random fixation of mutations occur. Among many such inadaptive lines one or a few are preadaptive, i.e., some of their characters tend to fit them for available ecological stations quite different from those occupied by their immediate ancestors. Such groups are subjected to strong selection pressure and evolve rapidly in the further direction of adaptation to the new status. The very few lines that successfully achieve this perfected adaptation then become abundant and expand widely, at the same time becoming differentiated and specialized on lower levels within the broad new ecological zone.

Simpson's thoughts on speciation derived in large part from Sewall Wright's. In fact, Simpson specifically discussed Wright's selective topographic fields, describing it as a landscape of hills and valleys. Simpson interpreted the hills as representing positive selection and the valleys as representing negative selection. He also saw each feature of this landscape as representing a different ecological subzone. In using this image to explain the evolution of various groups, including the horse, Simpson suggested that the course between valley and peak was a period of instability, the crossing of which could, in turn, lead to a rapid evolutionary shift, which would lead to the rise of a new group. Simpson identified this particular phenomenon as quantum, or explosive, evolution—explosive also in the sense that it could lead to increased rates of differentiation of a group. Subsequent to a spurt of quantum evolution and the rapid appearance of a group, more normal kinds of evolutionary rates and selective and adaptive pressures would prevail in the evolution of descendants within the newly evolved group. Although Haldane had earlier conceived of a similar model of evolutionary change—rapid evolution of a new species from a peripheral isolate, the establishment of a new group, and then slower evolution of the descendant species within the group—Simpson did not cite him in this context.

In a basic way, Simpson's scheme does not sound too dissimilar to Schindewolf's. The major difference is that Simpson invoked selection and adaptation in modeling his concept of speciation leading to the evolution of larger groups. But Wright was not pleased with Simpson's depiction of his selective topographic field idea. He thought Simpson was just plain ignorant of his other evolutionary proposals. In a review of Simpson's book, Wright took the paleontologist to task for relying too heavily on mutation as an instrument of change, whereas different genetic combinations could also have a similar impact; for dismissing intergroup competition or interaction as a possible source of change; for generalizing about rates of evolution at the level of the species when the data were not forthcoming from paleontology; for suggesting that speciation would occur with the complete, rather than partial, isolation of small populations; and especially for oversimplifying his selective topographic landscape so that peaks and valleys represented only ecological subzones, and only one of each, for that matter. In general, Wright was unconvinced that paleontology contributed much to the understanding of evolution at the lower levels of species and subspecies. Species, and especially subspecies, were, of course, the focus of population geneticists:

> Paleontology has little to contribute to the distinction between micro and macroevolution of Dobzhansky's definition in which the dividing line was that between the origins of subspecies and species. . . . For the paleontologist, the distinction between small scale and large scale evolution is rather that between the origins of genera and of families.

Since one of the possible contributions to evolutionary studies that stemmed from paleontology was an insight into evolutionary rates through the study of change as represented by the fossil record, Wright's review clearly removed paleontology from the most basic level of evolutionary inquiry: the origin of species. If the accuracy of paleontological resolution was restricted to higher taxonomic levels, it was only through a uniformitarian extrapolation of population genetic studies into the past that rates of species change and origin could be discussed.

The impact of Wright's review on Simpson is evident in the latter's chapter on rates of evolution in animals in the volume that he co-edited with Jepsen and Mayr, *Genetics, Paleontology, and Evolution.* Simpson did not mention Wright once. In his discussion of the different sources of information on evolutionary rates, he had this to say about genetics: "Genetic rates of evolution underlie most other rates and are more basic than most, but their direct study is so limited in scope that little will be said regarding them here." As far as Simpson was concerned, only morphologic and taxonomic rates of evolution could be studied with any meaning. Morphologic rates could reflect change "in a single character (e.g. length of skull), in a complex of characters (e.g. the dentition as a whole), or in the whole organism." As for taxonomic rates, he wrote:

> Taxonomic rates are concerned with the rate of origin of new taxonomic categories at any level, from subspecies to kingdom. . . . The taxonomic categories are almost always recognized and defined on a morphologic basis and are in the best modern practice, at least, assumed to have a genetic basis and made to correspond as nearly as possible with genetic groupings. The use of taxonomic rates may thus be viewed as a device for indirect study of genetic and morphologic rates in terms of the whole organism, rates not amenable to direct study by methods now available. This makes the taxonomic rates in many respects the most interesting of all.

The bulk of Simpson's article was concerned with how to calculate different evolutionary rates, which, he believed, were not always uniform. He continued to promote his theory of quantum evolution, and to allow for periods of speeding up as well as of slowing down in the emergence of and diversification within new taxonomic groups. One generalization that Simpson felt comfortable making was based on his sense that the smaller the taxonomic category (e.g., subspecies), the faster the rate of evolution. Conversely, the more inclusive the taxonomic category (e.g., family), the slower the rate of evolution. Of course, since taxonomic categories are at base arbitrary constructs, this kind of exercise must also be viewed with caution.

In spite of Walter Bucher's good intentions, paleontology and genetics would never mesh seamlessly. Although paleontologists would argue that they could identify species and follow evolution at this level, the upshot of the Synthesis was that there would always be a reliance on genetic models of evolutionary change and speciation. As such, even when it was allowed that there could be periods of more rapid or slower evolutionary rates, the change envisioned was based on the accumulation of micromutations and the natural selection of the small variations they produced. It was only a matter of how quickly or slowly these small changes accumulated. Since there would always be gaps in the fossil record, it would forever be suspect in providing support for gradual, or at least minutely cumulative, evolutionary change. Rather, it was the acceptance of slow, or at least minutely cumulative, evolutionary change that served to inform paleontology. With paleontology excluded from the realm of investigating the origin of species, Simpson and others focused more and more on the broader pictures of evolution, such as the appearance, diversification, and extinction of larger groups.

This, then, was the legacy of the Synthesis.

11

Toward a New Evolution

> *So far from a gradual progress towards perfection forming any necessary part of the Darwinian creed, it appears to us that it is perfectly consistent with indefinite persistence in one state, or with a gradual retrogression.*
>
> —Thomas Henry Huxley (1876)

After the Synthesis

Simpson, Haldane, and Wright represented a small group of individuals involved in the formulation of the Evolutionary Synthesis who were not unreserved proponents of some of its basic tenets. Of note was that these evolutionary biologists were not wedded to a picture of constant gradual change.

Haldane appears to have been made most peripheral by those in, as well as off center of, mainstream neo-Darwinism. His proposal that small groups of peripheral isolates should be the stuff of rapid speciation received only cursory mention by his contemporaries. His thoughts on gene-development–environment interaction, with a major component of heterochrony thrown in, also received little notice in the formulation of the Synthesis. And, certainly, his openness to the possibility that reduced selection pressure might be a factor in rapid character change and speciation was not in line with the role other evolutionary biologists assigned to selection. Generally, Haldane was cited for disagreeing with Fisher about whether change could occur within large, freely and randomly mating populations.

Simpson and Wright did not embrace the notion that the major cause of speciation was the imposition of a geographic barrier between a subspecies and the rest of its species. As for the genetic basis of evolutionary change, through his last publication in 1982, Wright steadfastly maintained that it was more than just the accumulation of small mutations and their effects. To the end, he was convinced that shifts in the balance of interactive gene combinations, as well as, for example, chromosome rearrangement, could lead to evolutionary change, both in transformation within a species lineage and in the divergence of a new species from its parent. Simpson stuck to the more neo-Darwinian geneticist's position that change was a matter of the accumulation of small changes. He did so, it seems, because of his continued rejection of Goldschmidt's and Schindewolf's theories, which invoked a single mutation as the basis of major organismal reorganization.

Although in his book of 1953, *The Major Features of Evolution,* Simpson took a swipe at Wright—in particular, at Wright's model of intergroup competition between subspecies within a large, widespread species as providing the impetus for evolutionary change of significance—the two scholars were, nevertheless, more theoretical bedfellows than archenemies. Both believed that shifts from one environmental situation to another could bring about relatively rapid evolutionary change. This, of course, was the basis of Simpson's theory of quantum evolution: A species that finds itself in a new ecological circumstance quickly adapts to it. Quantum evolution was actually a rewriting of Haldane's theory of peripheral isolates, but without the latter's emphasis on development. Wright's theories of shifting balances and selective topographic landscapes invoked the concept of rapid adaptation to new environmental circumstances as well as the possibility that different interactive gene combinations could be equally workable from an adaptive standpoint.

Simpson and Wright also serve as examples of the inability of the Committee on Common Problems of Genetics, Paleontology, and Systematics to meld these disciplines completely. Simpson worked hard in his writing to invest his paleontologically informed theories with what he thought were the salient aspects of Haldane's, Wright's, and especially Fisher's mathematical modeling of mutation rates and selective forces. But geneticists in general maintained a strong skepticism about the reliability of the fossil record in demonstrating speciation and, consequently, paleontologists' ability to understand how such evolutionary change occurs. A paleontologist could only point to fossils and remark on the fact that change had occurred. Beyond that, and depending on how good the fossil record of a particular group was, paleontologists might at best be able to address rates of evolutionary change and provide a rough picture of the times of origin and extinctions of major groups of fossilizable life.

The persistence of an intellectual hierarchy within evolutionary biology is reflected as recently as 1982, in one of Wright's articles:

> I am not in a position to discuss independently the data of paleontology and recognize that my field, genetics, bears directly only on microevolution, but I feel that we should explain phenomena at the higher levels as far as possible, as flowing from observed phenomena of genetics in the broad sense, including cytogenetics, before postulating wholly unknown processes.

Although Wright did not totally reject other disciplines, it is clear that he was reiterating the common belief that an understanding of evolutionary change at all levels ultimately derives from the studies of the laboratory-based population geneticists, who seemed to be able to demonstrate, and mathematically model, the effects of selection acting on the small mutations that arose. In the face of such assertions, what could the paleontologist do but accede to the majority view?

In the spirit of the Committee on Common Problems, Simpson accepted the premise that "[t]he essential picture of evolution under a Neo-Darwinian concept is that of a slow steady change through the gradual building up of minor modifications; over long periods of time animals become more perfectly adapted to a stable environment or change gently with slowly changing surroundings." This made sense, at least in terms of observations derived from population genetics. But, as a paleontologist, Simpson also saw rapid change recorded in the fossil record. However, because he was unable to embrace saltation of any sort, Simpson modified the typical neo-Darwinian scenario of evolutionary change to the following: "In rapid evolutionary changes in animal lines the process may have been a typically Neo-Darwinian one of the accumulation of numerous small adaptive mutations, but an accumulation at an unusually rapid rate." "Unfortunately," he had to admit, "there is in general little evidence on this point in the fossil record, for intermediate evolutionary forms representative of this phenomenon are extremely rare (a situation bringing smug satisfaction to the anti-evolutionist)." The gaps in the fossil record reared their ugly heads again.

It is to Ernst Mayr that most evolutionary biologists would probably turn as the one person whose notions of evolutionary change most affected the field in general. It was he, it seemed, who could bridge the gaps between the conclusions of geneticists and those of paleontologists. As a systematist trained in the study of living organisms, Mayr had access to an organism's entire biology. Paleontologists had only the typically fragmentary and incomplete bones and teeth of animals whose remains happened to end up in the right circumstances for becoming fossils, and then being discovered. As someone who studied whole, living organisms, Mayr was seemingly in the position of being able to apply the theoretical conclusions of geneticists to real-life, still-ongoing situations. As such, he was also someone who could translate the writings of Fisher and Wright for the vast majority of mathematically naive evolutionary biologists. Although Simpson would be recognized for bringing the rigor of statistical analysis to paleontology and systematics and making its applicability to fossil samples understandable, his discussions of the theoretical population geneticists' mathematical arguments were often beyond the grasp of his fellow systematists.

Mayr seemed to have a gift for explaining even the most complex ideas. He also published his evolutionary ideas through widely differing outlets, from the more rigorously argued *Animal Species and Evolution* to the simpler and more broadly accessible *Populations, Species, and Evolution*. Almost anyone interested in evolution could read the latter book, even if it required a bit more effort than the typical college or trade book. *Popula-*

tions, Species, and Evolution and Simpson's *The Meaning of Evolution* were the two major books on evolution that I had to read as a college student in the late sixties.

Mayr's basic message was the same throughout his work, whatever the actual publication: With the exception of chromosome multiplication in various plants and other aberrant phenomena, "[s]peciation proceeds through the gradual genetic modification of spatially segregated populations." The way that species formation most frequently took place was by the introduction of a physical or geographic barrier between a subspecies and the rest of its species. In general outline, Mayr followed Fisher's idea that a large population would be segmented into subsets, which, being separated from one another, would accumulate minute changes over a long period of time until they became reproductively isolated—that is, separate species. But Mayr did not put all his evolutionary eggs into one basket. He also allowed that, at times, rapid change was possible and that species could arise in ways other than those he favored. In spite of his insistence on speciation via gradual transformation, he even approached Haldane's model of species formation in his discussion of peripheral isolates, which, unlike Sewall Wright's peripheral populations, were truly separated from the parent species:

> Peripheral isolates, no matter how close to the main range of the species, almost always are noticeably different, in contrast to the essential uniformity in the contiguous range of the species. . . . The rapidity with which morphological changes take place in peripheral isolates confirms our conclusion that the genetic reconstitution permits or induces shifts in the previously existing developmental homeostasis. . . .
>
> The environment in the peripheral isolate is almost always rather unlike the optimal environment of the species in the center of its range. The biotic environment, in particular, is usually somewhat unbalanced at isolated locations. The new isolate will thus be exposed to a considerably changed selection pressure. . . . [T]he response of the isolated population to this selection pressure will be quite different from that of a population which is part of a contiguous array of populations held together by gene flow and [other] cohesive devices. . . . [T]he isolated population can respond to its local adaptive needs without having to compromise with the solutions found by other populations.
>
> The only answer that is possible in many cases is a shift into a new niche. Such a shift is greatly facilitated by a genetic revolution and the special properties of isolated populations. In particular, the genetic liability of such populations and the pronounced population fluctuations (in the absence of strong density-dependent factors) facilitate such shifts. In no other situation in evolution is there a greater opportunity for adaptive shifts or evolutionary novelties.

Mayr's concept of the role of peripheral isolates differed from Haldane's in that he did not specify rapid change as part of the process of species formation. Mayr also emphasized the importance of the environment much more than Haldane did, and he did not consider developmental processes at all. In fact, Mayr envisioned speciation as being possible and successful only when there were vacant ecological niches to be filled: "Most habitats are saturated with species at any given time, and there is room only for so many new species as are needed to fill newly opened niches." Without citing Haldane, Mayr concluded that, since "[m]ost species bud off peripheral isolates at regular intervals," their "continued presence . . . is a guarantee of the occurrence of speciation whenever the ecological situation is opportune." What is confusing at times when one reads Mayr's work is that he often used the more generic phrase "geographic isolate." Since he was a proponent first of geographic isolation as the key to the process of species formation, one cannot equate peripheral isolate with geographic isolate. Peripheral isolate is only one example of geographic isolate.

As for the process of evolution in general, Mayr reiterated the neo-Darwinian position that "evolution is a two-stage phenomenon: the production of variation and the sorting of the variants by natural selection." Following the lead of the population geneticists, Mayr accepted that variation was produced by small-scale random mutation. He also embraced Darwin's utilitarian argument for the selection of advantageous traits, as well as the latter notion that the advantageous traits selected in some way constituted an improvement over what had been the previous state. This is evident, for instance, in his response to those in the field who objected to the neo-Darwinian view of evolution:

> Whether function precedes a structure, or vice versa, gives rise to eternal argument. Did finches develop heavy bills because they ate seeds, or are finches able to eat hard-shelled seeds because they had developed heavy bills? The answer is, of course, that neither is correct. . . . The development of the heavy bill was a slow process, probably involving dozens of small mutational steps, each one surviving only if proving its usefulness in the actual test of selection. It must be admitted, however, that it is a considerable strain on one's credulity to assume that finely balanced systems such as certain sense organs (the eye of vertebrates, or the bird's feather) could be improved by random mutations. . . . However, the objectors to random mutations have so far been unable to advance any alternative explanation that was supported by substantial evidence.

Whenever Mayr spoke, evolutionary biologists, whether student or seasoned professional, listened. The weight of his pronouncements—such as the last sentence in the quote immediately above—was not inconsequential. However, proclamation and reality are two different matters. There was no substantial evidence to support the scenario of gradual finch-bill evolution. Nevertheless, Mayr's particular presentation of the evolutionary process was consistent with the uniformitarian approach of extrapolating from the laboratory experiments of population geneticists to the time-depth evolution required. But there is an underlying contradiction in Mayr's assertion. For, as he discussed elsewhere, there was a significant difference between the "closed" populations that population geneticists studied and the "open" populations that existed in nature:

> Population geneticists, who have worked all their lives with closed populations in which all genetic input is due to mutation, tend to underestimate the magnitude of genetic input in open populations. To be sure, it is immaterial for certain aspects of evolution whether mutation or immigration is responsible for new genes in a population. Yet it would be a great mistake to lump these two sources of variation in calculations of their effect, because they are of totally different orders of magnitude.

Although this statement was part of a criticism that Mayr levied at Wright's model of speciation via partially isolated peripheral subspecies—because he was critical of Wright's belief that partial rather than complete isolation can lead to species formation—it is not dissimilar in content to the latter scholar's insistence that interactive gene combinations play a major role in evolutionary change: That is, in addition to mutation (which population geneticists studied), new gene combinations also produced variation, if not also evolutionary change. And, as Wright envisioned it, one way in which new gene combinations could arise was for one subspecies to migrate into an area occupied by another subspecies of the same species. This, of course, would not lead to the emergence of a new species through a process of divergence. Rather, it would return an incipient species to the fold of the parent species. But it could, theoretically, lead to the generation of different gene combinations from the already existing gene pools of the subspecies.

The topic of human origins was not exempt from Mayr's evolutionary speculations. In 1950, he made the case that, of the many genera that had by then been proposed for various hominid fossils, only two were theoretically viable and taxonomically necessary. As a result, *Homo* and *Australopithecus* became the hominid genera of record, even though, Mayr admitted, he had his doubts about giving genus status to *Australopithecus*. By 1963, however, he was able to state: "I now agree with those authors who have since pointed out not only that upright locomotion was still imperfect but also that the tremendous evolution of the brain since *Australopithecus* permitted man to enter so completely different a niche that generic separation is definitely justified." The importance of the ecological niche in providing a space into which the first species of *Homo* could invade and become differentiated is evident in Mayr's thinking. As for not recognizing Robert Broom's *Paranthropus* as a distinct taxonomic entity above the species level, Mayr simply stated that this hominid "hardly shows the degree of difference from *Australopithecus* necessary to justify generic status." In 1963, and from their respective genetic and paleontological perspectives, Dobzhansky and Simpson also came to agree with Mayr's restriction of hominids to only two genera.

Not surprisingly, Mayr was convinced that the general course of human evolution had been a gradual one. In its details, bipedalism and the freeing of the hands evolved first. Then there was the reorganization of the pelvic girdle and the limbs. Toward the end of this phase came an increase in brain size and changes in the architecture of the skull.

As for the species of hominid at each of these perceived phases of human evolutionary change, Mayr was a bit ambiguous about the number of species that he thought represented *Australopithecus,* but the sense he gave was that it was probably two and, at most, three. Mayr mentioned by name no species other than Raymond Dart's *africanus,* which he suggested "might well have been one of the more extreme and aberrant races of the species." His taxonomic minimalism extended to the species of the genus *Homo* as well. He mentioned only in passing the new species that Louis Leakey and his colleagues had discovered at Olduvai Gorge and that they placed in the species *Homo habilis.* Although Mayr did not cite *H. habilis* by name, he accepted it as representing early *Homo.* From there, human evolution flowed directly into *Homo erectus* and then *Homo sapiens.*

The rationale for Mayr's view of human evolution lay in his belief that "since Recent Man is a polytypic species [a species of many varieties, or races] and since most species of mammals are polytypic, it can be assumed that the species of fossil hominids likewise were polytypic." Because, in Mayr's theoretical framework, the success of speciation depends on invading a vacant econiche, and hominids, in his view, had always been both widespread geographically and diversely adapted ecologically, there had been little opportunity for speciation. In addition, according to Mayr, "isolating mechanisms in hominids apparently develop only slowly." Consequently, although there may have been potential isolates, "isolation never lasted sufficiently long for isolating mechanisms to become perfected."

Clearly, these are not morphologically based conclusions about the taxonomy and evolutionary history of hominids. Rather, Mayr's theoretical constraints as to how he thought about evolution and speciation drove his interpretation of the morphology. Consequently, a diversity of often markedly morphologically dissimilar fossils were committed to the same species. Of course, this then created the impression that the degree of variation within a particular hominid species was so extraordinary that it surpassed the norm for all other mammals. But, as Mayr remarked, since present-day *Homo sapiens* is such a highly variable, or polytypic, species, "as concerns the fossil hominids, the simplest assumption would be that at any given time only a single polytypic species of hominid existed, and that the variety of observed types is merely a manifestation of individual and geographic variation." This particular scheme of human taxonomy and evolution came to dominate the field of paleoanthropology, and of anthropology in general.

The Post–Neo-Darwinian World

Although the paleontologists and systematists who contributed to the Evolutionary Synthesis were greatly influenced by the work of population geneticists, their concerns with evolutionary change were directed primarily at the level of the species. True, there was an awareness of processes within a species. But the focus was on the formation of species in terms of the questions "how come, and how fast?" The words *mutation, selection,* and *adaptation* were used, but usually in the context of "the organism versus the environment."

In 1966, George Christopher Williams, a professor of biology at the State University of New York at Stony Brook, published *Adaptation and Natural Selection: A Critique of Some Current Evolutionary Thought.* According to Williams, he began writing the book in 1963, which happened to be the year Mayr's *Animal Species and Evolution* appeared in print. As the title of his book suggests, Williams planned to address those aspects of Darwinian theory that had been omitted from the Synthesis. Clearly, while the leaders of the Synthesis were pursuing a solidification of their evolutionary positions, Williams and others like him were attempting to fill in the still-vacant lacunae in evolutionary theory. Although Williams's monograph was also written in part as a response to various evolutionary biologists who were not content to explain everything evolutionary in terms of selection arguments, it became the centerpiece of a new neo-Darwinian effort, which would become known as sociobiology.

The essence of Williams's thesis was that the organic world can be reduced to the level of genes alone, and that natural selection, pretty much as conceived by Darwin, operates in the realm of the available alternative alleles, or genetic character states. Williams's reason for focusing on genes rather than on the entire organism, or some subset of its anatomy or behavior, was that while organic bodies come and go, genes are essentially immortal. Genes are self-replicating and have the potential to be passed on through an endless number of generations of offspring. Admitting that "the gene" was still an abstract concept in population genetics, Williams conceived of it as the indivisible, or ultimately unfragmentable, unit of inheritance. As for chromosomal rearrangement, he argued that the rearranged chromosome fragment "behaves in a way that approximates the population genetics of a single gene." Putting this all together, Williams defined the gene as "that which segregates and recombines with appreciable frequency." He also reminded his audience that, as pre-Synthesis geneticists had learned, a gene could produce more than one effect. The sum total of the genes of an individual—its genotype—could be expressed differently in different individuals because, according to Williams, the genotype is interpreted by the "soma" of the individual.

Williams also reduced the concept of natural selection to the level of the gene. He did so through the simple mechanism of Mendelian inheritance. In sexually reproducing organisms, with each parent contributing half of its offspring's genes, natural selection would be faced with a choice between two different alleles of the same gene. By choosing from among the available, alternative alleles, there would be differential survival and, consequently, selective accumulation of genes. Ultimately, this selection of genes could have quite an impact on organisms. Although it is never stated fully, the implication throughout Williams's book is that even if the pace of accumulation were rapid, each allele so accumulated would have only a small effect. In this sense, the shift in gene frequencies would be a slow and gradual process. As for adaptation, Williams felt that "[N]atural selection would produce or maintain adaptation as a matter of definition. Whatever gene is favorably selected is better adapted than its unfavored alternatives. This is the reliable outcome of such selection, the prevalence of well-adapted genes."

Since a gene is ultimately expressed through the physical being, or phenotype, of the organism, in order for it to be selected it must be responsible for something that enhances the reproductive success of the organism. In other words—recasting concepts as they had first been proposed by Darwin in the light of population genetics—a well-adapted gene is one that contributes to the individual's reproductive success.

In dealing with Darwin's notion of competition, Williams admitted that while "natural selection works only among competing entities . . . it is not necessary for the individuals of a species to be engaged in ecological competition for some limited resource." In fact, he argued, selection could be most intense during the early phases of a population's expansion, when available resources might not be scarce. What is at stake, ultimately, is mean phenotypic fitness, which boils down to how many offspring an individual can produce and, consequently, how successful an individual will be in spreading its genes throughout subsequent generations. But it is not the absolute number of offspring an individual sires that is important. Fitness is judged on the basis of how many offspring an individual produces relative to the average number of offspring produced by the individual's population or species. An individual that sires more offspring than the average for its population would be doing well within this view of selection and adaptation. If, then, adaptation is thought of in terms of reproductive success, with genes being the ultimate focus of selection, the phenotypic traits of importance would be those that enhanced the gene's chances of being transmitted to as many offspring as possible. Although physical characteristics can be accommodated in this model, so, too, can behavioral, including social, attributes.

In Williams's formulation, natural selection serves to maintain stability. This would make sense in a context in which natural selection chooses among alleles that are already present within the species and available to selection through the phenotypes of individuals. On the other hand, mutation, according Williams's theory, acts to disrupt this stability. This suggestion seems to imply that a mutation is expressed via the phenotype shortly after it has arisen. Nevertheless, Williams had to admit, "[m]utation is . . . a necessary precondition to evolutionary change," which "takes place, not so much because of natural selection, but to a large degree in spite of it."

Williams's speculations about how evolution would be enacted over a period of a million years is exemplified in the following passage:

> The important process in each such period was the maintenance of adaptation in every population. This required constant rectification of the damage caused by mutation, and occasionally involved gene substitutions, usually in response to environmental change. Evolution, with whatever trends it may have entailed, was a by-product of the maintenance of adaptation. At the end of a million years an organism would almost always be somewhat different in appearance from what it was at the beginning, but in the important respect it would still be exactly the same; it would still show the uniquely biological property of adaptation, and it would still be precisely adjusted to its particular circumstances. I regard it as unfortunate that the theory of natural selection was first developed as an explanation for evolutionary change. It is much more important as an explanation for the maintenance of adaptation.

Williams was very clear about how he read the work of paleontologists and systematists, and even population geneticists who were interested in processes that went beyond the level of the individual. As he saw the situation, these evolutionists were discussing only speciation, long-term morphological change, and environmental or ecological adaptation. Williams believed that these evolutionary biologists were not using "[t]he principle of natural selection . . . in an adequately disciplined fashion." As far as he was concerned, "[m]ost of the conclusions on patterns of speciation would be much the same whether based on Lamarckian, nineteenth-century Darwinian, or modern genetic concepts." Just because a publication on speciation contained the words *selection, mutation,* or *gene flow* did not mean that it was "conceptually much advanced beyond what Lamarck or Darwin might have written." In fact, Williams continued, "Darwin's or even Lamarck's concepts form a perfectly adequate basis for explaining most of the phenomena of systematics."

Williams adopted the term *teleonomy* for the discipline he was developing: the study of adaptation as informed by a particular theory of gene selection. Although *teleonomy* was later replaced by *sociobiology,* the focus remained the same: "Its first concern with a biological phenomenon would be to answer the question: 'What is its function?' " The answer to this question came by defining natural selection as "the differential survival of alleles." The simpler the explanation to the question, the stronger, Williams argued, would be the strength of this theory of natural selection.

Although the goal of simplicity may lead to a better, or at least a clearer, explanation, it does not constitute a test of the theory of natural selection itself. Nonetheless, it is important to realize that if, according to Williams, natural selection strives to maintain stability, then the picture of evolution that emerges is one in which nothing of significance happens until it is disrupted by mutation. Natural selection may be accommodating individuals to the vicissitudes of daily existence by choosing among the already available alleles, but it is not creating anything new. This does indeed appear to be what Williams meant when he so clearly distinguished between individual adaptation and the processes of evolutionary change that lead to speciation. In fact, he was even clearer on this point in a more recent book:

> The microevolutionary process that adequately describes evolution in a population is an utterly inadequate account of the evolution of the Earth's biota. It is inadequate because the evolution of the biota is more than the mutational origin and subsequent survival or extinction of genes in gene pools. Biotic evolution is also the cladogenetic [branching] origin and subsequent survival and extinction of gene pools in the biota.

This is certainly a different approach to neo-Darwinism than the one the predecessors and founders of the Evolutionary Synthesis propounded. They—from Morgan to Dobzhansky and Mayr—were invested in the applicability of observations on laboratory population genetics to microevolutionary processes and the expandability of microevolutionary processes to explain all aspects of evolution, from the individual to the largest of evolutionary groups. Following Fisher, in particular, the mottoes of the neo-Darwinism of the Synthesis was "variation furnished by random micromutation," and "change enacted by natural selection on this low-level variation." Because microevolution was slow and gradual, so, too, must be macroevolution.

But then came Williams, a neo-Darwinian who suggested that microevolution and macroevolution should be decoupled. Granted, Williams was no Bateson or early Morgan, or a Goldschmidt or Schindewolf. But he was not a promoter of the kind of neo-Darwinism that became incorporated into the Synthesis. Nonetheless, the intellectual offspring Williams sired often cite him as suggesting that his adaptationist program is explanatory of all levels of evolution. Perhaps this overgeneralization of Williams's intent is due to a confusion that arises when the word *evolution* is used to discuss two very different phenomena: the maintenance of adaptation versus the introduction of change. But, clearly, as the quotes above demonstrate, Williams was not promoting the kind of uniformitarianism that was followed by the neo-Darwinians who preceded him; namely, that of expecting the processes of micromutation within species to be of equal validity for understanding the origin of species.

Back to the Fossil Record

In 1972, a different approach to distinguishing between processes that are relevant at the within-species level and those that result in the origin of species came from two graduate students in invertebrate paleontology at Columbia University, Niles Eldredge and Stephen Jay Gould. The impetus for this particular joint effort came from the studies on a group of fossil invertebrates—the primitive, shrimplike trilobites—that Eldredge had been conducting.

A marine environment provides a better setting for a more reliable picture of what happened over a period of geologic time than does a terrestrial environment. Being in water facilitates fossilization. A dead body will probably sink to the bottom, and become covered by sediment, which will protect it. And small marine organisms tend toward large populations, so there will be a greater number of individuals contributing to the potential population of fossils. Consequently, marine invertebrates, such as trilobites, would have a much better chance of demonstrating the details, over geologic time, of their comings and goings geographically, and of their evolutionary changes, than, for example, would elephants or even most insects.

By following the representative fossils of different groups of trilobites over time, Eldredge was struck by how stable a particular group remained in its morphology. Moreover, it appeared that one population of trilobites with, typically, fewer lenses quickly replaced another population that, typically, had a greater number of lenses. The fossil history of trilobites did not conform to a model of gradual evolutionary change, with transformation via all possible permutations from one lens count to another. Rather, these fossils presented a totally different picture: long periods of relatively no change, and occasional rapid episodes of morphological change. To Eldredge, this paleontological observation suggested a particular mode of allopatric speciation, or speciation by way of geographic separation: allopatric speciation via peripheral isolates. It turned out that Gould, too, had been impressed by a similar picture of the fossil record of the fossil Bermuda snails he had been studying.

In their joint article, Eldredge and Gould introduced the concepts of phyletic gradualism, stasis, and punctuated equilibria. They used the phrase "phyletic gradualism" to encompass the kind of slow change via a succession of intermediates that, especially in paleontology, had become the image of how evolutionary change actually occurs—in spite of the frequent gaps in the fossil record that intrude upon this ideal. In this framework, gradually accumulating change can produce a continuum of transformation that does not increase species diversity, but it can also be involved in the origin of a species. In contrast, the picture of evolution that derived from the study of trilobites and Bermuda snails was that of long periods of stasis—no morphological change beyond minor individual variation—that were infrequently punctuated by abrupt or rapid episodes of population replacement. In addition to summarizing their work on trilobites and snails, Eldredge and Gould cited the fossil record of the horse, the extinct oyster *Gryphaea,* and another extinct invertebrate, the echinoid *Micraster,* as other examples of rapid, or punctuated, evolution.

By rapid evolution, Eldredge and Gould were not suggesting that a new species arises instantaneously, or even within a matter of generations. Rather, in everyday life—that is, in terms of ecological time—the process of speciation might not appear excessively rapid, even if one knew that it was taking place. However, in terms of evolutionary or geologic time, a new species would be seen to arise much more abruptly than could occur through a process of gradual transformation. Granting that there would always be cases in which the completeness of the fossil record of a particular group would be questionable, Eldredge and Gould suggested that the time had come to see the field of paleontology as contributing to an understanding of evolution. If the fossil record was not a woefully fragmented representation of a once-gradual picture of change, perhaps the gaps in the fossil record were real. If so, the gaps were consistent with a model of rapid speciation via peripheral isolates. That is, an isolate would not occupy the central region of the species; the emergence of a new species from a peripheral isolate could be rapid; and a species arising from a peripheral isolate that invaded the core territory of another species would create a discontinuous sequence of succession. If uniformitarianism can be characterized by the phrase "the present is the key to the past," then Eldredge and Gould's fossil approach could be summarized as "the past is the key to the present."

In suggesting that a model of punctuated equilibria might replace one of phyletic gradualism in representing the picture of evolution, Eldredge and Gould were not claiming that their speculations were correct. To the contrary, they were very open about their position:

> We readily admit our bias towards it and urge readers, in the ensuing discussion, to remember that our interpretations are as colored by our preconceptions as are the claims of the champions of phyletic gradualism by theirs. We merely reiterate: (1) that one must have some picture of speciation in mind, (2) that the data of paleontology cannot decide which picture is more adequate, and (3) that the picture of punctuated equilibria is more in accord with the process of speciation as understood by modern evolutionists.

Another model, or bias, of course, is gradual, or phyletic, evolution.

The reason Eldredge and Gould gave for focusing on a particular model of allopatric speciation—allopatric speciation via peripheral isolates—was that "[t]he central concept of allopatric speciation is that new species can arise only when a small local population becomes isolated at the margin of the geographic range of its parent species." Although Haldane had first provided an all-encompassing argument for viewing peripheral isolates as central to speciation, this was not Wright's or Mayr's first source of speciation, nor was it considered likely by such geneticists as Fisher and Dobzhansky. Interestingly, Haldane had formulated the essentials of punctuated equilibria—including using *Gryphaea* as an example—back in 1932. But, obviously, Haldane's suggestion became side-shelved during the Synthesis. When Eldredge and Gould revived this model, however, it could no longer be ignored.

In discussing the implications of their model of punctuated equilibria, Eldredge and Gould explored the property of homeostasis, or self-regulation, which had been discussed by the geneticist Isadore Michael Lerner. Lerner had distinguished between ontogenetic self-regulation, or developmental homeostasis at the level of the individual, and genetic homeostasis, or the self-regulation of populations. In the realm of developmental homeostasis, the heterozygous, or "different allele," condition tends to be more stable ontogenetically than the homozygous, or "same allele," state. As for genetic homeostasis, Lerner argued that natural selection would tend to favor the more or less average phenotypes for the group rather than the extremes. Consequently, the normal situation for a species would be the maintenance of stability, precisely because, Eldredge and Gould pointed out, "the basic property of homeostatic systems, or steady states, is that they resist change by self-regulation." "That local populations do not differentiate into species," they continued, "even though no external bar prevents it, stands as strong testimony to the inherent stability of species in time." But, Eldredge and Gould reiterated, peripheral isolates tend not to represent the average of their species. As such, peripheral isolates are more susceptible to the disruption of genetic stability, which, in turn, could lead to speciation. They concluded with the following:

> The norm for a species or, by extension, a community is stability. Speciation is a rare and difficult event that punctuates a system in homeostatic equilibrium. That so uncommon an event should have produced such a wondrous array of living and fossil forms can only give strength to an old idea: paleontology deals with a phenomenon that belongs to it alone among the evolutionary sciences and that enlightens all its conclusion—time.

Since Eldredge and Gould were not engaging in a discussion of adaptation and natural selection, it is not surprising that they did not cite G. C. Williams's 1966 book. But it is interesting how close to certain aspects of Williams's stated position Eldredge and Gould actually came, especially considering their very different perspectives as paleontologists. But from his particular perspective Williams had also argued for stability. The theory behind his argument, of course, was that natural selection acts to maintain stability. Although Williams did briefly discuss Lerner's suggestion that the heterozygous condition may actually serve to resist rapid evolutionary change, in the end he rejected any interpretation of heterozygosity—leading either to rapid evolutionary change or to stability—as providing an example of potential biotic adaptation. Nevertheless, even though Eldredge and Gould found Lerner's conclusions of potential importance for understanding evolutionary change while Williams did not, it is perhaps more important that the paleontologists and the adaptationist came to the same conclusion from totally different theoretical and methodological perspectives: that the major picture of evolution at the species level is that of stability. For Williams, mutation interrupts the state of stability that natural selection is trying to maintain. Also for Williams, natural selection and mutation alone do not explain the origin of species. But this is where Eldredge and Gould's resuscitation of Haldane's model of speciation via peripheral isolates comes into play, providing the extraordinary circumstances necessary for the emergence of something new.

The response to Eldredge and Gould's model of punctuated equilibria was mixed in the sense that it was either immediately embraced or adamantly rejected. One would have thought that many paleontologists, faced with the diminished importance in evolutionary studies that such geneticists as Wright claimed for their discipline, would have rallied around the potential synthesis provoked by the model of punctuated equilibria. But this was not the case. There were quite a few paleontologists, including some who studied invertebrates, who thought that their fossils demonstrated phyletic gradualism. Among the most forceful defenders of gradualism was Philip Gingerich, a vertebrate paleontologist at the University of Michigan.

From his, as well as from other, published studies of various fossil land mammals from deposits in the American West, Gingerich was convinced that the sedimentary record was sufficiently complete, and the stratigraphy reasonably undisturbed, to enable him to arrange the fossils contained in each deposit chronologically. After doing so, he suggested that the resultant sequence of fossils reflected the phylogenetic history of the species. Although he reported morphological changes, such as in tooth cusp presence or size, his primary focus was change in overall body size, which, he argued, was correlated with the surface area of the first lower molar, which he could measure. He presented his results in the form of diagrams that gave the mean and maximum range of first molar crown areas for the fossils at each stratigraphic level. Although his plotted data were not continuous, Gingerich thought that, overall, this fossil record revealed a picture of phyletic gradualism, both in the unilineal transformation of a species and in the splitting of species during speciation. Consequently, Gingerich concluded that "[t]his 'punctuated equilibrium' picture of phylogeny suggested by the previous studies is now seen to be an artifact of methodology."

In 1977, five years after the publication on punctuated equilibria, Gould and Eldredge responded to the numerous publications that had come out either against or in support of their model. They pointed out that, in virtually every example of supposed phyletic gradualism, including Gingerich's, the evidence did not exclusively or conclusively demonstrate this claim. In some cases, when the data seemed complete enough, the model of punctuated equilibria was supported. Interestingly, Gould and Eldredge pointed to the human fossil record—which had long been considered a bastion of phyletic gradualism—as sufficiently complete and geographically sampled to provide evidence of punctuated evolution. Only one study—on fossil foraminifera, which are microscopic, marine, shell-encased, single-celled organisms—satisfied Gould and Eldredge's criteria for testing punctuated equilibria: The samples were large enough to rule out randomness and to allow them to study geographic variation, and the stratigraphic range was extensive and detailed enough to demonstrate a possible correlation between morphological and environmental change.

In their second publication, Gould and Eldredge also discussed the possible genetic background of punctuated change, and the possible role of selection in speciation. With regard to the latter, they chose to endorse the notion of species selection, which had been formally proposed by their colleague, the geologist and paleontologist Steven M. Stanley. Inasmuch as this concept involved selection at the level of groups of individuals, rather than at the level of individuals, many neo-Darwinians had little time for it. Beginning with G. C. Williams in 1966, group selection, which had even been part of Wright's formulations for a while, was not a valid evolutionary consideration, in large part because selection was envisioned as choosing among different alleles, and this could occur only at the level of the individual. With regard to genetics, Gould and Eldredge suggested that knowledge of the relatively newly discovered regulatory and structural genes could provide some insights into the tempo of speciation.

By the late 1970s, it had become known that organisms that appear as different from one another physically or phenotypically as chimpanzees and humans could also be almost identical in terms of their structural genes: the genes that transcribe the proteins that contribute to the bits and pieces of hard, soft, and molecular morphology that make up a complete, functioning organism. In fact, it was Mary Claire King and Allan Wilson, working in the latter's laboratory at the University of California at Berkeley, who discovered that chimpanzees and humans share about 99 percent of their structural genes. But the reason that chimpanzees and humans look so different from each other lies in their regulatory genes, which control the timing and interaction of the structural genes and, consequently, the shapes of structures themselves. As King and Wilson concluded: "A relatively small number of genetic changes in systems controlling the expression of genes may account for the major organismal differences between humans and chimpanzees." Interestingly, and reminiscent of Goldschmidt, they also suggested that "[s]ome of these changes may result from the rearrangement of genes on chromosomes rather than from point mutations." Point mutations, which seem to occur most frequently in the third base of a triplet codon in the DNA sequence encoding a specific molecule, tend to have no expressed effect on the molecule's structure and its function in the organism. Consequently, a point mutation is essentially silent. Yet the idea of point mutation, representing a form of micromutation, was something that could be invoked as leading to gradual change.

Gould and Eldredge saw potential in the regulatory aspect of genetic information. This is not surprising, perhaps, considering Gould's long-term interest in ontogeny and heterochrony. In their article, they turned to Emile Zuckerkandl, who was one of the first scientists to explore the theoretical and practical aspects of molecular genetics, especially with regard to hemoglobin evolution. On a more global scale, Zuckerkandl had suggested that "[r]eproducible morphogenesis depends on constancy of genic regulation to a larger extent than on constancy of genic structure." Consequently, if one can discuss adaptive evolution at the genetic level at all, it would be in terms of regulatory genes and the effects of genetic regulation on determining the shapes of features, rather than in terms of structural genes, which produce the raw materials from which specific shapes are created. An appreciation of the potential significance of regulatory genes made inescapable the need to understand the role of ontogeny in evolution. As Gould and Eldredge put it:

> We applaud the burgeoning emphasis on change in regulatory genes as the stuff of morphological evolution . . . if only because one of us has written a book to argue that the classical, and widely ignored data on evolution by heterochrony should be exhumed and valued as a primary demonstration of regulatory change. We do not see how point mutations in structural genes can lead, even by gradual accumulation, to new morphological designs. Regulatory changes in the timing of complex ontogenetic programs seem far more promising—and potentially rapid, in conformity with our punctuational predilections. The near identity of humans and chimps for structural genes, and the evidence of major regulatory change indicated by human neoteny provides an important confirmation.

Clearly, even though they share virtually their entire suite of structural genes, humans and chimpanzees are members of different species. In fact, even if they are closely related, as some people believe, they are still widely separated from each other evolutionarily. If nothing else, the whole of the human fossil record—with what we now know to be numerous species of *Australopithecus*-like forms, as well as an increasingly recognized number of species of *Homo*—intervenes between them. This, in turn, suggests that all extinct hominid species would have shared most of their structural genes with humans and chimpanzees. But humans, chimpanzees, and the species of extinct hominids are remarkably, and in some cases almost unbelievably, different from one another morphologically.

The simple explanation for the morphological differences between such obviously different species is that they are due to regulatory gene differences—if not in entire regulatory genes, at least in their activities. This suggestion makes sense in light of the fact that the same kinds of structural building blocks are found among a wildly diverse array of organisms—from yeasts to humans—that have fashioned the resultant structures differently. For instance, all mammals have prismatic enamel, but the number and shapes of teeth are very different among different groups of mammals. All four-footed animals, or tetrapods—amphibians, reptiles, birds, and mammals—produce a fibrous protein known as keratin, which gives hardness to such seemingly diverse structures as claws, nails, horns, and feathers, all of which also come in a variety of shapes and sizes. On another level of inquiry, the differences that can be seen from one individual to another within the same species are probably due to different alleles within the domain that controls the structural genes. This suggestion is certainly consistent with

Williams's theory, in which natural selection acts on different alleles expressed in individuals within a species.

Among the geneticists who responded to Gould and Eldredge's second article on punctuated equilibria was Sewall Wright. As had become his style, Wright's discussion was mediated through his shifting balance theory of the selective topographic landscape, which he adamantly maintained could explain all evolutionary phenomena, including a picture of punctuated equilibria. On the grounds that morphological change and speciation are genetically different phenomena, he suggested that Gould and Eldredge's comment—"Punctuated change dominates the history of life: evolution is concentrated in very rapid events of speciation (geologically instantaneous even if tolerably continuous in ecological time)"—should be changed to read: "Punctuated change dominates the history of life: evolution is concentrated in very rapid events of morphological change. . . ." In seeming contradiction to this declaration, however, Wright later stated that the origin of a new group, via speciation, usually accompanies rapid change. As he envisioned it, the likelihood that a new species will emerge is dependent on the availability of a vacant econiche. He pointed to the well-worn paleontological example of mammals existing for millions of years without differentiating into a diversity of species until the dominant land animals, the dinosaurs, became extinct.

But although Wright was fixated on his own theory of speciation, he appeared to be rather adventurous in considering how, from the mutation side of things, rapid morphological change leading to a new species, and then to a new group of organisms, could occur. He commented: "Of special importance . . . are the cases in which evolution along a restricted line happens to lead to an adaptation that happens to lead to an adaptation that turns out to open up an extensive new way of life." His first example was the conversion via neoteny of the larval stage of an invertebrate—the tunicate, or sea squirt—into an animal that we would think of as a chordate, a primitive vertebrate. As additional examples of morphological change, he offered the modification of the first gill arch into a functional jaw apparatus, the fins along the lower aspect of a fish's body into weight-supporting structures, and the fish's swim bladder into organs of respiration that functioned out of water.

Wright realized that by suggesting that major organismal reorganization could occur very rapidly, he could be mistaken for making a case for "hopeful monsters." But he headed off any attack based on this presumption by stating that although "[i]t may seem that mutations with impossibly drastic effects would be required for the origins of the higher of the taxa [evolutionary groups] . . . Such origins . . . probably all occurred from species, the individuals of which were so small and simple in their anatomies that mutational changes, that would be complex in a large form, were not actually very complex." For example, he speculated that a snail-like animal with a single shell could have evolved into a two-shelled mollusk-like one by a simple mutation. But in spite of these provocative speculations, Wright was content to consider, as he always had, that chromosomal rearrangement and shifting gene combinations were at the base, genetically, of change. In the end, he concluded that "the evolutionary processes indicated by the fossil record can be interpreted by the shifting balance theory without invoking any causes unknown to genetics or ecology."

The population geneticists Brian Charlesworth of the University of Sussex in England, Russell Lande of the University of Chicago, and Montgomery Slatkin of Washington University in St. Louis took a hard-line neo-Darwinian stance with regard to punctuated evolution. They cited Simpson's 1953 book *The Major Features of Evolution* as authority that "there is nothing in the fossil record that compels one to adopt a saltational interpretation." Furthermore, they stated, "[a]lthough geologically rapid, from a genetic or ecological point of view, major adaptive transitions are probably slow and gradual, involving many thousands of generations." They criticized the model of allopatric speciation via peripheral isolates because "[t]his idea is, in essence, a special case of Wright's theory of random genetic drift as a mechanism for triggering shifts from one stable equilibrium (adaptive peak) to another." They rejected the role of chromosomal rearrangement in the process of speciation—a role that, they claimed, had been stressed by punctuationists—because they were not convinced that this phenomenon created anything new. They stated that Goldschmidt had argued that "the genetic basis of variation between species differs qualitatively from that of variation between populations of the same species, or that between individuals of the same population." And then, citing articles by Gould and by Stanley (the author of species selection) as having been based on this incorrect notion, they commented that "it is difficult to understand why such a claim should be resurrected except in support of preconceived notions about evolutionary mechanisms." Goldschmidt did not make this claim. He suggested that the genetic basis of variation among individuals of the same species was not the same as was involved in the origin of species.

Charlesworth and his colleagues continued their assault on punctuation by associating its advocates—Gould and Stanley, specifically—with Goldschmidt, and now also Schindewolf. In this case, they were defending "whether microevolutionary theory is adequate to account for the evolution of major new adaptations" by rejecting the idea that a single mutation could result in a major new adaptation. They cited Darwin as having emphasized that "such adaptations . . . involve a set of mutual adjustments of the parts of a complex structure (the vertebrate eye is the classic example)" and that "complexes of characteristics are gradually built up by selection of the best individuals in each generation with respect to some criterion (e.g., speed in running among racehorses and greyhounds)." They also cited Fisher's argument "that small, successive steps are required, since a small random change in a complex structure has a reasonable chance of being in the direction of increased adaptation, whereas a large change is almost certain to be out of adjustment in some respects." Charlesworth and his colleagues did admit that "[t]he only genetically credible alternative theory to the Darwinian process of step-by-step evolution under the guidance of individual selection is Wright's shifting balance theory." But even this theory, they concluded, relying as it does on a large population of partially isolated local populations, cannot be embraced as "a general model for the evolution of adaptation."

Not unexpectedly, perhaps, Charlesworth, Lande, and Slatkin concluded:

> In contrast to this [Wright's] model, and to saltational [punctuational] models for the evolution of adaptations, Darwinian natural selection is an almost inevitable process that requires only a supply of genetic variability to work in virtually any type of population structure. Its operation in present-day populations has repeatedly been demonstrated. Since plausible scenarios can be constructed to account for the step-by-step evolution of even the most complex structures, there is no a priori reason to deny individual selection a major role in such evolutionary events.
>
> [P]unctuationists claim that macroevolution is "decoupled" from microevolution, and deny that gene frequency changes within populations are the foundation of major morphological changes. This argument seems to neglect the fact that every living or fossil organism owes its existence to a continuous line of descent going back generation by generation into the remote past. . . . [T]here is no evidence suggesting the need for qualitatively new mechanisms to account for macroevolutionary patterns.

In 1994, however, support—genetic as well as morphological and paleontological—for punctuated evolution came in a totally unexpected way. Beginning in 1986, Alan Cheetham, an invertebrate paleontologist at the National Museum of Natural History of the Smithsonian Institution who could not embrace Eldredge and Gould's model, set out to disprove it through his studies on bryozoans. Bryozoans are tiny, colonial, filter-feeding, coral-like marine invertebrates that deposit around themselves an external "skeleton." By the end of his study, however, which he published in 1995 with a colleague, Jeremy Jackson, Cheetham had become a convert to punctuated equilibria.

Jackson and Cheetham found that their identifications of living species of bryozoan based on morphology (aspects of the external skeleton) and genetics (gene-correlated protein sequences) were identical. This made determination of fossil species of bryozoan on the basis of morphology more reliable. Their morphological analysis of the fossils clearly demonstrated that the species had existed essentially unchanged for at least 1 and up to 16 million years. In addition, after figuring out the possible evolutionary relationships between the fossil and living species of bryozoan and looking at all their data, Jackson and Cheetham discovered that changes in morphology and proteins occurred together and at the same time during speciation. They concluded that "[m]orphological stasis and the episodic pace of speciation thus may imply an episodic pattern of molecular evolution, rather than a constantly evolving [gradual] protein clock." Jackson and Cheetham did not, however, discuss the process of speciation that would have produced the punctuation that they described or the role, or lack thereof, of selection of whatever sort. But they did demonstrate the pattern.

The Return to Development

At the same conference at which Wright and Charlesworth, Lande, and Slatkin presented their papers addressing punctuated equilibria, Gerald Oster, a biophysicist at the University of California at Berkeley, and Pere Alberch, a comparative morphologist at Harvard's Museum of Comparative Zoology, discussed development and evolution. In the publication that derived from this conference they stated: "The idea that development can tell us something about the organization of the diversity of life predates the establishment of Darwinian evolution." The problem, however, was that "despite its recognized significance, the role of morphogenetic interactions in controlling phyletic transformations remains to be successfully integrated into the body of evolutionary theory." This state of affairs was reflected, for example, in the fact that "modern textbooks on evolution make only scant reference to developmental biology." They were going to try to fill this lacuna.

The theme of Oster and Alberch's paper was epigenesis: the interaction of the constituents of the different levels of cellular organization and development that emerge, or unfold, after fertilization of the egg and the union of parental genes. The eighteenth-century German embryologist Caspar Wolff had first coined the term for his "theory of epigenesis," which derived from his observations that rather than being a miniature version of a fully formed adult, the developing embryo passed through ontogenetic stages in which its cells first organized into layers and then differentiated into organs. A more sophisticated theory of epigenesis was put forth in 1974, more than two hundred years after Wolff's, by the Swedish developmental and experimental biologist Søren Løvtrup.

The reason, according to Løvtrup, why we need to think in terms of epigenesis is that development is not immutably preformatted. If it was, he said, its course could not be altered experimentally, which, of course, happens in the laboratory all the time. Consequently, he concluded, "it appears unavoidable that *everything, without exception,* which happens in the embryo after fertilization must be classified as *epigenetic events.*" In the eyes of Løvtrup and others, and very reminiscent of Haldane's ideas, the unfolding of ontogeny is an ongoing process of interaction among the genetic and the cytoplasmic contents of the cell, differentiating cells, and the external environment. The course of embryonic development can be altered at the genetic level by mutations, in a typical neo-Darwinian fashion. Or, as Løvtrup also suggested, by changes in other properties of the fertilized egg that alter the chronology or spatial organization of patterns of cellular diversification.

Since most multicellular organisms possess the same kinds of cells, as well as the same chemical substances that form the immediate environment of the cells, Løvtrup concluded that variation in the spatial and chronological organization of cellular differentiation must be the key to evolution.

Løvtrup began his case for epigenesis with the commonly accepted generalization that, in the course of development, one set of cellular events is determined by the events that immediately went before. For instance, he pointed out, an egg can come in more than one shape. If it is spherical, it will establish an axis with a front end and a rear end, but it can do so in one of two ways, so that the axis is oriented differently in each case. Once this axis of polarity is determined, then other developmental decisions come into play depending upon whether it is going to be a bilaterally symmetrical animal with right and left sides, such as a mouse, or a radially symmetrical organism, such as a sea urchin. If it is going to be a bilaterally symmetrical animal, axes specifying up and down and right and left must be established. But there is also more than one way in which these axes can be sorted out, which may also depend on how the front-rear axis was oriented.

The multiplication of cells, Løvtrup pointed out, is not a straight-line process, either. Cells need not divide into two daughter cells of equal size. The creation, by way of meiosis, of one large egg and three tiny polar bodies serves to demonstrate this. If cells do divide equally, there are certain paths of differentiation they can follow and spatial relationships they can achieve. But if a cell divides into daughter cells of different sizes, there are other outcomes—for instance, packaging and spatial relationships—that become available. Even the position of the chromosomes relative to the center of the cell or the outer cell membrane at the time of cell division leads to differences between daughter cells in their sizes and, consequently, their spatial relationships.

As cells divide and multiply, different kinds of cells differentiate, creating cell lineages. For example, cells from the margins of the invaginating neural tube of the early central nervous system escape and migrate throughout the head region, lodging in areas that will develop into membranous tissue, sense organs, and cartilage, with cartilage cells eventually differentiating into bone-forming cells. As another example, the cells in bone marrow can differentiate into an array of cell types of different attributes, ranging from cells that produce or destroy bone to a variety of blood cells. As cells differentiate into more specific roles, they interact with one another and the different products they produce in the creation of tissues and organs. Since, Løvtrup suggested, the materials that an animal cell produces are synthesized from that cell's DNA, yet the products of cells derived from the same stem parent can differ, cell differentiation must involve the differential transcription of the DNA code. That is, although daughter cells retain the entire DNA package of the original cells of the lineage, they use only some of that information to produce proteins or other products. Because the rest of the DNA remains unread, it will not produce or transcribe a protein. The use of

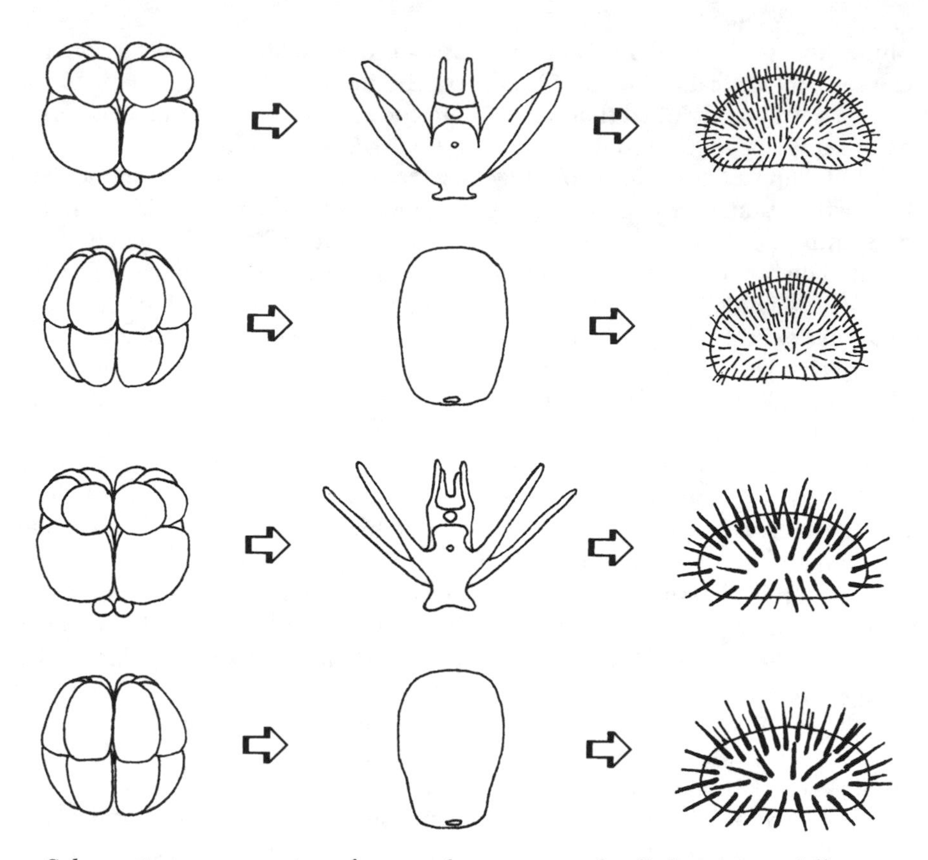

Schematic representation of unequal versus equal cell division in different species of sea urchin. Although the adult stages may superficially look similar, as in the two sea urchins with many small spines and the two with fewer large spines, the cellular paths to these ends were different. Cell division after fertilization can be equal or unequal (left column), and the larval stages may be simple or complex (middle column). (Adapted from Swalla and Jeffrey)

part of the total DNA in a cell's nucleus is correlated with the particular stage of division and differentiation the cell lineage is in as its cells assume more specific roles and functions.

But the mere presence of cells does not mean that each cell is necessarily capable of performing a specific function. As experiments in grafting cells from embryos of certain ages to other embryos have demonstrated, cell lineages have to acquire the competence to produce structures. For instance, a clump of cells taken from the presumptive jaw of a mouse embryo and transplanted into an older individual's back might merely become incorporated into the back of the recipient. But if a clump of cells from the presumptive jaw of a mouse embryo even twelve hours older than the first were transplanted into the back of another individual, a tooth and all of its surrounding tissues could form and erupt. In the case of the younger embryo, the cells in the presumptive jaw had not become competent to produce teeth, whereas in the older embryo they had. In the younger embryo, the cells that could have produced dental structures if they had remained in place became submerged or incorporated into the cell lineages that normally produce non-dental structures. In the case of the slightly older embryo, the cell lineages in the presumptive jaw had "lived" long enough—that is, gone through enough cellular divisions—that the older generations of cells had become committed to a specific role in the production of a given structure.

In their discussion of epigenesis, Oster and Alberch pointed out that there are major themes in development. For example, after fertilization, cell division gives rise to a more or less hollow, ball-shaped structure called the blastula. Depending upon the organism, one area of the blastula's cell wall will either fold inward, or invaginate, or it will buckle outward, or evaginate, as the cells proliferate to form the next stage of development, the gastrula. On a higher level of cellular organization, the neural tube forms as a result of infolding, or invagination, whereas the cusp of a tooth arises through a process of buckling outward, or evagination. In general, Oster and Alberch envisioned "the construction of an organism in ontogeny as a hierarchically organized sequence of processes." The interactions among cells—which cells need in order to communicate and from which ontogeny emerges—are, they suggested, "far removed from the province of direct genetic control." Morphology, or the phenotype of an organism, "emerges as a consequence of an increasingly complex dialogue between cell populations, characterized by their geometric contiguities, and the cell's genomes, characterized by their states of gene activity."

Oster and Alberch offered the generalization that the basis for the development of many structures in a complex organism lies in the interaction of two cell types that become differentiated early in embryogenesis: epithelium and mesenchyme. This interaction had been well documented through experimental studies on tooth development. Epithelium constitutes an organism's external layer of cells. Mesenchymal cells lie beneath the epithelium. An organism is encased in a "glove" of epithelium, which continues into the mouth, or oral cavity, and carries right on through to the anus. The mesenchymal cells in the head and in some regions of the rest of the body derive from neural crest cells that migrate from the edges of the invaginating neural tube. In other areas of the body, mesenchymal cells result from local cell division and differentiation. The development of many organs, such as teeth, scales, hair, and feathers, results from the interaction of epithelium and the underlying mesenchyme. The interaction is followed by the phases of cell proliferation that ultimately create a structure that has its particular characteristics.

A major reason why the same two kinds of cells can produce structures as different-looking as hair and feathers is that there are different developmental paths their interactions can take early in the process of structural differentiation: The proliferating cell mass either invaginates or evaginates. If the cells invaginate, a hair in a follicle will develop. Skin glands also develop through a process of invagination. If, on the other hand, the cells evaginate, the resultant structure will be a feather. Scales also form through a process of evagination. The final shapes of these structures—which all derive from the interaction of the same kinds of cells and the deployment of the same fibrous protein, keratin—are due in large part to two factors: the mechanical and elastic properties of the cells and the uptake of water.

In the world of development, hydration is behind almost everything. Cleft palate, for example, occurs when the two halves of the palate do not

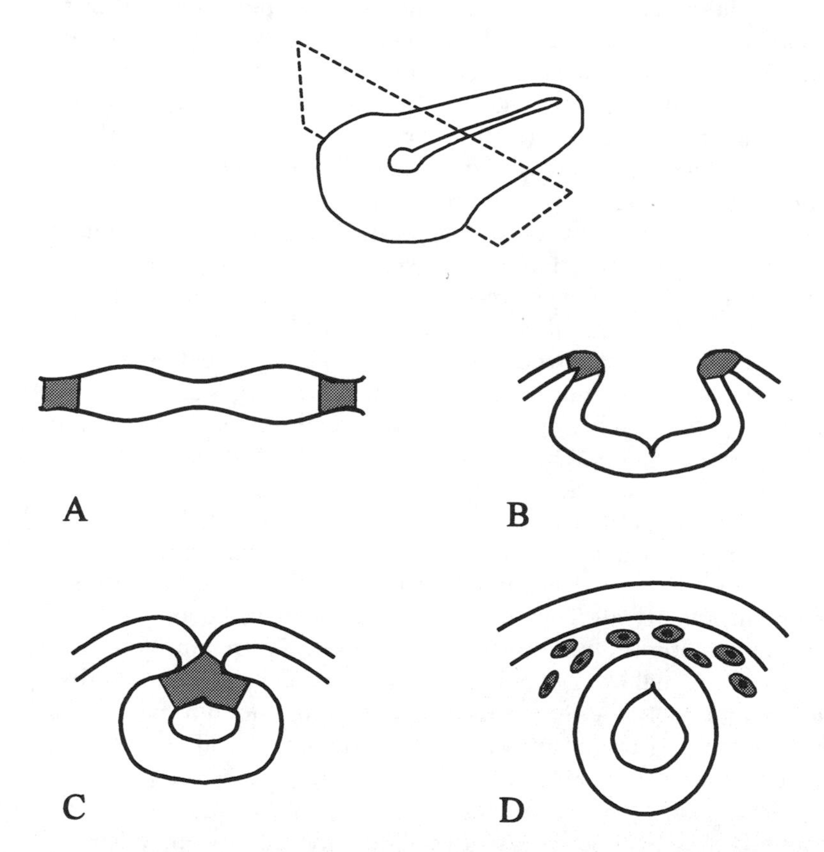

Development of the neural tube and the origin of neural crest cells. A: A cross section through the upper surface of a developing embryo. B. The invagination of this surface to form the neural groove, from whose edges—the neural crests—neural crest cells will eventually separate and migrate, particularly into the developing head region. C: The closure of the neural groove to form the neural tube, or early spinal cord. D. The complete separation of the neural tube from the upper, or back, surface of the organism. The uppermost drawing of a chick embryo shows the plane of section of the other diagrams. The neural crest and migrating neural crest cells are darkened.

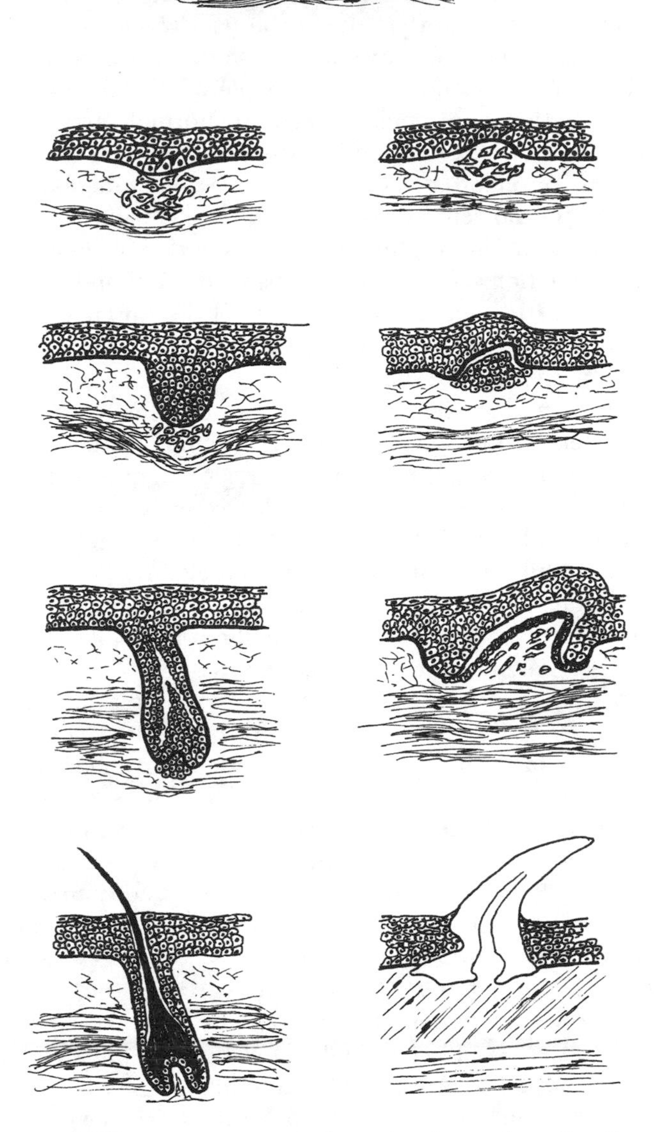

Development of different structures by invagination and evagination. The top, or external, layer of cells (uppermost diagram) consists of epithelial cells. Next down, mesenchymal cells have migrated into the region below and are interacting with the epithelial cells. In the left-hand column, the epithelial cell layer then thickens and invaginates down into the underlying region as the product of the interaction between epithelial and mesenchymal cells—in this case, a hair—takes shape. Sweat glands also develop by invagination. In the right-hand column, the direction of the interacting cells is outward, with the product—in this case, a scale—developing through a process of evagination. Feathers also develop by evagination. (Adapted from Oster and Alberch)

swing up from their original vertical position and close over the roof of the oral cavity. The mechanism that allows the palatal shelves to change from the vertical to the horizontal is hydration—water swells the cells, which, in turn, stretch or contract and become capable of moving larger structures. The key to hydration and, consequently, to normal development, is the accumulation, in the right cells and at the right time, of a certain class of large molecules, or macromolecules, called acidic mucopolysaccharides, which attract water molecules.

One of Oster and Alberch's main points was that their model of a hierarchy of cell interactions—their dialogue between cells and their products—precluded the need to invoke other models of development. One of these models proposed that the information to which a cell responded was correlated with the timing of differentiation of cells within a cell lineage. Another model was based on the notion that a chemical substance that provoked the formation of a series of repeated structures, such as teeth along the jaws or bones within a limb, spread throughout the region in question as it enlarged. These models were not necessarily mutually exclusive, and both were indicated by experimental work led by the British experimental embryologist Louis Wolpert on the developing wing of the chick embryo. From their studies, Wolpert and his colleagues formulated the theory of pattern formation, which had been foreshadowed by one of William Bateson's preoccupations: the development of meristic, or repeated, parts.

The investigation of pattern formation grew out of a realization that there was an order to the sequence of bones in a limb or vertebral column, or teeth along the jaw. These repeated or serially arranged parts formed a gradient of size and shape, and, in the case of the limbs, a gradient in number of bony elements. In the forelimb of a typical mammal, for instance, there is one large upper-arm bone, or humerus, two slender and slightly smaller lower-arm bones, the radius and the ulna, and then a greater number of smaller bones tacked onto the end of the limb. If one did not consider the very small wrist, or carpal, bones, this was a perfect gradient of repeated parts. In terms of development, the blobs of cartilage that represented the presumptive individual bones emerged as the limb was elongating and changing from a mere swelling of the body wall to the full appendage.

Wolpert and his colleagues sought to understand the mechanism behind the development of the limb. The vertebrate they worked with was the chick embryo. Although a chick, like all birds, has fewer bones in its forelimb or wing than reptiles and most mammals, it is a practical experimental animal because chicks are plentiful and they can be manipulated surgically and kept alive. In a mammal, the mother would have to be sacrificed and the embryos would not survive. By experimentally cutting out pieces of the elongating limb bud and transplanting them to other parts of the limb bud, Wolpert and his co-workers could discover when cells reached certain levels of competence and became committed to producing certain structures. After many experiments, they determined that the proliferating mass of mesenchymal cells at the elongating end of the limb bud "assigned" information at certain positions that specified the formation of the specific bones. Different generations of the proliferating cell lineage would produce the different bones. Once cells were assigned positional information, that was that. Even if these committed cells were transplanted back onto the body wall, they would produce, for example, the finger bones they would have done had they remained on the end of the limb. The cells of these positional sites would be activated by a chemical called a morphogen, which spread from the body end of the limb and diffused throughout the limb.

The impact of the pattern-formation model on the field of developmental biology was profound. An experimental embryologist at Guy's Hospital Medical School, Andrew Lumsden, pursued studies on developing mice and demonstrated that the cells that eventually produced the first molar and all of its differentiated cellular products—such as enamel and dentine—proliferated and gave rise to the cells that produced the second molar and its components, and that, in turn, these cells proliferated and gave rise to the cells that produced the third molar and its components. It was undeniable that all the information necessary to form all the molar teeth, and all the bits and pieces of each tooth, was present in the initial cell mass. Since mouse molar teeth form a gradient of decreasing size and morphological complexity, these repeated structures conformed well to a model of pattern formation in which the cell mass diminished in structure-producing potential as it proliferated.

Jeffrey Osborn, a dental embryologist who had also been at Guy's with Lumsden but is now ensconced at the University of Alberta in Edmonton, Canada, proposed the clone model to describe the emergence of repeated structures, whether teeth or limb bones, from the proliferation of, and subsequent differentiation within, cell lineages. The clone model relied on the notion of increasing or decreasing potentials of proliferating cell masses to produce repeated structures that either increased or decreased in relative size and shape throughout their series. The clone model also helped to explain what Bateson had suspected: that so-called anomalous conditions affecting series of repeated parts—such as additional or supernumerary structures, as well as the lack of expected structures, as in number of fingers, toes, finger and toe bones, vertebrae, and teeth—can be understood in the context of normal development. In terms of the language of the clone, or pattern-formation, model, differences in series of repeated parts result from varying degrees of potential of proliferating and differentiating cell masses. In other words, the same developmental process that gives rise to the expected condition in a species can also be responsible for an anomalous expression of its potential, either on the "too much" or "too little" side of normal.

It turns out that the information that signals cells to produce a certain bone of a particular size and shape within a patterned, complex structure, such as a limb, comes from the differential expression of a particular class of regulatory gene, the homeobox gene. In fact, homeobox genes are ultimately implicated in the differentiation of all aspects of an organism.

Homeobox genes, which were first identified in the early 1980s and are currently the best-known group of regulatory genes, are necessary for proper development. Homeobox genes tend to come in clusters, or complexes. Homeobox genes code for proteins that do not necessarily go immediately and directly into the formation of a particular structure. Instead, the protein one homeobox specifies may travel to another stretch of DNA, corresponding to another homeobox gene, and bind with it. In turn, this protein may activate or turn off that homeobox gene. Consequently, one homeobox gene can initiate a cascade of effects of gene-protein-gene interactions that can lead to the unfolding of an ostensibly complex, but nonetheless exquisitely simple, organized pattern such as the coloration of a butterfly's wing or of a brilliant tropical fish. Adding to the realm of possibility in development is the demonstration that more than one homeobox gene can be active at the same time. As a result, the domains of homeobox genes can overlap and their joint products can then have regional effects as well as more widespread, systemic effects. And, since all body cells contain the same DNA, different stretches of these long molecules, corresponding to different homeobox genes, can be activated at different times. Consequently, homeobox genes that are "recruited" at different times will have different effects on the development of the organism.

In chick limbs, there are five related genes in a homeobox gene complex—originally identified as the *Hox-4* complex, but later renamed the *Hoxd* complex—that are involved in wing and leg formation: *Hox-4.4, -4.5, -4.6, -4.7,* and *-4.8.* Their domains of expression are determined almost immediately after the limb bud begins to swell, and they remain active throughout limb elongation. The cells on the "little finger" side of the wing express all five of these homeobox genes. But the cells on the "thumb" side of the wing express only the *Hox-4.4* gene. The same pattern of expression occurs in the leg. All five homeobox genes are expressed on the side of the leg that bears the "little toe," which is the outside front toe. However, on the side with the "big toe," which is the small, backwardly directed toe that cocks use in fighting, only the *Hox-4.4* gene is expressed. Experimentally "turning on" the *Hox-4.6* gene in all cells of the leg affects the normally tiny and backwardly twisted "big toe." It grows longer through the addition of an extra segment and comes to lie forward, in alignment with the other three toes. Curiously, when the *Hox-4.6* gene is experimentally activated throughout the cells of the wing, an additional "thumb" is produced. The formation of the tibia and fibula in the leg, and the radius and ulna in the wing, results from the interaction of the *Hox-4* gene domains and the domains of another homeobox gene complex, *Hox-1* (later identified as the *Hoxa* complex).

Studies on limb development in the mouse reinforced the conclusions drawn from the studies on homeobox genes and pattern formation in the chick limb. Mice are more typical of vertebrates than birds because they have more bones in the arms (forelimbs) and legs (hind limbs). They also have many wrist, or carpal, bones, many ankle, or tarsal bones, more fingers and toes, and, especially in the "hand," or forefoot, more individual bones overall. As in the chick wing, though, the domains of the *Hoxd* (the old *Hox4*) complex are established early in the development of the mouse forelimb, and they are activated in an orderly sequence. Also as in the chick wing, more homeobox genes are active in cells on the posterior, or "little finger," side of the mouse forelimb than in cells on the anterior, or "thumb," side. As had been strongly suggested by experiments on chick limb buds, there is a gradient of activation of individual homeobox genes that proceeds from the cells on the posterior side of the mouse forelimb to cells on the anterior side. Interestingly, when the very last gene in the *Hoxd* complex—the *Hoxd-13* gene—is deactivated experimentally, most of the finger bones are affected, becoming stunted in size. This result suggests that in limb formation some homeobox genes have a more specific effect on development, while others have a more widespread effect. It also illustrates, at least for the limb, that pattern formation is influenced by specific positional information as well as by shifts in concentrations of homeobox gene products.

The discovery of homeobox genes involved in limb development was also important for understanding the evolution of the limbs of tetrapods, or four-footed animals. Since tetrapods must have evolved from a fishlike ancestor, the question was: How could forelimbs and hind limbs, with many supportive bony elements in each limb, be derived from the cartilaginous fins of fish? Most living fish—teleosts, such as a trout, or chondrychthyans, such as a shark—are not relevant to unraveling this mystery because they have numerous cartilaginous rays arranged more or less parallel to one another or in a fanlike pattern in each fin. But lungfish and coelacanths are notably different from the typical fish. They have fins that are fleshy at the base and are supported by a series of bones that extend the length of the fin's long axis. The fins of lungfish and coelacanths do contain small, thin fin rays, but these structures adorn the anterior and posterior flanks of the main axis of fin-supporting bones. Fossil relatives of lungfish and coelacanths had similarly configured fins.

Since the large bones of the tetrapod limb are also arranged in a linear fashion, paleontologists at first thought that the evolution of a limb from a fin involved the simple addition of forefeet and hind feet to the end of this structure, complete with wrist and ankle bones and toes. But in 1986, Neil Shubin, a biologist at the University of Pennsylvania, and Pere Alberch proposed a different evolutionary scheme. They suggested that what had probably happened in the "conversion" of a simple finlike appendage to a tetrapod limb was that the end of the fin had become flexed developmentally, so that it was bent over to the "thumb" side of the limb. Consequently, the toes at the end of a tetrapod's limb are merely derivatives of the rays that would have extended along the posterior side of the ancestral fin.

Confirmation of the bent-fin theory of pattern reorganization in the formation of the tetrapod limb came in 1995 from comparative studies by Paolo Sordino, Frank van der Hoeven, and Denis Duboule, all of the University of Geneva, on homeobox gene expression in the fins of zebra fish and the limbs of mice. They found that zebra fish and mice express the same *Hoxd* genes—*Hoxd-11, Hoxd-12,* and *Hoxd-13*—plus one particular *Hoxa* gene during the early phases of fin and limb development. This finding suggests that the large bones of the tetrapod limb derive from the same homeobox genes that produce the bones that link a fish's fin to its body. But in later phases of development in the mouse, the limb bud expresses *Hoxd-11, Hoxd-12,* and *Hoxd-13* gene activity across the front of the elongating of the limb, whereas the fish's growing fin bud does not. In the mouse, there is not only early homeobox gene activity along the posterior side of the limb but there is also later homeobox gene activity that spreads anteriorly at a right angle to the previous phase of activity. As Sordino and his colleagues pointed out, this difference between the fish and the mouse was due to an asymmetry in cellular activity at the proliferating end of the mouse limb, not necessarily to a physical bending of the original fin axis. Nevertheless, this new cellular field at the end of the developing mouse limb—coursing across the

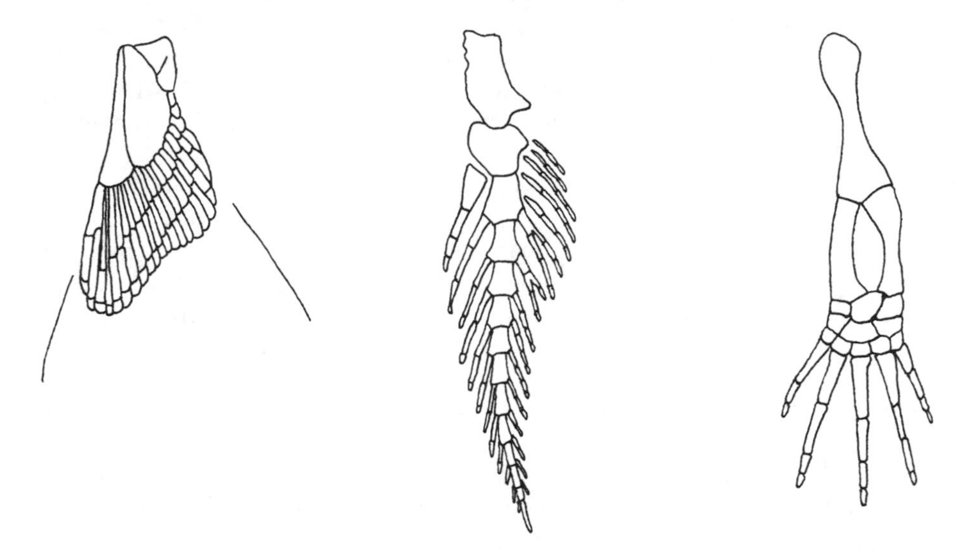

The major supporting structures in the fin of a fish (left), the fin of a lungfish (middle), and the forelimb of a typical four-legged animal. In the fish (represented here by a fossil shark), there are numerous cartilaginous rays, arranged in fanlike fashion, that give stability to the flare of the thin fin. In the lungfish, the fin is thick and fleshy and is supported by a central series of bones, on each side of which is a series of smaller bones. The tetrapod limb is composed of a single large upper bone (in the case of the forelimb, the humerus) and two smaller and thinner lower bones (here, the radius and the ulna). In addition, the end of the limb bears toes. In between the toes and the lower-limb bones is a series of small bones (wrist in the forelimb and ankle in the hind limb). The bent-fin theory posited that the lower part of a lungfish-like fin turned in at a right angle and the outer, smaller set of bones became the toes of a tetrapod.

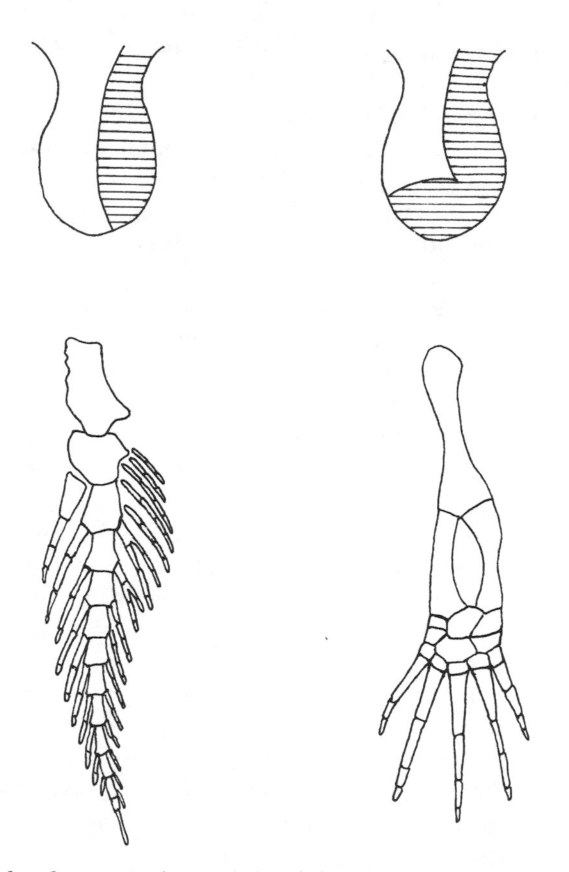

Diagram of a developing and an adult left fin of a lungfish (left column) and a developing and an adult left forelimb of a typical tetrapod (right column). In fish, the Hoxd *genes* 11–13 *are active only along the back side of the developing fin. In tetrapods, these genes are active along the back side as well as across the front of the developing limb bud. (Adapted from Shubin, Tabin, and Carroll)*

front of the limb bud from the posterior to the anterior side—ultimately leads to the formation of cartilaginous condensations that, in turn, become the wrist and finger bones. Clearly, as represented by the mouse, a novelty responsible for the emergence of the tetrapod foot at the homeobox gene or homeobox gene product level had been introduced in the evolutionary emergence of this group of vertebrates.

The exact nature of this molecular novelty was inadvertently resolved the following year by a team of collaborators at Harvard University. These scientists—Yasuteru Muragaki, Stefan Mundlos, Joseph Upton, and Bjorn Olsen—investigated one of the anomalies in humans that William Bateson often discussed: polydactyly, the development of more fingers and toes or more bones per finger or toe than is normal. This condition is a dominantly inherited trait, as Bateson had suggested. Upon studying individuals who had the condition, Muragaki and his colleagues discovered that those with multiple fingers, toes, or bones in fingers or toes had longer repeat sequences of the amino acid alanine than normal individuals do. Since the involvement of fingers and toes in this developmental anomaly implicated the *Hoxd* complex, and the *Hoxd-13* gene is the last homeobox gene to be expressed during limb development, Muragaki and his team focused their analysis on this gene and its activity. Not surprisingly, perhaps, they discovered that the *Hoxd-13* gene in humans contains a segment that encodes a sequence of repeated alanines. When they looked at data on other animals, they found that the chicken also has an alanine-encoding stretch in its *Hoxd-13* gene. The chick differs from the human, however, in that its *Hoxd-13* gene specifies a much shorter sequence of repeated alanines. The zebra fish lacks an alanine-encoding component of its *Hoxd-13* gene altogether.

Clearly, the presence of an alanine-encoding segment in the chick and in the human *Hoxd-13* gene indicates that this is one of the keys to understanding hand and foot development, as well as evolution, in tetrapods. Muragaki and his colleagues suggested that the greater number of alanine repeats in the human, who normally has a greater number of toes and toe bones than the chick, reflects the evolutionary increase in the alanine-encoding stretch of the *Hoxd-13* gene in mammals. But this theory is probably the wrong way around: Since birds have fewer toes and toe bones than many other vertebrates, including humans, there had been a decrease in the number of alanine repeats in the former group.

Muragaki and his colleagues found that human polydactyly is correlated with an increase in the number of repeated alanine. Individuals that were heterozygous for the mutant *Hoxd-13* condition had extra fingers and toes as well as pathological development of the cartilaginous precursors of the various bones themselves. An individual who was homozygous for the mutant gene had stunted finger bones that looked more like wrist bones, which are typically short. Muragaki and his colleagues suggested that polydactyly is a result of the excess amount of alanine, which interferes with the coordination between the *Hoxd-13* gene and the *Hoxd-11* and *Hoxd-12* genes that is necessary for the proper formation of fingers and toes. Specifically, the excess alanine produced by the *Hoxd-13* gene binds to the other *Hoxd* genes and inactivates them for a longer period of time than is usual. In these cases, excess alanine correlates with extra numbers of digits as well as of individual bones.

Shortly after Muragaki and his colleagues reported their findings on human polydactyly, Jószef Zákány and Denis Duboule, both of the University of Geneva, published their study on polydactyly in mice, in which they experimentally inactivated the *Hoxd-11, Hoxd-12,* and *Hoxd-13* genes in embryos. The result was the development of polydactyly, with some mice having seven instead of five toes on the forefoot and six rather than five toes on the hind foot. Most recently, a research team at the Jackson Laboratory, in Bar Harbor, headed by Kenneth Johnson, found that some of their experimental mice had spontaneously developed polydactyly. After much study, they determined that the mutation, which was in the recessive state, involved an expansion of the alanine repeat sequence encoded in the *Hoxd-13* gene.

Clearly, manipulation of the various homeobox genes is implicated in the generation of extra fingers and toes. Also evident from the studies on humans, mice, and chickens is the normal, nonexperimental role of alanine in regulating the activity of its associated homeodomain. The greater the number of alanine repeats, the greater the number of digits and bones within digits. The reverse also appears to be true: The shorter the repeated alanine series, the fewer the number of digits and bones within digits. In the chick, which has a severely reduced number of wrist bones as well as digits and bones within digits in its wing, and four instead of five toes on its feet as well as fused ankle bones, the number of alanine repeats is 40 percent less than it is in humans and mice. When we put this all together, it appears that

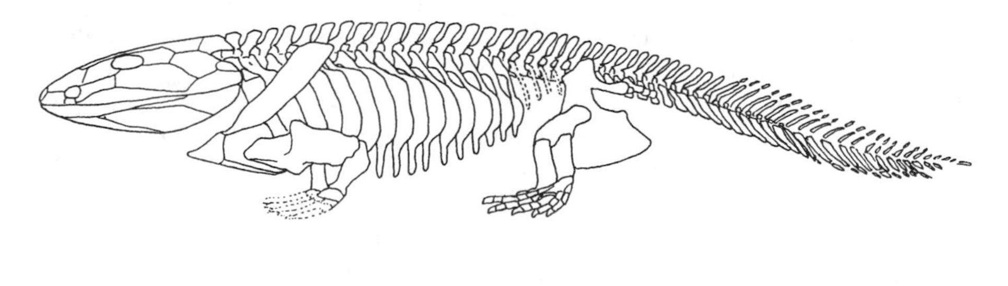

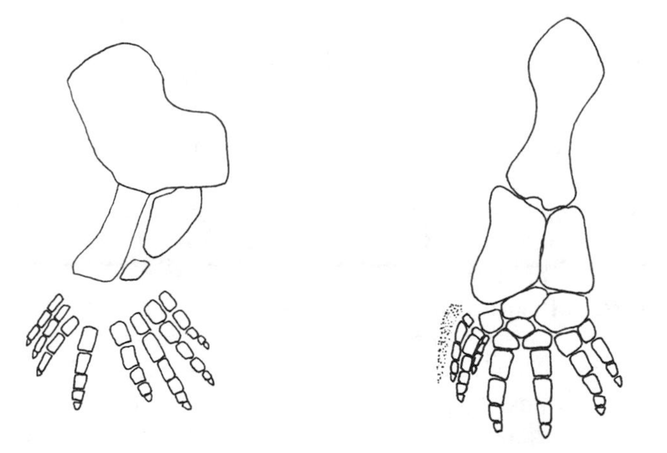

Diagram of the skeleton of the early tetrapod Ichthyostega *(above) and the forelimb and hind limb bones of the equally early tetrapod* Acanthostega *(below, left and right). The hind limbs of both fossils bore at least seven bony toes; an eighth toe may have been present (stippled), but it may have been cartilage that did not become transformed into bone. In the forelimb, as known from* Acanthostega, *there were at least eight toes. (Adapted from Carroll)*

the evolution of four-footedness is correlated with changes in homeobox genes and their functions: the expansion of homeodomains across the end of the developing limb bud and the encoding of alanine by the *Hoxd-13* gene.

Since the earliest known tetrapods—*Ichthyostega* and *Acanthostega*—had at least eight digits on the forefeet and at least seven digits on the hind feet, with as many as six bones in each digit and a number of wrist and ankle bones, it is likely that their *Hoxd-13* genes produced high levels of alanine. Alanine production would no doubt have been decreased in the last common ancestor of the group that includes living reptiles, birds, and mammals. The greatest number of digits a reptile or mammal develops on its forefeet or hind feet is five. Since mammals have fewer bones per digit than do primitive reptiles, the length of the alanine repeat series would again have become diminished, probably close to the number of alanine repeats we see in humans. Humans are not different from the ancestral mammal in terms of its forefeet and hind feet: five digits on hands and feet, the maximum number of bones for a mammal in each digit; and many separate wrist and ankle bones.

Within groups of tetrapods, there were multiple parallel cases of reduction in alanine repeats that were in some way correlated with reductions in forefeet and hind feet. For instance, living amphibians, such as frogs and salamanders, typically have only four digits on their forefeet and five on their hind feet. Among reptiles, some groups of dinosaurs developed only three digits on their forefeet and hind feet. In forms like *Tyrannosaurus rex,* not only were the digits reduced in size and number but the entire forelimb was also severely reduced in size, which suggests that the expression of other homeobox genes was altered as well. In *T. rex,* as indicated by studies on chicks, the *Hoxd-9* gene was probably also affected. Birds, whose closest relatives are to be found among the reptiles, have a reduced alanine repeat series, and reduced digits, particularly in the wing. And snakes, which are reptiles, are the most extreme of all vertebrates in that they lack limbs and digits altogether.

Digit and limb reduction has also occurred independently in a variety of groups of placental mammals. For example, among artiodactyls, or even-toed mammals, pigs have two functional digits flanked by the somewhat reduced "trotters," while the majority of this group, such as deer, sheep, and cows, have only two toes. In these two-toed, or cloven-hoofed, animals, the two bones of the "palm" and "sole," which articulate with a reduced number of wrist and ankle bones, fuse together early in life to form a single bone. These animals also lack a fibula, one of the two lower-leg bones. Among odd-toed animals, or perissodactyls, some, such as tapirs, have only three toes on each foot, while digital reduction has been most severe in horses, which also do not develop a fibula. Three- and two-toed sloths represent another case of digital reduction. Regardless of whether whales are most closely related to an extinct group of carnivores or to a group of artiodactyls that includes cows, they are the snakes of the mammals as far as lacking hind limbs is concerned. The earliest fossil whales, which are at least 45 million years old, are truly whales in skull, tooth, and most of their body plan. They have hind legs, but these appendages are very diminished in size and number of bones.

It is common knowledge that, among tetrapods, there has been a general decrease in limb and foot complexity from the condition seen in the earliest amphibians. As was recently reviewed by the molecular biologists Neil Shubin and Cliff Tabin of Harvard University, and the paleontologist Sean Carroll of the University of Wisconsin at Madison, these morphological changes resulted from changes in the timing of expression and, consequently, of phases of interaction of homeobox genes that were already present in vertebrates. Clearly, alteration in the timing of the expression of interactive sets of homeobox genes represents a major way of introducing morphological change—even profound morphological change with, at the molecular level, little change at all.

Other, perhaps more remarkable morphological innovations appear to have occurred by way of altering the regulation of established homeobox genes. For instance, one of the questions in the realm of invertebrate evolution is: Where did the insect wing come from? One of the two competing hypotheses was that insect wings somehow evolved from limblike structures that functioned as gills. The other theory was that wings simply grew out of the body wall, and have no counterpart in non-winged, aquatic invertebrates. It turns out that two homeobox genes—dubbed *nubbin* and *apterous*—are expressed in a similar pattern in the development of the wings of the fruit fly (representing insects) and in the limbs of *Artemia,* a simple brine shrimplike crustacean. Ergo the structures that bud from the body wall, whether they are wings or legs, are controlled by the same genes.

Although the vertebrate eye and the insect eye are morphologically different—for instance, the former has a deformable, single lens while the latter is multilensed—they are both light-sensing and -gathering organs. In insects, a homeobox gene involved in eye formation is called *eyeless,* and its counterpart in vertebrates is *Small eye.* Since loss of function of these genes results in reduction in size or even in absence of the eye, they clearly have a role in eye development. This role is further indicated by experiments on fruit flies that provoked *eyeless* gene expression in typically non-eye-forming regions. The result was that miniature eyes developed on the appendages—legs as well as antennae—of these genetically manipulated insects. The reason that insects and vertebrates have the same homeobox gene for eye development, and yet have eyes that are so different in final form, may lie in another homeobox gene, the *Rx* gene.

The *Rx* gene is required for normal eye development in insects and vertebrates. In insects, as judged by the fruit fly, the *Rx* gene is expressed very early in ontogeny in the presumptive eye disc itself, as well as in a small area of cells at the anterior end of the embryo. In vertebrates, the *Rx* gene is expressed early in ontogeny in the region of the presumptive forebrain. In

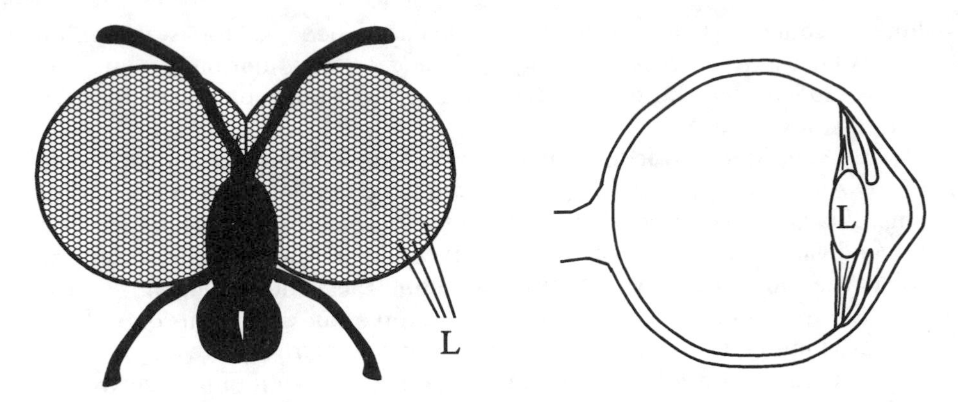

Schematic representation of the eyes of an insect with multiple lenses (left) and a cross section of an eye of a vertebrate, such as a human (right). The lenses (L) of an insect are fixed, whereas the shape of the lens (L) of a vertebrate can be changed or deformed by the muscles that are attached to it.

frogs, representing amphibians, the *Rx* homeodomain splits into two, giving rise to the first stage of eye formation, the optic cups. Later on, the *Rx* gene is also expressed in the region of the amphibian retina. In mice, representing mammals, the *Rx* gene is also expressed early in development in the forebrain. But in mice there is also early expression of the *Rx* gene in the presumptive eye itself. This, I suspect, may explain why the eyes of amphibians and mammals are essentially the same structure but the proteins in their lenses differ. Fish, such as the zebra fish, have an extra *Rx* gene, which presumably arose by gene duplication. Two of these genes are expressed in the fish retina and the third in the fish forebrain.

The team of molecular biologists that did this study also investigated what would happen when the *Rx* gene was overexpressed as well as when it was turned off. In the experiments on frogs on overexpression of this gene, extra retinal cells were found in the region between the presumptive eyes and the neural tube, or early spinal cord. In mice that had been bred for loss of *Rx* function, the results depended on whether the offspring were homozygous or heterozygous for this experimentally induced mutation. In mice that were heterozygous for the mutation, there was normal eye development. This means that the mutant condition was the recessive state and was not expressed because the normal condition was dominant. As would be expected, mice that were homozygous for functionless *Rx* genes did not develop either an eye or its bony socket.

When the *Rx* gene is functional, there is an eye. When it is deactivated, there is no eye, even though other genes involved in eye formation, such as the *Small eye* gene, are still present. Clearly, the interaction of more than one homeobox gene is necessary for eye formation. And equally clearly, the positions in the head region where the genes are expressed, and the times at which these genes are active, are responsible for the formation of not only an eye but of eyes that can be of great morphological difference.

Insects and vertebrates are similar in that their head regions consist of two main segments, anterior and posterior, which are specified by different combinations of homeobox genes. The posterior head segment in vertebrates corresponds to the hindbrain, and the homeobox genes involved in its formation are also differentially expressed along the rest of the body. The anterior head segment in vertebrates corresponds to the forebrain and the midbrain. Although insects have neural clumps, called ganglia, which serve as sensation- and information-integration or -processing centers, they do not have a brain, not even a simple one.

A variety of regulatory genes, such as *Sonic hedgehog,* are expressed in specified regions along the segmented vertebrate forebrain, mirroring the regional activation of *Hox* genes that create segmentation in the hindbrain and the trunk. Some thirty of the homeobox genes expressed in the vertebrate forbrain have counterparts in homeobox genes that control head development in the fruit fly. Obviously, in spite of their sharing numerous homeobox genes involved in the patterning of the head region, insects and vertebrates are different. Vertebrates develop a neural tube, which sloughs off neural crest cells that are necessary for structural development, and they have a tripartite brain. In this regard, it may be of some significance that all vertebrates so far studied express the *Rx* gene in the forebrain and develop a single-lens eye.

When there is change at the level of the homeobox gene itself, it may come in one of two ways. As in the case of alanine production by the *Hoxd-13* gene, a homeobox gene can be altered by the insertion of a molecular sequence. (One can conclude that the deletion of a small molecular segment could also occur.) The other way in which change within homeobox clusters can take place is by the duplication of genes.

Insects and vertebrates have similar homeobox genes for regional organization or segmentation for, essentially, Bateson's repeated parts. In insects, segmentation is noted, for example, in the development of body segments, as reflected in the plates that course along their abdomens, as well as in their jointed legs and antennae. The vertebrate brain, trunk musculature, and vertebral column are also examples of segmented structures. When homeobox gene control of segmentation was first discovered in insects, the gene cluster was identified as *Antennapedia.* In vertebrates, there are four *Hox*-gene clusters of the *Antennapedia* group. These four *Hox*-gene clusters must have arisen through a process of duplication. The development of anterior and posterior pairs of fins and their limb homologues in tetrapods may also have resulted from gene duplication (although it has been suggested that paired fore fins and hind fins and limbs are due merely to the activation of the appropriate *Hoxd* and *Hoxa* genes in these positions along the body). Gene duplication is not restricted to animals. Gene duplication in plants, creating new species within the span of a generation, had been known to plant breeders as well as to the early geneticists and those of the Evolutionary Synthesis. Gene duplication certainly counts as a kind of mutation.

The insertion or deletion of a molecular sequence encoding a particular protein in a homeobox gene, as well as gene duplication, can be easily embraced as representing different types of mutation. They conform more to the traditional notion of what we learn in school as constituting the basis of change. But there are even simpler mechanisms for producing change: the activation of homeobox gene expression in an area or areas in which these genes had not been active, and the deactivation or inhibition of homeobox gene expression in a certain area or areas. The former would have been the case in the expression of *Hoxd-11, Hoxd-12,* and *Hoxd-13* genes, which are otherwise silent across the front of the developing fish fin, in the same region of the tetrapod limb bud. The activation of homeobox gene expression in novel positions or in novel combinations at different times certainly produces significant changes.

It is, however, one thing to identify potential categories of genetic change, based on the delineation of homeobox genes, their domains of expression, their fields of interaction, and the structures they ultimately create, and, from this, to have a sense of the evolutionary background of major groups of multicellular animals. But it is another to try to understand the picture of evolution by means of homeobox gene regulation.

12

The New Evolution

Until the Utopian day when the processes of evolution are really well understood, we cannot afford to close our minds conclusively to any factors that might conceivably prove to be at the root of the many mysteries still remaining.

—G. G. Simpson (1949)

The Marvels of Homeobox Genes

With each passing day, another group of articles on homeobox genes appears in scientific journals. Oftentimes the focus of the research is on the identification of a new homeobox gene or homeobox gene cluster or the products of homeobox genes and their effects on development. The diverse levels of growth and differentiation—from the basic organization of an organism's body axes to the emergence of a specific toe—that can be investigated are staggering. Research has even begun to disclose the very beginning phases of cellular organization, such as in the determination of which end of the oocyte would become the tail, or posterior end, and which the head, or anterior end, of an organism. Since the oocyte is the cell that will eventually give rise via meiosis to the mature ovum and the three abortive polar body cells—and it is the ovum that might or might not eventually be fertilized—its axial organization does indeed take place at a very early phase of cellular differentiation.

It is developmentally imperative that an anteroposterior axis become established, because the body plan of a bilaterally symmetrical animal, with its right and left sides and head and tail, is orchestrated around it. As has been demonstrated in the fruit fly, an oocyte achieves its anteroposterior axis by first positioning itself against the cells of the posterior wall of the ovarian follicle, or egg chamber. The oocyte induces the cells of the egg chamber to become committed to inducing a posterior polarity. Cells of the egg chamber then send a protein, called Gurken, back to the oocyte. Gurken signals to a specific protein in the oocyte, called Torpedo/DER, which, in turn, proceeds to define the posterior end of the oocyte. The anterior end of the oocyte is determined secondarily, by default of the posterior end's having been determined first. Once the anteroposterior axis of the oocyte is established, the same messenger RNA that transcribed the protein Gurken, which is identified as *gurken* mRNA, or *grk* mRNA for short, is recruited to determine the dorsoventral, or up and down, axis of the cell. So it is, at least in the fruit fly, that well before an oocyte is transformed into a mature egg, the fields of right and left, front and back, and up and down are in place. Afterward, the oocyte goes through the phases of meiosis that will produce the ovum that will await fertilization—which may never occur—to initiate the unfolding of a bilaterally symmetrical organism whose body axes had been predetermined two cell generations earlier.

Since this sequence of cell-cell and cell-product interaction yields the body axes of animals with heads and tails, right and left, and up and down, one might expect that a different set of homeobox genes, messenger RNAs, and proteins would be involved in programming the body plan of a radially symmetrical animal, such as a starfish or a sea urchin, in which the appendages of the organism are arranged like the spokes of a wheel around a central axis. This, however, is not the case.

Starfish and sea urchins belong to a particular group of invertebrates, classified as echinoderms, which develop an internal calcite skeleton with externally protrusive spines and a vascular system whose fluid is plain water. Although the adult may be arranged in a radial pattern, an echinoderm begins its developmental life in the same way that a bilaterally symmetrical animal does, with cell organization reflecting up-and-down, side-to-side, and front-to-back axes. But after starting off life with the cellular axes of a bilaterally symmetrical animal, an echinoderm then becomes transformed into an organism with a specific, five-part, or fivefold, radial body plan.

The shift to becoming a fivefold, radially symmetrical echinoderm is tied to the recruitment or co-opting of the very same regulatory genes that perform totally different roles in arthropods, such as crustaceans and insects, and chordates, a group that includes the eel-like lancelets and all vertebrates, which are all bilaterally symmetrical animals. The specific homeobox genes are *distal-less, engrailed,* and *orthodenticle.* In the bilaterally symmetrical arthropods and chordates, the *distal-less* gene plays a role in organizing the position of the growing limb along the organism's anteroposterior axis. This gene was nicknamed "*distal-less*" because, when it was experimentally induced into a mutant state in fruit flies, it produced legs and antennae that lacked their terminal, or most distal, structures. In bilaterally symmetrical animals, the *engrailed* gene is involved in the growth of the central nervous system along the body's anteroposterior axis. And the *orthodenticle* gene is critical for the proper differentiation of individual structures in the head region of bilaterally symmetrical organisms.

In echinoderms, however, these regulatory genes do not perform the functions they do in bilaterally symmetrical animals. Instead, rather than one gene contributing to the formation of paired limbs, another to the formation of a linear central nervous system, and the third to details of the head region, as in arthropods and chordates, these three homeobox genes are expressed simultaneously during the larval and juvenile phases of echinoderm development and consequently produce the pattern that will yield the fivefold radial symmetry of the adult. The result is the loss of the initially established body plan of bilateral symmetry and the establishment of an entirely different scheme of symmetry. In various subgroups of echinoderms, some of these genes are further co-opted to perform more specialized functions. For example, the *distal-less* and *engrailed* genes are expressed together during the formation of the tubelike feet of all starfish. And, at a finer level of morphological detail, the *engrailed* gene of the brittle starfish is active in those cells that span the gaps between the calcite plates of this echinoderm's internal skeleton.

These observations on echinoderms and on fully committed bilaterally symmetrical animals point out the fundamentals of development from the standpoint of homeobox genes. First, the basic body plan of an organism is laid out. Then it is subdivided into regions. And then the more detailed aspects of morphology emerge. Throughout all of this, the same homeobox genes and products can be involved, but their roles in the development of an organism can differ in degree of morphological generality or specificity. The idea of a hierarchy of cellular and structural differentiation that proceeds from the general to the specific had been proposed first by the eighteenth-century German embryologist Caspar Friedrich Wolff. Since then, the reality of such a hierarchical arrangement of increasingly restricted cell function has been steadily documented. In the late 1970s, Andrew Lumsden, an experimental embryologist at Guy's Hospital Medical School in London, made two critical observations during his study of molar development in mice: The cells that give rise to the first tooth of the molar series clone and give rise to the second and then the third molar, and these cells also differentiate into the cells that produce the different cellular layers of the crown and the root of each tooth. More recently, developmental geneticists have discovered that the *distal-less* gene is not only involved in the development of an insect's legs, antennae, and other appendages but that it is also recruited to perform even more detailed and localized activities, such as the formation of wing spots in butterflies. While further documenting, but to a much finer degree than ever before possible, the different levels of cell fate, the beauty of regulatory gene studies is that they also provide the genetic backdrop to the processes of cellular and systemic differentiation.

The realization that all multicellular animals share many, if not most, of their homeobox genes has not escaped the notice of those who are interested in evolution. In a fascinating book, *The Shape of Life,* the Indiana University developmental biologist Rudolf A. Raff waxed eloquent on the implications of homeobox genes for understanding the myriad versions of body form that animal life has evolved over millions upon millions of years. There is, of course, a limit to the kinds of basic shapes an animal might assume. As the Harvard evolutionary biologist Stephen Jay Gould has remarked, animals have not evolved wheels.

The essence of Raff's thesis is that evolution of different animal body shapes should no longer be thought of as merely the result of de novo mutations introducing something entirely new to the gene pool, from which, in an equally unexpected manner, diversity then emerges. Rather, within certain limits to the shapes and parts that are available to all organisms—wheel-like structures not being among the latter—the differences we see among multicellular animal life in body plan are due to the recruitment, or, as he phrased it, the co-opting into new locations and in different combinations, of homeobox genes that are already present in an individual's cells.

All multicellular animals are united by their shared homeobox genes. This commonality reflects the persistence for more than half a billion years of the genes that are basic to the development of organisms as distinctive in their own right, and blatantly morphologically different from one another, as worms and humans. When particular genes are turned on for certain lengths of time and in certain regions, a worm may emerge. If the same or other genes are expressed for different lengths of time and in different regions, a more complex organism may develop. With the co-opting of specific genes and their expression in virgin cellular territory or in novel combinations, a worm may become segmented or a vertebrate may develop a tripartite brain, replete with fore-, mid-, and hindbrain components. Recruitment of yet other genes may create a wing spot on a butterfly's wing or an eye spot on a fish's tail.

The fact that the evolution of complex shapes and structures is not due to the gradual accretion of minute changes but, instead, results essentially from a shifting of when and where various homeobox genes are turned on or off has captured the imagination of a growing number of evolutionary biologists. In a recent column in *Natural History* magazine, Stephen Jay Gould ultimately discussed in terms of homeobox genes the suggestion, put forth by the early-nineteenth-century French comparative anatomist and natural historian Étienne Geoffroy Saint-Hilaire, that invertebrates were clearly relatable to vertebrates if one only perceived an invertebrate, such as a lobster, as being an inverted version of a vertebrate, such as a dog. The major difference, according to Geoffroy Saint-Hilaire, is that the skeleton of one organism surrounds its musculature, while, in the other, it is the reverse. Gould then summarized the attempted refutation of Geoffroy Saint-Hilaire's theory by the late-nineteenth–early-twentieth-century developmental anatomist Walter H. Gaskell. Gaskell asserted that the invertebrate nerve and gut, which lie near the undersurface of, for example, a worm's body, had actually migrated upward and become transformed into the structures we know as a vertebrate's brain and central nervous system. Gould praised both scientists for what were seen in their day as crazy propositions but that, in light of the amazing demonstrations of studies on homeobox genes, could now be appreciated as having a kernel of reality. On a very simple level of understanding homeobox genes, it really is possible to conceive of an invertebrate's being "converted" into a vertebrate without invoking much, if any, addition to the already available genetic information. The seemingly distantly related and very dissimilar groups we call invertebrates and vertebrates are, in their genes, much closer than scientists even ten years ago could have imagined.

But while it is certainly appropriate to marvel at the possibilities of producing different body plans merely by the recruitment of extant homeobox genes, with, on occasion, a bit of genetic spice being added by way of duplication of a homeobox gene cluster or by insertions or deletions from a particular homeobox gene, we are still at a loss to identify the mechanisms by which organismal change can emerge and become established in groups of organisms. We might know that *x, y,* or *z* homeobox genes are shared by a fruit fly and a mouse. And we might be able to delineate with reasonable precision the spatial and interactive domains of genes in the formation of a certain structure—as, for instance, in knowing the role of the *eyeless* gene in the formation of the multiple-lensed insect eye and the role of its mammalian counterpart, *Small eye,* in the development of a single, deformable eye lens. Yet, as far as relating the astonishing impact homeobox genes have on the development of structures to the fundamental mechanisms of evolution, there remain important unanswered questions.

One such question is: "How does more than one individual come to have a novel structure?" Goldschmidt, for example, was on the right track in terms of recognizing that significant morphological reorganization could occur. But he was ensnared by his reliance on major chromosomal rearrangement as the basis of significant morphological reorganization, and he fell back on the idea of the "hopeful monster": the chance emergence of an individual bearing major genetic and morphological novelty waiting and hoping for another individual like it, but of the opposite sex, to appear on the scene. Hugo de Vries also did not propose a mechanism for how a genetic mutation and its associated physical characteristic would arise in more than one individual. Instead, he relied on his observations of the unexpected appearance in the same generation of a few "monsters" of the evening primrose to argue that, indeed, it could happen in any species.

A second question is: "How will a novelty look when it does appear?" This question is basic to the debates on whether gradualism or punctuation is the only, or at least the more prevalent, mode of evolutionary change. Paleontologists have long argued over the reality of documenting graded series of morphological intermediates in the fossil record. In the wake of decades of failed attempts to demonstrate truly continuous sequences of fossils representing unquestionable ancestor-descendant transitional forms, there is still a great deal of disagreement on the reality of the gaps in the fossil record. Are representative fossils missing because of preservational complications, or because of a more rapid and saltational process of speciation?

Inheritance Lost

The history of the discovery of the fundamentals of inheritance and its aftermath embodies a curious historical twist. After the rediscovery and refinement of Mendel's principles, and the recognition of how simply the genetic underpinnings of morphology can be transmitted from one generation to the next, the direct discussion of inheritance ceased to be part of the literature on evolution. I do not mean to suggest that scientists, including such geneticists as Thomas Hunt Morgan, Theodosius Dobzhansky, and Richard Goldschmidt, as well as such naturalists as Ernst Mayr and such paleontologists as George Gaylord Simpson, were ignorant of genetics. To the contrary, their writings were very much dependent on the reality of Mendelian principles. But these scholars, and most evolutionists since, have relied less on the application of Mendelism to evolutionary problems than on the assurance that Mendelian principles of heredity provide the connectivity of life, whether that continuity takes place on a grand scale over vast expanses of time among a multitude of ancestors and descendants or only during the course of a few generations, depending on one's model of speciation.

This historical curiosity is probably at the heart of why Richard Goldschmidt failed to convince the leading evolutionists of his day that his observations of major morphological alteration in gypsy moths should be taken seriously in the formulation of a model of evolutionary change. Goldschmidt simply could not couch his argument in understandable terms of inheritance. When Darwinism was finally merged with Mendelism, population geneticists were invoking small mutations, or micromutations, as the basis of eventual, significant morphological change. The simplicity of Mendelian inheritance became tied to the transmission from parent to offspring of simple and minor genetic, and, consequently, simple and minor morphological, changes. Since Goldschmidt was not convinced of the relevance of minute variation to the origin of species, he was forced to seek a different kind of genetic revolution as the source of evolutionarily significant morphological novelty. He thought he had found his answer at the level of the chromosome. Although chromosomes had been accepted as conforming to a general picture of Mendelian inheritance, the profound reorganization—the large mutations, or macromutations—that Goldschmidt postulated for the origin of species seemed uninheritable. Gradualism appeared to make sense because the underlying notion of small genetic changes could be accommodated by Mendelian genetics. Macromutations, especially of the sort Goldschmidt proposed, did not.

Within the constraints of Darwinian evolution, natural selection could operate only in a physical realm composed of continuously graded characteristics. In this context, intrapopulational variation, on which natural selection acted, could be altered only in equally trivial ways. For the Synthetic Theory of Evolution, being founded on a particular coalition of Darwinism and Mendelism, the same processes that worked on individuals within species were perfectly reasonable forces in producing new species. All that was needed was sufficient time and selection pressure.

For Goldschmidt, however, speciation was not merely an extension of the daily struggles of existence of individuals within a species; it was something more profound, and it needed an equally profound genetic revolution in which to enact the process. Unfortunately, Goldschmidt was done in both by the general lack of knowledge at the time of the interplay of genetics and development and by his own "hopeful monsters." This in spite of the irony that no one on the side of gradual change leading to speciation had ever observed, much less studied genetically, the emergence of a new species. It was simply assumed that the processes that were supposed to occur among individuals and between generations would work at levels of evolution beyond the individual.

A similar dilemma befell Stephen Jay Gould and Niles Eldredge's second article on their model of punctuated equilibria, which was published in 1979. During the seven years that followed the initial publication of their theory, regulatory and structural genes had been discovered. Correctly, Gould and Eldredge saw in these different classes of genes the potential for understanding evolution. For example, even though great similarity between humans and chimpanzees in the proteins that are coded by structural genes had been demonstrated, which, in turn, implied that the structural genes were essentially identical, Gould and Eldredge were not convinced that this was the proper genetic level at which one should seek answers to questions of evolutionary change. For, in spite of sharing well over 90 percent of their proteins and structural genes, humans and chimpanzees are blatantly dissimilar generally and in detail in many aspects of their reproductive physiology, as well as in their hard- and soft-tissue morphology. In this regard, Gould and Eldredge argued, one should scrutinize the regulatory genes, which are responsible for orchestrating the expression of and the interactions between the structural genes themselves. Even a small difference in only one regulatory gene could make a huge difference in terms of whether an organism developed into a chimpanzee or a human.

But here Gould and Eldredge stumbled into the same kind of trap that snared Goldschmidt. Although they had good reason to be smitten by the implications of regulatory genes, Gould and Eldredge did not provide a model of how changes in regulatory genes could happen, or how such changes would arise in more than one individual and subsequently spread throughout the population. Perhaps because of the common knowledge that genes provide the basis of what is inherited and are themselves inheritable, Gould and Eldredge felt that it was sufficient to leave their argument at the point they did: recognizing that not all genes are alike in their effects and that this difference could indeed have tremendous evolutionary import. In keeping with the tradition of the gradualists, who sensed that studies on individual selection would work at the level of producing new species, Gould and Eldredge left their readership with the feeling that the answer to significant evolutionary change—which to them meant novelty and speciation—lay not in the genetics underlying individual variation but in the regulatory genes that control the entire developmental spectrum.

Gould and Eldredge's plea for considering the role of regulation in producing evolutionary novelty was obviously not satisfactory for such population geneticists as Brian Charlesworth, Russell Lande, and Montgomery Slatkin. In their refutation of the model of punctuated equilibria, Charlesworth and his colleagues argued that there was no reason to invoke any mechanism to explain change other than the basics of population genetics that had been documented by Morgan, Fisher, Wright, Dobzhansky, and geneticists who followed. From this position, Charlesworth and his collaborators appealed to the effects of Darwinian selection as it acted on available genetic variability, which was kept churning at low levels by the incorporation of the occasional micromutation. Whatever the ultimate new feature, they argued, the way in which it was acquired was by the slow accumulation of minor changes at both the genetic and the morphological levels, which, in the end, added up to the novelty in question. In their presentation, Charlesworth, Lande, and Slatkin kept alive the notion that the path to evolutionary change was through the marriage of Mendelism and Darwinism and could be fulfilled only within the realm of the kind of micromutation that was responsible for individual variation. But at the same time they were extolling the virtues of Mendelian inheritance, they failed to apply its principles to the transmission of mutation.

Although it was not stated outright, the gist of Charlesworth and his colleagues' case for Darwinian evolution rested on the implication that not only would mutations be small but their effects would be virtually instantaneously available to natural selection. This is odd, because these three geneticists sought support in the work of the mathematical population geneticist Ronald Fisher, who had himself struggled with the fundamentals of gene transmission and its theoretical implications. While these four geneticists would have agreed that evolutionary change was and had to be gradual, Fisher would not have embraced the notion that the effects of a gene change would be expressed and, consequently, made available for selection soon after the mutation. Fisher knew all too well, and wrote about in detail, the mechanism of Mendelian inheritance and its impact on evolution.

George Williams bypassed morphology altogether and went straight for the gene, which he also redefined from its being solely a small segment of a chromosome to its being anything at the genetic level—even, theoretically, an entire chromosome—that behaved as a unit of inheritance. But Williams's discussion of selection on genes via the morphologies and behaviors they generate carried with it a certain implication. As with Charlesworth, Lande, and Slatkin's rejection of punctuated equilibria, Williams's argument was imbued with the idea that the effects of change at the gene level, though anathema to his model of stability, actually become quickly accessible to natural selection.

The general theme of genetic change becoming expressed, in short order, in the morphological variability of a population was not new with either Williams or Charlesworth and his colleagues. It had begun to crop up decades earlier, especially in the writings of Ernst Mayr and Theodosius Dobzhansky. It was never articulated as such, but it is implied in the language these evolutionists used. Perhaps this assumption is a natural outcome of the historical process of going from discovery to acceptance. When Mendelian principles were first being explored in detail and applied to evolution, it was necessary and still important to spell out the basics of inheritance, including the nature of dominant versus recessive alleles. When it became abundantly clear how inheritance worked—how different allelic states were expressed and inherited—it was no longer necessary to repeat the essentials; they were implicitly understood. Everyone knew that genes were at the base of morphology, how genes were passed on from generation to generation, and that a change in a gene produced a change in morphology. Armed with knowledge of the continuity of life, evolutionists could proceed directly from a hypothesized mutation to its being acted upon by natural selection without going through the exercise of figuring out how that would happen and what its effects would look like—which is what such geneticists as Bateson, Morgan, Fisher, Haldane, and Wright would have done.

It is, however, more imperative now than ever that we do return to the basics of inheritance; that we rethink the debates that have ensued throughout most of this century on the congruence of genes and morphology and how they fit into a picture of evolution. Goldschmidt may have relied too exclusively on chromosomal rearrangement to explain the profound morphological differences he documented in gypsy moths. But this does not mean that his, or anyone else's, observations of morphological disjunction and discontinuous variation were incorrect.

Back to the Basics

Perhaps the best way to begin piecing together a fresh look at evolution is to start simply. Following in the footsteps of Mendel, we need to think first about genes as coming in the form of one of two possible allelic states: recessive or dominant. When an individual is heterozygous, the dominant state will be expressed and the recessive masked. As had been known to early population geneticists, including Bateson, Morgan, Fisher, Haldane, and Wright, and has recently been confirmed by the Jackson Laboratory group in mice that spontaneously developed polydactyly, most of the mutations that cropped up in whatever experimental organism they were studying manifested themselves in the recessive state. Mutations that emerged in the dominant state were less frequent and typically injurious, if not plain lethal, to their hosts. But since organisms do possess a number of dominant alleles that are not deleterious to their bearers, a major question that had to be confronted was: "How is a recessive state, in which nondisruptive genetic novelty is safely introduced, converted into a dominant one?"

Fisher, Haldane, and Wright tried to answer this question by invoking what they called a "modifier," a term that is now applied to a totally different concept of gene action. Fisher thought that the only way that evolution by natural selection could work was if a recessive mutation were quickly modified into the dominant state. Haldane and Wright could get behind neither Fisher's dominance theory nor the idea that there was a specific modifier for the task of converting a recessive allele into a dominant one. Haldane and Wright's modifiers occupied a vaguer role in the grand scheme of recessives being altered to dominant status. To this day, no one knows how the transformation of a recessive allele into a dominant one occurs.

The introduction of genetic novelty via the recessive state was an important piece of information that Fisher, Haldane, and Wright incorporated into their speculations on the tempo of evolutionary change. All three theoretical population geneticists believed that the heterozygous state was preferable to the homozygous state. Heterozygosity was supposed to confer greater selective advantage on an individual than homozygosity because the former state maintained the genetic variability that would be tapped from one generation to the next. Novelty could reside in the recessive allele, while heterozygosity was maintained by the dominant allele.

Because Fisher thought of the evolutionary process in terms of the effects of natural selection on large populations, he concluded that the spread of the recessive mutant would be slow, as would also be the spread of the dominant state once it had been achieved. Small populations, however, were key to Haldane and Wright's scenarios of evolutionary change. As these two geneticists envisioned it, mutation arising in the recessive state would, through the production of heterozygotes, spread quickly within and across generations until there was a critical mass of heterozygotes sufficient to produce homozygotes for the recessive state. Along with the emergence of homozygotes for the new recessive allele would come the physical or behavioral expression of the mutation. The smaller the size of the population, the more rapid would be the increase in number of heterozygotes for the recessive genes that would eventually give rise to a population of homozygotes. For Wright, the small population was the incompletely severed peripheral isolate that could feed genetic novelty back into the parent population. For Haldane, the peripheral isolate was totally genetically independent of its original species. Wright and Haldane were in agreement, however, on the point that, once it was expressed in the homozygote, the morphological change resulting from the expression of the homozygous recessive allele could be acted upon by selection.

Fisher was the most adamant of the three mathematical population geneticists in maintaining that the only way that evolutionary change could come about was if the recessive mutation, once it was available in a small number of individuals, was quickly converted into the dominant state. For only when a particular morphology or behavior is governed by an allele that is in the dominant state can natural selection act on the novelty. If, according to Fisher, a mutation arose in the dominant state at the very beginning, it would probably be eliminated from the population by selection immediately, even if its effect was not deleterious to its bearer. Since, theoretically, this is a mutation that affected only one individual, the novelty would most likely have no relative selective advantage. But even if it did, its presence in a single individual would not be conducive to its spreading within the species. Given these problems, Fisher arrived at a compromise: Because genetic change is introduced in the recessive state, it has to remain in that state for at least a few generations in order to spread somewhat within the species. But then it would have to be converted into the dominant alternative because that would allow for the possibility of an individual's having one dominant and one recessive allele—and heterozygosity was critical to Fisher's view of what was genetically best for a species. If selection, which could now act on the expressed dominant state, found the change adaptive, it would conserve it and its population of bearers would pass it on to their offspring.

The types of mutations that geneticists typically discussed in the context of the constant introduction of new, but minor, twists on continuous variation are those that affected such features as height and eye color in mammals or bristle number and relative wing length in fruit flies. Indeed, what convinced Morgan and his colleagues that evolutionary change was gradual was Muller's experimental manipulation of generation after generation of fruit flies in which he painstakingly altered the predominant wing length from being longer, to being a bit shorter than, body length. Muller was able to do so because there were a few individuals with short wings, although most individuals of the parental population had long wings. Occasionally, as the experiment proceeded, individuals would emerge with slightly shorter wings than usual and Muller selected them to breed for the next generation. Eventually, wings that were shorter than body length became the norm for the population.

In considering the significance of Muller's experiment, it is important to realize that the genes for short wings were already present in the experimental population and that even when he managed to shift the tendency of the population from long to short wings, there were still individuals with long wings. Muller had certainly changed the character of the population in terms of average wing length, but he had not manipulated the genes that were responsible for the presence or absence of wings. Morgan assumed from Muller's experiment that, if the process of selection were to go on long enough, wing length would eventually become so diminished that the wingless state would eventually result. But in doing so, Morgan was conceiving of the state of winglessness as the result of a progression in the expression of variation in wing length. This extrapolation is, however, not valid.

The question that must be addressed is: "Is observation of change in minor aspects of a structure equivalent to the change that brought about the structure itself?" The answer, we now know, is no. All fruit flies had wings in spite of the fact that someone experimentally shifted the average wing length of the population. Homeobox genes code for the development of wings and for the location of wings in specific places. Differences between individuals in wing size result from differences in the expression of the genes that affect wing length. As such, whether a fruit fly is homozygous or heterozygous for long or short wings is of no consequence to the alternative states of "having wings" or "not having wings." Developing a major structural feature and having a nuance in its size or shape are two different things.

Tetrapods have digits—fingers and toes—on forelimbs and hind limbs not because they have a set of homeobox genes that no other organism has but primarily because an existing series of homeobox genes is activated in a region of the early limb bud where it is otherwise not expressed in the growing fish fin. The only molecular difference lies in tetrapods' having a short alanine-coding sequence in the very last of these homeobox genes, the *Hoxd-13* gene. These, then, are the two reasons why tetrapods have limbs with digits at their ends. True, some tetrapods have more digits than others, and some have shorter digits than others, just as some have more or shorter bones in a limb. But as we know from studies on alanine repeat sequences in chickens, which have three vestigial fingers, versus humans and mice, which have five almost equally sized digits, the difference in the number of digits results not from different numbers or kinds of homeobox genes but from differences in interactions between alanine sequences of different lengths and the *Hoxd-13* gene. In the case of a shorter alanine series, an individual will have fewer and shorter fingers and toes. A greater number of alanine repeats has the opposite effect. Indeed, humans with a mutation that expands the alanine-coding sequence in the *Hoxd-13* gene develop more fingers, toes, wrist, and ankle bones, as well as longer fingers and toes, than is typical not only of our species but of mammals in general.

Physically normal humans are pentadactyl, having five fingers on each hand and five toes on each foot. The presence of pentadactyly in humans is not, however, specific to being a human, or even a primate, for that matter. Pentadactyly is certainly a retention from the last common ancestor of all mammals, and probably a retention from an even earlier vertebrate ancestor. Even though some humans have longer fingers than others, the third finger of *Homo sapiens* is usually the longest of the set. But this configuration is not unique to *H. sapiens.* All anthropoid primates—humans, apes, Old World monkeys, and New World monkeys—have third digits that are to some extent longer than the other fingers. But within each species of anthropoid primate, some individuals have longer or shorter, or thicker or thinner, fingers. Clearly, the general pattern of digit length among anthropoid primates reflects a different and more specific level of gene expression than that which defines pentadactyly. Individual variation in digit size or shape is yet an even lower level of gene expression, one that is far removed from the origin of the tetrapod "hand," the establishment of pentadactyly in the ancestor of a large group of organisms, and the patterning of digital proportions in a group of primates.

Since the Synthetic Theory of Evolution is predicated on a synergistic relationship between Mendelian inheritance and Darwinian selection, the expectation is that the only way in which novelty of the sort that distinguishes one species from another arises is by the gradual accumulation of minute structural modifications that result from small changes at the genetic level. According to this model, one major feature can eventually be transformed into another. The evolution of the vertebrate eye comes to mind as an organ that, under these constraints, had to have evolved piecemeal. Although Darwin's own doctrine forced him to believe this, the thought of how such a complicated structure, which is useful only in its complete state, could have been built biological grain by biological grain was definitely troublesome to him. More recently, the sociobiologist and promoter of the notion of "selfish genes," Richard Dawkins, has championed a proposal that estimated how many generations it would take to go from having no eye almost to having the kind of eye that vertebrates and octopuses have.

The scenario that Dawkins embraced had been formulated by two biologists, Dan Nilsson and Susanne Pelger. They began with their hypothetical eye being already in a formative stage that consisted of a patch of light-receptive cells at the skin's surface that was sandwiched between a transparent protective layer of cells above and a layer of darkly pigmented cells below. By way of mathematical modeling, Nilsson and Pelger calculated, conservatively, that it would take only 400,000 generations for this non-eye region of skin to be transformed into an organ of sight. Even when only a 1 percent change per generation is invoked over what is really a relatively short period of geologic time, Nilsson and Pelger predicted that selection would be able to cause the skin to invaginate, bringing the presumptive retinal layer down with it, and then to fill with fluid of a very low refractive index. Still no functional eye, but then a lens begins to emerge and eventually it achieves a refractive index sufficient to provide sight. The maximum number of tiny steps required to go from the flat to the invaginated structure was estimated at 1,033. It took only 529 more steps to make a lens and put the eye into its final, semi-flattened shape.

Absent from this simulation was consideration of the origin of the patch of layered cells in the right place, the development of a variable iris and controlled focusing, the creation of the nervous optic chiasma that is the region in which the optic nerves cross over, and the innervation of the eyeballs by the optic nerves, which constitute one of the twelve pairs of primary nerves that emanate from the brain itself. Curiously lacking, as well, was any discussion by Dawkins of the selection pressure that would have set the process in motion and of the selective advantage members of more than 399,000 generations of their species would have enjoyed as they served as conduits for this ever-invaginating, liquid-filled pair of pockets in their head region. But, once the process had taken off on this trajectory, there was, as Dawkins saw it, no turning back. For, in his view of evolution, "[u]nlike human designers, natural selection can't go downhill, not even if there is a tempting higher hill on the other side of the valley." It is, however, one thing to model how such changes might have occurred seamlessly and gradually, and another to have a basis for doing so.

Do we actually need to invoke such an elaborate thought experiment in order to understand the origin of the vertebrate eye, or any eye, for that matter? I think not. And the reasons lie in knowing that there are homeobox genes for eye formation and that when one of them, the *Rx* gene in particular, is activated in the right place and at the right time, an individual has an eye. When something goes awry with this gene, the other homeobox genes involved in eye development cannot do their job, and an eye does not form. Clearly, the difference between having or not having an eye is a different proposition altogether from the gradual accretion of the bits and pieces that make up an eye. At the genetic level, major morphological novelty can indeed be accomplished in the twinkling of an eye. All that is necessary is that homeobox genes are either turned on or they are not.

But having marveled, yet again, at the importance of homeobox genes in helping us to understand the basics of evolutionary change, we must deal with the question of how innovation can both arise and be inherited. Only after addressing this subject can we speculate about how such inheritance would manifest itself evolutionarily.

When lack of function in the *Rx* gene is experimentally induced, the heterozygous bearer of this mutation is normal in both eye development and morphology. This is because eye formation is controlled by the dominant state. A heterozygote is potentially capable of passing on the dominant state for eye formation as well as the experimentally induced recessive state for eyelessness. If such an individual were to mate with an individual that was homozygous for normal eye development, all of their offspring would have functional eyes. Some of these offspring would, of course, be genetically homozygous for the dominant state of normal eye development, while others would be heterozygous for the abnormal condition. The latter situation would not, however, make any difference to the individual itself, because the dominant allele would allow eye formation to proceed apace. Through the mating of homozygotes for the normal state with heterozygotes, the recessive state would be able to infiltrate the species. Individuals that were homozygous for the recessive mutation would be eyeless. Since the *Rx* gene experimenters knew the pedigrees of their laboratory animals, they could produce eyeless homozygotes quickly, in a matter of a single generation. But this is not how it would work in nature.

Whether you adopt Fisher's, Haldane's, or Wright's calculations for estimating how many generations it would take before there were sufficient numbers of recessive alleles available in a population for homozygotes to begin appearing, it would seem, at first glance, to be relatively significant: hundreds, if not thousands. However, for tiny organisms like the fruit fly, in which the turnaround time between generations is a matter of days, and for many small mammals, such as shrews, where it can be less than a year, even a few thousand generations would be a mere hiccup as far as the age of the earth is concerned. If there was some amount of inbreeding—which Haldane pointed out would occur naturally in not-so-small groups, such as peripheral isolates, and would not be genetically injurious to offspring over long periods of time—the emergence of homozygotes for the recessive allele would be more rapid and their emergent numbers greater.

The upshot of this exercise is that while the number of heterozygotes for a recessive allele was increasing generation to generation, no one in the species would be expressing the genetic condition it represented. Only with the appearance of homozygotes for the recessive state would that allele be realized morphologically. This, it would seem, is precisely what happened in Thomas Hunt Morgan's laboratory colony of fruit flies. For generations, all fruit flies had normal eyes. Then, one day, he and his colleagues came to work and discovered that a number of eyeless mutants had hatched. By selectively breeding these individuals, Morgan and his collaborators were able to increase the eyeless population. No doubt, without knowing it, they were inadvertent observers of a mutation in a homeobox gene that was similar to the mutation that was produced in mice by experimentally targeting a specific gene in order to alter it.

In the case of experimentally induced mutation in the *Rx* gene, the recessive state would be eyelessness. And if it, as in Morgan's fruit flies, were to spread by way of heterozygotes from generation to generation, eventually individuals that were homozygous for the eyeless mutation would emerge. Since, however, as has long been known to geneticists from William Bateson to the present, most nonlethal mutations arise as recessive alleles, it is likely that the *Rx* gene also made its first appearance in this state. Only sometime after this new allele had become established in the species would it then have been converted to the dominant state in which we now know it.

If we consider that the genetic basis for eye formation must have presented itself first in the form of a recessive allele, we can predict—as Fisher, Haldane, and Wright would have done—how it would have spread throughout a species. Even one heterozygote for the recessive state for eye formation could produce a small number of heterozygotes for this allele. These heterozygotes would, in turn, produce more heterozygotes, and so on, from generation to generation. Once there was a critical number of heterozygotes for the recessive, eye-forming state, homozygotes for this allele would probably appear. Following Mendel's prediction of a 3 to 1 ratio of offspring expressing the dominant allele versus the recessive allele when hybrids for these two states mate, there would be a one in four chance that heterozygotes for the recessive eye-forming allele would produce an offspring that had eyes. In addition to there being a 25 percent chance that these

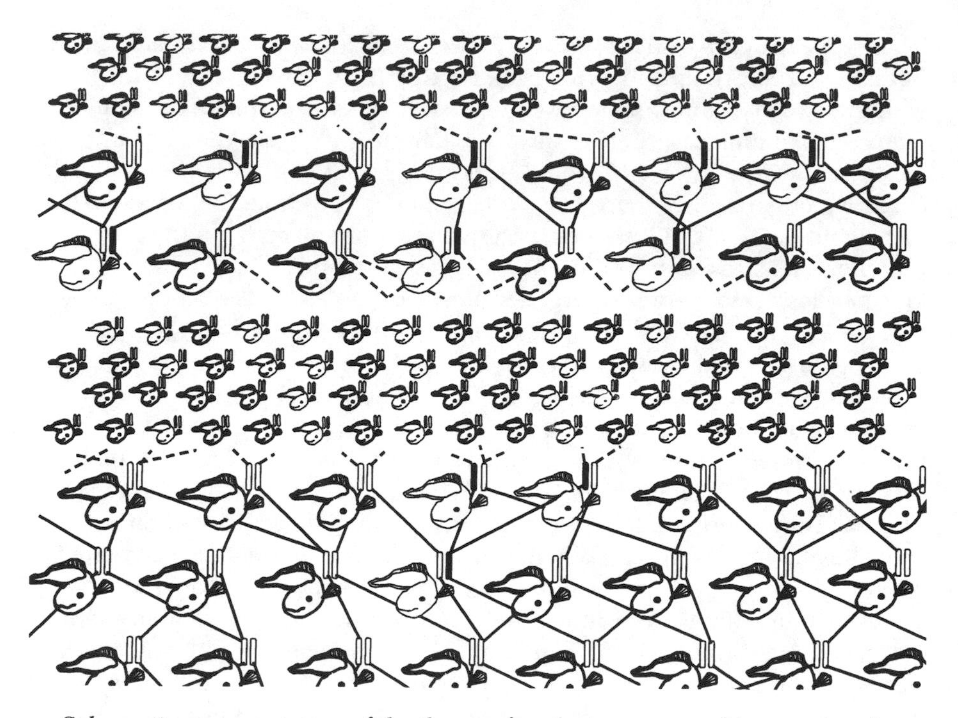

Schematic representation of the theory of evolution proposed here, using the emergence of the first toothed vertebrate species as an example; the large images give a magnified view of what is going on, and the lines connecting parents with their offspring indicate the transmission of alleles. Proceeding from the bottom of the page to the top, we begin with a species lacking teeth. Some individuals of this species would be homozygous for dominant alleles and others for recessive alleles. Others would be heterozygous. These conditions are not indicated in the chromosomes beside each image. The second line contains an individual in which a mutation arises in the recessive state (indicated by the darkened chromosome). Next, other individuals pass their

heterozygotes would produce a sighted offspring, the available pool of heterozygotes would probably ensure the production of more than one sighted individual in the next generation.

From an evolutionary perspective, once recessives could pair up with recessives, morphological novelty would appear abruptly, even though it would have taken a number of generations for the critical mass of heterozygotes to increase to the point of producing homozygotes. In this model, more than one homozygote for the recessive—and potentially many more than one—would emerge in the same generation, or within several generations of each other. Rather than a single individual presenting itself as a "hopeful monster," waiting for the chance occurrence of an individual of the opposite sex as morphologically and genetically extraordinary as it, the homozygotes produced by heterozygosis would be in the company of a number of poten-

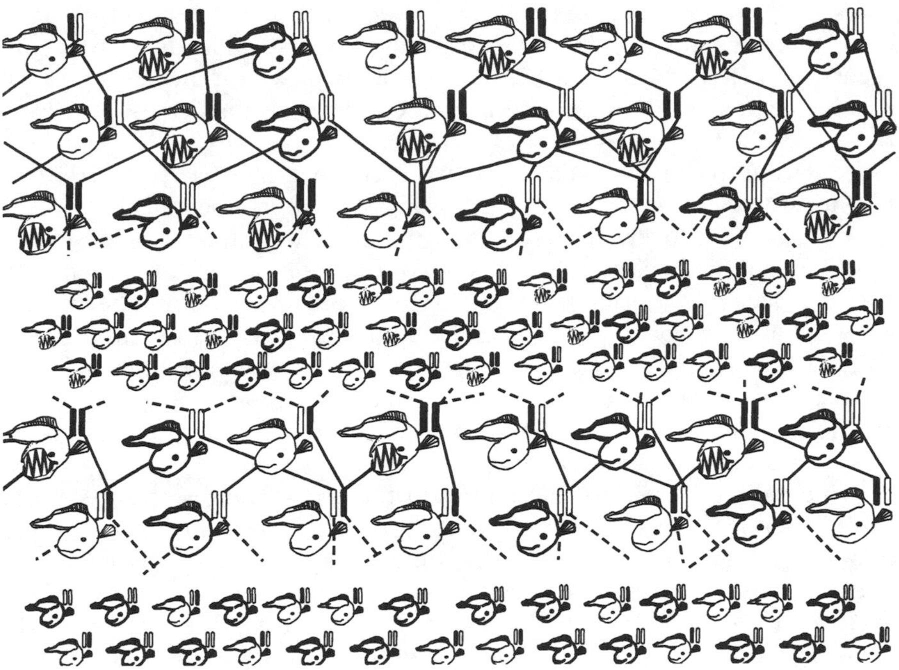

unmutated alleles on to the next generation, and this individual passes the mutation on to two individuals in the next generation. As this process proceeds for a number of generations, the mutated recessive allele spreads throughout the population but is unexpressed because it is in the heterozygous state. At some point, the population becomes so saturated with heterozygotes that they produce homozygotes for the mutated recessive allele, which, in this example, results in a number of individuals with teeth. These toothed individuals will mate with each other and produce more of their kind while heterozygotes continue to add to the number of toothed individuals by producing homozygotes for the recessive state.

tial mates, both from their own generation and from those that would follow. In addition, these homozygotes would all be phenotypically different from individuals of their original species.

Being different, these mutants would probably not be sought as mates by members of their parental species. Nor would they be likely to seek the latter as mates. On the basis of whatever the novel feature was—morphological, physiological, behavioral—the homozygotes for it could be perceived as constituting a different species from its original population. And, certainly, avoidance of mating, or choosing not to mate, would constitute a barrier to reproduction—which is another criterion that evolutionary biologists have used to distinguish species from one another. But even in the face of the novelty, and the likelihood that the homozygotes bearing it and members of the original population might not mate, this does not mean that the homozygotes for the mutation could not have mated with members of the parent population and produced fertile offspring—because individuals with the recessive allele had been mating without problem with individuals bearing the dominant allele during the period of heterozygosis that led eventually to the emergence of homozygotes.

Knowing that individuals of a newly emergent species apparently can but, for whatever reason, may not reproduce with members of the species from which they diverged allows us to understand better two phenomena that have plagued systematists in their pursuit of a clean definition of a species. Occasionally, animals that do not breed with one another in the wild and are physically different enough for systematists to identify them without hesitation as separate species (such as horses and zebras and panthers and tigers) will breed successfully in captivity, producing offspring that are not sterile. How could this happen if, according to the popular view, species are supposed to be genetically incompatible entities?

The other perplexing situation lies in the occasional occurrence in the wild of what biologists call hybrid zones. A hybrid zone is an area of overlap between populations of two seemingly different species in which individuals from each species freely interbreed. Often there is a gradation, for instance in physical and behavioral attributes, from one species, through the hybrid zone, to the other species. Although the characteristics of the species at each end of this continuum may be consistent with what a systematist normally uses to distinguish species from one another, the maintenance of genetic continuity through the hybrid zone is inconsistent with the popular biological species definition, which is based on its genetic discontinuity and incompatibility with other species. But while a hybrid-zone situation constrains some systematists in recognizing two distinct species, it is also used as an example of speciation in progress: The distinct populations at each end of the continuum will be distinct species once gene exchange through the hybrid zone is disrupted.

For example, the New York University geneticist and primatologist Clifford J. Jolly has for years studied the hybrid zone between the yellow and the Anubis baboons in Kenya and that between the Hamadryas and the Anubis baboons in Ethiopia. Although pure yellow and Anubis baboons away from the hybrid zone are easily distinguished by the color of their fur alone, and Hamadryas baboons by the males' manes, Jolly and others involved in baboon systematics have suggested that all of these baboons should be considered members of the same species. If a species is defined according to the ability of its members to breed and produce viable and fertile offspring, this is a reasonable conclusion. More recently, biologists have discovered that a local species of Indiana crayfish, the blue crayfish, are producing hybrids with an invading Kentucky species, the rusty crayfish. The aggressive hybrids are able to produce offspring and, in turn, are driving out the original species of crayfish. So far, the blue and rusty crayfish are still recognized as separate species.

Although these are documented examples of hybrid zones, it is quite possible that scientists are studying potentially real species, but species that are not so different from one another that they cannot recognize each other as potential mates. Given the model I am proposing for the spread throughout a species of the genetic basis of an evolutionary novelty, we should expect mate recognition to be a primary factor in determining whether members of closely related species will breed with each other. In the absence of species-specific cues, whether morphological, physiological, or behavioral, hybrid zones would exist, and since the spread of a mutant allele and its eventual expression in homozygotes does not mean that individuals of a new species cannot breed successfully with members of the original population, hybrid zones would naturally occur.

If, however, the difference between the mutants and others of the original population were of a certain magnitude or kind morphologically, physiologically, or behaviorally, these individuals probably would not recognize each other as potential mates. The novelty could also affect the structure of the reproductive organs or cells or interfere with reproductive physiology or embryonic viability, which would certainly isolate the mutants from the parental population. If the mutant and original populations became geographically separated from one another, there would be another, but still significant barrier to mating.

Following Hugh Paterson's model of a species mate-recognition system, we would predict that these mutant homozygotes would certainly respond to each other as potential mates because they would all possess the same novelty, whether it be behavioral, physiological, or morphological. If the bearers of the expressed mutation managed (or "chanced," as Thomas Hunt Morgan would have put it) to survive in the circumstances in which they found themselves, and they produced offspring, which, in turn, produced offspring, and so forth, the population of novelty-bearing individuals would increase. At some point, the recessive would convert to the dominant state.

A New Way to Look at Evolution

The expectations of this model differ from those of most discussions of mutation and natural selection. As originally portrayed by Thomas Hunt Morgan and emphasized in the Synthesis, mutations of the kind that contribute to evolutionary change are typically thought of as small and of little phenotypic consequence. Natural selection picks and chooses from among the resultant multitude of morphological and behavioral variants that arise through the constant motion machine of mutation. The time between the appearance of a mutation and its availability to natural selection is often implied to be short. As in the writings of, for example, George Williams and Ernst Mayr, the reader is often left with the impression that the mutation in a parental population will be fully expressed and will become part of the pool of variation of the next generation. But, obviously, the basics of Mendelian inheritance, and the fact that most mutations emerge in the recessive state, to which natural selection is blind, do not yield the traditional picture of evolution by mutation and natural selection.

It is certainly true that mutations, if they are of the kind usually studied by population geneticists, will not be of major consequence to the overall morphological or behavioral repertoire of the individuals bearing them. But whether or not I have an allele for blue eyes, although I actually have brown eyes, matters little to the homeobox genes that correctly orchestrated the formation of my eyes. Similarly, the fact that I have small hands and feet and a head whose size does not conform to Leonardo da Vinci's formula that an individual's height must be equivalent to the height of eight heads stacked is irrelevant to the role of the homeobox genes that produced my hands and feet, my head, and my tripartite brain.

The simple model of introducing evolutionary change within the constraints of Mendelian principles of inheritance that I am proposing does not depend on a particular kind of mutation. It only makes use of the fact that most mutations arise in the recessive state. If a mutation were of a sort that affected the controlling region of a homeobox gene cluster, or a specific homeobox gene, and it altered the interactions of homeobox genes and their products and, ultimately, the final morphology of the organism, then that would be something of significance. But it would still be a mutation, and it would still behave within the parameters of Mendelian inheritance. It is just that the consequence of such a mutation would be profoundly different from the genetic change that would alter eye color, height, or any similar kind of variation on an already established genetic and morphological theme.

The nature of a mutation, its impact on the individual, its mode of inheritance, and its effects on offspring are critical to a workable theory of evolution. It is not sufficient just to demonstrate that regulatory changes can affect individuals in profound ways. For instance, in recent experiments on fruit flies, changes in temperature were thought to have altered their morphology drastically and abruptly. In actual fact, however, mutation of the gene in question is typically spontaneous and of unknown origin, but can be, and was in this experiment, chemically induced. Since the homozygous condition for the mutation is lethal, these and other geneticists have kept the mutation alive through lineages of heterozygotes. The gene in question, *Hsp90,* stabilizes cell cycle and developmental regulators that are crucial for normal development and physiology. Fruit flies with the mutated gene (*Hsp83*) displayed a variety of severe deformations that were asymmetrically expressed if bilateral structures (such as truncated wings or legs or multiple vestigial eyes) were involved, or disorganized if linear structures (such as body segments) were affected. These heterozygotes were bred with normal laboratory colony fruit flies as well as with normal wild-caught fruit flies with the result that their heterozygous offspring developed a similar range of deformities. Experimental selective breeding produced generations of deformed individuals, but did not increase the population of individuals with a particular anomaly. Extremes of temperature acted to enhance the expression of an anomaly, but did not serve to increase the population of similarly deformed individuals, either. Rather than demonstrating the role of the environment (in this case, temperature) in producing a mutation that yields viable and evolutionarily significant morphological change, this study actually points out how important regulatory control of normal development really is.

Another aspect of the model I am suggesting is that it demonstrates how a mutation involving the expression of homeobox genes can produce a morphological, physiological, or behavioral novelty that would emerge in a full-blown and viable state. Consider again the consequences of the experiment in which the *Rx* gene is altered from the dominant eye-producing state to a functionless recessive state. Individuals that are homozygous for these mutated, recessive *Rx* genes do not develop eyes or eye sockets at all. With two dominant alleles, an individual has functioning eyes in bony sockets. With one dominant and one recessive allele, an individual still has functioning eyes and bony sockets. With two recessive alleles, an individual has no eyes, not even the sockets in the skull for eyes. Since it is likely that the mutation that produced the original allele for having eyes arose in the recessive state, the genetics and the morphology produce the following evolutionary picture: with two dominant alleles, no eyes; with one dominant and one recessive allele, no eyes; but with two recessive alleles, fully functional eyes housed in bony eye sockets. Although it may have taken a number of generations for the *Rx* gene to spread throughout the population in which it appeared, once homozygotes for it were produced, they would have had completely useful and fully formed eyes—not shallow depressions in the front of the head, or even half eyes, but actual eyes. Consequently, as Haldane had long ago suggested was the case, there would not have been a string of graded morphological intermediates between the "before" and "after" states of "not having" and "having" eyes. Given the potential of homeobox genes to be fully rather than partially expressed, we can appreciate why "missing links" are so elusive in the fossil record. They probably did not exist.

The lack of transitional evolutionary stages between adult invertebrates and chordates—animals that as adults have a stiffening rod and a central nerve cord along the back, a brain anteriorly, serially arranged trunk muscles, and a tail—is a case in point. The only potential links between these groups are found in forms like the tunicate, but only in their larval stages.

The larvae of one species of tunicate are free-swimming and have all the basic features of a chordate. It turns out that a particular homeobox gene, identified as *Manx* (which is named after the virtually tailless Manx cat, which comes from the Isle of Man, hence Manx), is responsible for the development of chordate features in the larval form of this species of tunicate. In another species of tunicate, whose larvae do not develop chordate features, the *Manx* gene is unexpressed. When the *Manx* gene is experimentally activated in the latter species, this typically non-chordatelike larval tunicate actually develops chordate features. In tunicates in which the *Manx* gene is normally expressed, it is active only during the early phases of ontogeny. When it shuts down, the larva metamorphoses into its non-chordatelike adult form.

If Walter Garstang's suggestion is correct, and chordates did arise from a tunicate-like form by retaining the chordate features of their larval stage of development into adulthood, then the first chordates must have been affected by a regulatory mutation that kept the *Manx* gene activated for a longer period of time. The result of this change in homeobox gene regulation—which probably involved more than just the *Manx* gene—was the retention of childlike features: in this case, paedomorphosis by way of the retention of chordatelike larval tunicate features in the adults of true chordates. Whether the impetus for such a regulatory change came, as Haldane, de Beer, and others proposed, through a chain of interactions from the genetic level to the environment and then back again, or from a simple mutation, is unknown. But whatever the ultimate basis for the change, it happened. And it would have occurred without leaving a trail of fossil species caught in the act of accumulating the features of being a chordate.

The sudden appearance of novelty is not, as Otto Schindewolf emphasized, an unusual aspect of the fossil record. For instance, feathers, which have recently been documented in the fossil dino-birds *Protarchaeopteryx* and *Caudipteryx,* which preceded *Archaeopteryx,* are not the result of a gradual transformation of scales. Rather, as seen in the older fossils, downlike and regular feathers are all there, although the latter do not have the asymmetry typical of modern flighted birds and *Archaeopteryx* as well. Although it is unlikely that any of these dino-birds flew, they were still the bearers of a regulatory change that produced feathers. Even before *Protarchaeopteryx* and *Caudipteryx,* the earliest tetrapods so far known not only had forefeet and hind feet but their feet were fully developed and full of bones—more bones than any living tetrapod possesses. And even earlier, the first fish were armored forms, which were not just patchily covered with thick dermal scales but were fully surrounded by them.

With regard to footedness in general among subgroups of tetrapods, the picture of change has been one of structural reduction after the novelty is introduced in its fully expressed state. As in various dinosaurs, feet and limbs became reduced in bone number and size. Among mammals, the ancestors of two- and three-toed sloths, single-toed horses, and even-toed animals, such as pigs and deer, reduced the number of bony elements in their feet. In horses and even-toed animals, the ulna of the lower forelimb and the fibula of the lower hind limb are vestigial, if not absent. In the ancestors of modern whales, limb bones became diminished in length, essentially leaving the foot skeleton as the sole supporting structure in the paddlelike flipper.

A similar phenomenon of evolutionary structural reduction happened with teeth. The first known jawed fish had both a lifetime of teeth that multiplied in number as they "cloned" horizontally along the jaws as the animal grew and a lifetime of successional teeth that replaced previously erupted teeth. Reptiles, such as alligators, and most fish, such as sharks, have retained the full genetic potential for developing multiple tooth generations. Mammals, however, produce only two functional sets of teeth, the milk, or deciduous, teeth and the permanent replacement teeth. The most extreme example of dental reduction comes from birds: Modern birds do not produce any teeth at all, even though the epithelial cells of their oral cavities can be induced to do so.

The observation of reduction in complexity in many groups of organisms after a structure has emerged in a full-blown state makes sense in terms of homeobox genes. If a mutation were to activate a homeobox gene or gene cluster, it would already possess the potential to create a complete structure. Although, of course, there could be an increase in output of the homeobox gene from the time of its activation, this, as indicated by the fossil record on the emergence of new kinds of organisms, was not the case. A homeobox gene or gene cluster is capable of orchestrating maximal development from the very time it is turned on, and it does so.

Fully expressed homeobox gene activity must be the reason why the earliest tetrapods had more bones in their feet than do more recent tetrapods. From what we know about limb and digit formation in living vertebrates, we can conclude that it was in the first tetrapods that the *Hoxd-11, Hoxd-12,* and *Hoxd-13* genes were expressed across the front of the enlarging limb bud and that mutation in the *Hoxd-13* was in the form an alanine-encoding insertion. From this point onward in tetrapod evolution, there were multiple instances of decreases in *Hoxd*-gene output, especially in *Hoxd-13*-derived alanine repeat sequences, with the result that digit and limb-bone number and size decreased in many groups. In snakes, limbs and feet were inhibited from developing altogether.

The potential for complete homeobox gene expression at the time of activation is also reflected in the fact that the earliest jawed fish developed a lifetime of successional teeth. Since the homeobox genes that are active during tooth development—the *Dlx* gene family—are also present, but silent, in tunicate larvae, the first species of toothed vertebrate would have been the bearer of a mutation that merely activated this gene cluster. Once activated, however, early jawed fish had the competence to, and did, develop a limitless number of successional teeth. As seen in descendant fossil and living fish, and also in the vast majority of reptiles, this tooth-forming propensity was retained. But in mammals the competence to produce teeth was truncated to the development of maximally two sets of functional teeth, while in birds it was completely inhibited.

In general, then, after a new, or mutant, interactive system of homeobox genes and their products become established, the length of time during which the system is turned on and is interactive with other genes, and the levels of the proteins they produce, can be reduced. The reduced number of alanine repeats in the chick, with fewer bony elements in its legs and especially in its wings than most tetrapods, certainly appears to bear this out.

A model of fully expressed emergent homeobox genes also contributes to our understanding of the origin of groups of individuals that systematists would refer to as species. As Richard Goldschmidt argued, a new species would arise, not through the gradual accumulation of minor variations but through the sudden appearance of a major genetic and morphological reorganization. The genetics of everyday life, in which natural selection fine-tunes individuals by picking and choosing from among the available variation, Goldschmidt referred to as microevolution and the genetic changes underlying minor individual differences as micromutation. But because the founders of the Synthesis had based their model of gradual species change on the accumulation of micromutations, Goldschmidt was forced to invoke a different genetic term and concept to accommodate the model of major organismal reorganization that he thought lay behind the origin of species. Consequently, as the counterpart to micromutation, Goldschmidt proposed macromutation as the mechanism behind speciation.

One can see the bind Goldschmidt was in. However, because we now know that a small molecular change can produce an effect as profound as having or not having an eye, we also know that such a micromutation can produce what Goldschmidt would have described as macromutation leading to macroevolution. Since mutations in homeobox genes are inherited in the same way that earlier fruit-fly population geneticists understood the inheritance of the alleles for wing length or bristle number, we can appreciate that evolutionarily significant novelty—which, in turn, could result in the emergence of a new species—can be passed on from parent to child as easily and simply as eye color.

Although there is a real and important distinction between the concepts of microevolution and macroevolution—the former involving natural selection and the refinement of adaptation, and the latter the origin of species—they can both be understood within a framework of micromutation. The major difference between the concepts of microevolution and macroevolution is the effect on the individual of the genes that mutated. Perhaps because the word *macromutation* could be used to connote regulatory mutations of the sort that would yield new species, there might be a conceptual reason to retain it in evolutionary language. But, inasmuch as micromutation is perfectly adequate to explain the phenomenon, retention of the term would only promote confusion. In retrospect, we must admit that Goldschmidt's insistence on recognizing macroevolution as something different from microevolution was on track even though he was not in a position to tackle its genetics.

The other side of the evolutionary coin was eventually represented by Thomas Hunt Morgan. In his second professional life, as a fruit-fly population geneticist, Morgan argued that Darwinian evolution and Mendelian inheritance were compatible. He promoted the relationship between continuous variation and micromutation and the role of natural selection in guiding evolutionary change. In his anti-Mendelian and anti-Darwinian days, however, he was a fierce proponent of discontinuous variation. He denied that natural selection had anything to do with the origin of species. Instead, he thought that Darwin had erred in the way he had portrayed the role of natural selection in evolution. For the early Morgan, natural selection was relevant only, if at all, to the origin of adaptation among the individuals that comprise a species, not to the origin of a species itself.

Decades later, this was precisely the argument that George Williams made in his discussion of natural selection. Although Williams's followers may extrapolate from his writing that he also argued that the origin of a species is a natural extension of the process of refining interindividual adaptation from generation to generation, this was not his treatise at all. Williams was concerned only with the role of natural selection in choosing from among different alleles via their differential morphological or behavioral expressions and, consequently, in keeping the species alive. In his view, mutation disrupts the stability of a species that natural selection works to maintain. Speciation is not a factor in Williams's adaptationist program. Adaptation via natural selection is an aspect of the daily lives of the individuals of a species. And that is all.

As had been suggested by Thomas Hunt Morgan during the early phase of his career, we should expect that novelty—of whatever sort—would arise via random mutation that then affects gene regulation and expression. If an interaction with something in the individual's environment, such as a toxin, radiation, or temperature, causes a mutation, the ultimate expression of that mutation—genetically or morphologically—would not necessarily have anything to do with the circumstances under which the mutation occurred. And because it would take a number of generations before the mutation, which would probably be in the recessive state, was expressed in homozygotes, it is quite unlikely that it would have been an adaptive response to the initial provocation. Consequently, instead of evolutionary change—certainly of the sort that produces species—being the sum total of innumerable adaptations to the environment, it is more likely that those individuals that just happened, or chanced, to find themselves with certain attributes in a particular environment or situation would either survive or die off. The evolution of life is probably strewn with the carcasses of failed species.

Although evolutionary biologists try to guess what the adaptive significance of a feature is or was, they can do so only in retrospect and with reference to what they think is the adaptive significance of similar features in other organisms. It might just be the case that, if a novelty doesn't kill you, you retain it, whether or not it will ever do you or your descendants any good. As Huge de Vries saw it, if you need to invoke selection at all, it would be in the context of eliminating a feature by way of eliminating its bearers. Death is certainly a strong selective force. But if natural selection does work as Darwin proposed, as opposed to just eliminating the unfit, it would be involved in fine-tuning some of the features or adaptations of those individuals that did survive the unexpected inheritance of a morphological, physiological, or behavioral novelty.

An individual may have one or more characteristics that are unique to it, but it exists as a functioning organism primarily because it possesses a combination of features that it inherited from a string of ancestors, from the very distant to the very recent past. A species can be thought of in the same

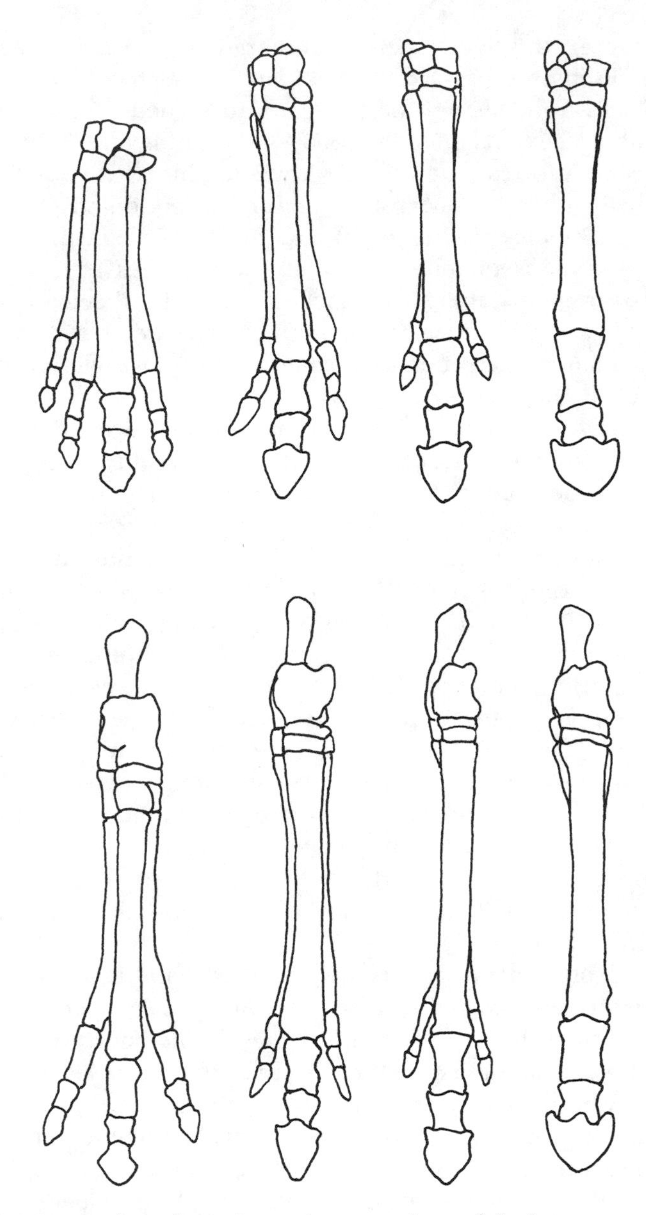

Reduction in number of toes in various members of the horse group, as seen in the forelimb (the illustration on the right and upper row of the illustration on the left) and the hind limb (lower row of the illustration on the left). The earliest horse had four toes in the forelimb and three in the hind limb. In horses that had three toes in both the forelimb and the hind limb, there was a

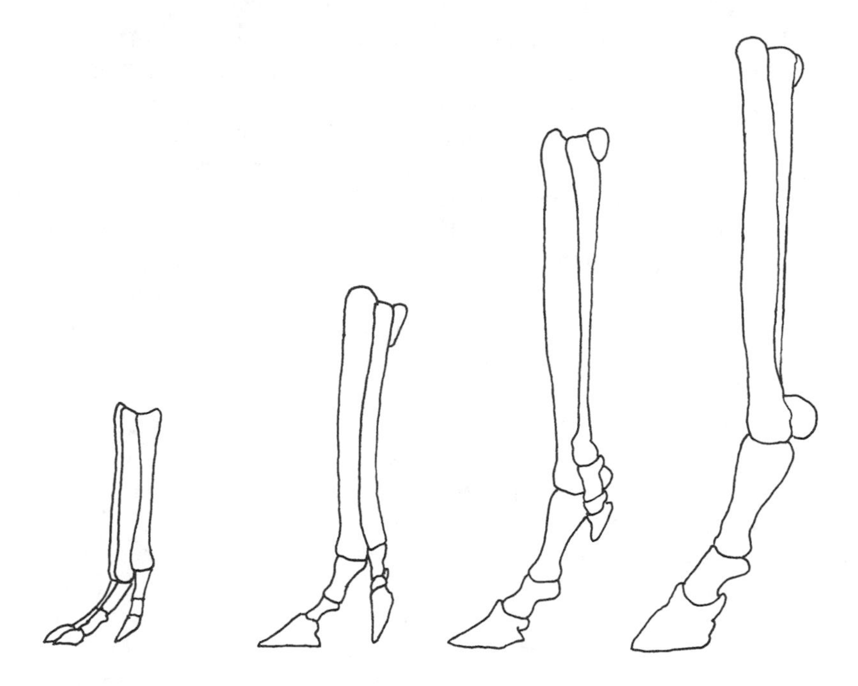

reduction in the size of the bones of the lateral toes and an increase in the size of the central toe. In modern horses, the lateral toes are absent altogether. The genera of the horses in the drawings, from right to left, are Hyracotherium, Miohippus, Merychippus, *and* Equus. *(Adapted from Schindewolf)*

way: Its members may be distinguished by one or a few unique characters, but they mostly retain features from a string of ancestors. Mammals are a unique group of vertebrates. Vertebrates are a distinct group of chordates. And chordates are a particular group of animals. In each case, each subgroup is retaining the regulatory mutation, or mutations, that occurred in various ancestral species. But the nature of the regulatory mutation need not always be the production of profound novelty. Sometimes it can be within the realm of differences within the same regulatory domain.

For example, the body plans of horses and bears are pretty much the same, and both can be distinguished from a reptile by these features they share because they are mammals. But horses and bears are also different from each other in obvious ways. During the evolution of the horse, there were changes in two major skeletal systems: the teeth and the limbs. The teeth became higher-crowned, with more complicated folding of the chewing surfaces; and the toes were reduced to one functional digit, with a decrease also in the number of other foot bones. In contrast, the skeleton of a bear has changed relatively little from the basic mammalian body plan.

George Simpson thought that the fossil record of horses presented a smoothly transitional picture of evolution. In stark contrast, Otto Schindewolf looked at the same fossil record and thought that the evolution of the horse had been saltational. A model of change that is based on developmental regulation as well as on developmental thresholds supports Schindewolf's, not Simpson's, interpretation of the evolution of the horse.

Experiments on *Hoxd* gene-cluster expression in mice, as well as studies on expression of the *Hoxd-13* gene in birds and underexpression of this gene in humans and mice, clearly demonstrate that wrist-, finger-, ankle-, and toe-bone size and number are controlled by these genes. Overexpression results in an increase in bone size and number, sometimes causing what would normally be single bones to bifurcate. Reduction in gene activity reduces bone size and number. But, as was demonstrated as long ago as the 1930s by the German experimental geneticist Hans Grüneberg, neither structural reduction or loss, nor structural addition, is a completely gradual process. Instead, the loss or gain of a structure is developmentally constrained to be more step-wise, or saltational.

Grüneberg was able to show, especially by studying tooth formation, that if a structure does not have the developmental competence to reach a certain critical size threshold, it will not continue to grow. It will be resorbed. If a structure has the developmental competence to reach that critical size threshold but to go no further, it will be stunted. In the case of a tooth, it has to reach a certain size before it can erupt. Since the realization of a structure is tied to crossing that critical threshold, we can appreciate that the appearance or disappearance of structure, even if stunted in size, would be abrupt.

Mammals have maximally five digits per foot and three bones in each digit of each foot. The earliest horses lacked the two lateral toes in the hind foot and one lateral toe on the forefoot. It would seem that these toes had been inhibited from forming by regulatory mutations in the *Hoxd* cluster. Next during the evolution of the horse, the central toe increased in size and another lateral toe on the forefoot was inhibited from developing. As the central toe on each foot became larger in various fossil species, the two smaller toes on either side of it became slightly smaller. Eventually, the central toe was very large and the smaller lateral toes were positioned well off the ground. But even in the fossil species with the smallest lateral toes, there were always three bones in each toe. From a species of this fossil configuration emerged the horse as we know it, with a single large toe on each foot. The evolution of the horse was not characterized by a structural reduction that went from three bones in the lateral toes to two, one, and then none. In species with lateral toes, each digit was made up of three bones. And in the first modern horse there were no lateral toes at all.

Grüneberg's critical-development threshold model provides the only reasonable explanation for toe loss in horse evolution: Once homeobox gene expression and cell potential were insufficient to produce lateral digits of a minimum requisite size, they did not continue to develop or begin development in the first place. Parts of the sequence of horse-foot evolution may appear to have taken place gradually, such as the enlargement of the central toe and the overall diminution of the last pair of lateral toes. But the entire process—especially the loss of digits—was not gradual at all. There were inevitable leaps due to inescapable developmental constraints.

Putting Humans Back into Evolution

The evolution of our own group, hominids, has never been exempt from the vicissitudes of random mutation and the emergence of novelty that affects all other organisms. Nevertheless, it has been commonplace in paleoanthropology for humans, and even hominids more broadly, to be thought of as a special case because hominids can both alter their environment and create an environment around themselves. If one adheres to the notion that an organism changes in response to its environment, then hominids have been directing the course of their own evolution. This in spite of the fact that humans have long been invoked as classic examples of the particular form of paedomorphosis called neoteny (caused by the prolongation of juvenile growth rates) and that heterochrony in general (the differential rates of physical versus sexual development) was considered by prominent embryologists and comparative anatomists such as Louis Bolk, Julian Huxley, and Gavin de Beer to be a potential source of rapid and major evolutionary change. Perhaps even more significant is the fact that, whether it was Charles Darwin, Alfred Russel Wallace, Hugo de Vries, or Thomas Hunt Morgan who did the speculating, humans have long held a central position in a diversity of theoretical formulations of evolutionary change.

Because the field of paleoanthropology has not taken seriously the possibility of punctuated evolution (although the model has received increasing support in other areas of evolutionary biology), it has remained the most resistant to the possibility that evolution may not be gradual. Clearly, however, if present-day humans are neotenic animals, this implies the advent of regulatory changes, and this, in turn, is consistent with the great morphological differences that exist between *Homo sapiens* and fossil hominids. The similarities that humans share, to varying degrees, with extinct hominid species reflects merely the recency or antiquity of past common ancestors. For instance, the presence in all hominids, including the earliest species, of various, almost fully modern, skeletal features that are associated with bipedalism—such as the curve of the lower spine and knee joints that angle in under the body—would reflect the regulatory changes that had occurred in the ancestor of the entire group.

Since hominid evolution is as much as history of regulatory changes as is the evolution of any other group of animals, we should not expect to find a trail of intermediates proceeding slowly from one morphological state to another. But because the role of gradualism in hominid evolution has been a mainstay of paleoanthropology, the literature is replete with scenarios of how and why features changed from an apelike to a humanlike state. Darwin set the stage in many ways, including his discussion of the transition of the apelike foot, with its divergent big toe, to the aligned big toe of the most civilized of modern humans. Armed with this expectation, we are faced with the specimen from the South African site Sterkfontein, which Ronald Clarke and Phillip Tobias nicknamed Little Foot. Surprisingly, this hominid had a divergent big toe. And when Clarke and Tobias compared these foot bones with the foot bones attributed to *Homo habilis,* they found that this supposed biped had also had a divergent big toe. I predict that we will never find a hominid with a halfway divergent big toe. And the reason lies in the way in which feet, and hands, for that matter, develop.

The human hand and foot emerge first as paddle-shaped ends on the limbs. Within these paddles, cartilage cells concentrate in specific locations to form the precursors of the bones of the hands and feet as they will be in the adult. Later, cell death in the regions between the developing fingers and toes frees them up and they become separate digits. The same developmental picture characterizes all tetrapods that have been studied. The divergent big toe of most primates and the more aligned big toe of *Homo sapiens* develop in the same positions they have in the adult foot. The Australian physical anthropologist Arthur Abbie was very much mistaken when he surmised that the difference between a divergent big toe and one that is aligned with the other toes was due to differential degrees of clefting between the digits.

Because of what we know about development, we should not expect to find a series of intermediate fossil forms with decreasingly divergent big toes and, at the same time, a decreasing number of apelike features and an increasing number of modern human features. If Little Foot and *Homo habilis* had some features possessed by apes and others possessed by humans, this means that they retained the apelike features from a much earlier ancestor than the one from which these fossil forms retained the features also seen (and retained) in *Homo sapiens:* After hominids and apes diverged from their common ancestor, some hominids were affected by regulatory changes that influenced the architecture of their feet.

The discovery of modern humanlike elbow and knee joints among the 4-million-year-old fossils from the Kenyan sites of Allia Bay and Kanapoi suggests that the same genes that produce these characteristic morphologies in *Homo sapiens* were also then available, and that they were expressed and they interacted in a similar way. If this similarity indicates a close evolutionary relationship between these ancient fossils and other hominids who had these features, then the underlying regulatory event would have occurred in an ancestor that these fossils and *H. sapiens* shared. If we do not think that the Allia Bay and Kanapoi specimens are closely related to other hominids with similar elbow and knee morphology—because we have some other features in mind that support other relationships we find more tenable—then we would consider these similarities to be evolutionary parallelisms.

This is the frustration a systematist faces: trying to sort out similarity that is reflective of evolutionary relationship from similarity that is not.

Although a systematist may have a preferred theory of who's related to whom, there will always be alternative theories of relationship, because there will always be different sets of shared features among the species one is studying. And this, we can now appreciate, is because all organisms share many of their genes, which means that there is always the possibility that the same genes will be turned on or off in the same way and in the same place in different, but unrelated, organisms. What this also suggests is that it might not be correct to try to sort out the evolutionary relationships of organisms on the basis of overall genetic similarity, especially of structural genes. As Stephen Jay Gould and Niles Eldredge pointed out decades ago, the fact that such wildly anatomically different primates as chimpanzees and humans share well over 90 percent of their structural genes does not shed any light on the regulatory underpinnings of these differences.

Understanding the role that homeobox genes—and the larger class of regulatory genes of which they are a part—play, and can play, in the origin of novelty leads to a hierarchical view of evolution. The often heated and sometimes nasty debates that have taken place between gradualists and punctuationists, or between micromutationists and macromutationists, have been generated by the perception that there is only one evolutionary question, for which, in turn, there can be only one correct answer. The result is an arena in which the sentiment is "We both can't be right, so you have to be wrong." But if we take a different approach, and assume that both sides of a typical evolutionary debate have something valid to offer, then the theoretical and methodological disagreements between different schools of thought may just be a matter of having the right answer to a different question.

There is room in evolutionary biology for the investigation of both the roles of natural selection and adaptation and the roles of regulatory gene interaction and expression. But their levels of significance with regard to what is generally referred to as evolution are worlds apart. The former relates to the survival of a species over time, whereas the latter provides insight into the origin of species. Far from the expectations of the Synthesis—that we know enough of the basic outline of how evolution works that we can concentrate on the minutiae of details—we are only now beginning to understand the broad picture. As such, there is the very real need to return the study of comparative morphology, and especially development, to the fore of evolutionary biology. But developmental and comparative morphologists will need to embrace the new insights derived from genetics every bit as much as developmental geneticists will need to embrace the complexities of comparative morphology. In addition, all should be conversant in systematics, which, though central to evolutionary biology, has too long been ignored or misunderstood. In short, we need to resynthesize the Synthesis.

More than one hundred years ago, William Bateson suggested that studying the regulation and timing of development was the key to understanding evolutionary change. He was right.

References and Notes

Chapter 1: A Rash of Discoveries

The quote from Kaufman is on page 31 of *At Home in the Universe: The Search for the Laws of Self-Organization and Complexity* (New York: Oxford University Press, 1995).

The opening paragraph and much of this section of the chapter refers to T. D. White, G. Suwa, and B. Asfaw, "*Australopithecus ramidus,* a new species of early hominid from Aramis, Ethiopia," *Nature* 371: 306–312 (1994); and also G. WoldeGabriel et al., "Ecological and temporal placement of early Pliocene hominids at Aramis, Ethiopia," *Nature* 371: 330–333 (1994).

The note by T. D. White, G. Suwa, and B. Asfaw introducing the genus *Ardipithecus* is in *Nature* 375: 88 (1995).

Examples of W. W. Ferguson's taxonomic publications on fossil hominids are "An alternative interpretation of *Australopithecus afarensis* fossil material," *Primates* 24: 397–409 (1983); and "Taxonomic status of the hominid mandible KNM-ER TI 13150 from the Middle Pliocene of Tabarin, in Kenya," *Primates* 30: 383–387 (1989).

The material on Little Foot derives from the article by R. J. Clarke and P. V. Tobias, "Sterkfontein Member 2 foot bones of the oldest South African Hominid," *Science* 269: 521–524 (1995).

Information on the arrangement of foot and ankle bones, and on the human skeleton in general, can be found in J. H. Schwartz, *Skeleton Keys: An Introduction to Human Skeletal Morphology, Development, and Analysis* (New York: Oxford University Press, 1995).

For specific reading on the Olduvai foot (OH 8), see M. H. Day and J. R. Napier, "Fossil foot bones," *Nature* 201: 967–970 (1964).

Specific information on the Laetoli footprints can be found in M. D. Leakey and J. M. Harris, eds., *Laetoli: A Pliocene Site in Northern Tanzania* (Oxford: Clarendon Press, 1987); see section 13, "The hominid footprints," pages 490–523, with contributions by M. D. Leakey, L. M. Robbins, and R. H. Tuttle.

For a historical review of features that have been used to distinguish *Homo sapiens* from other mammals, see J. H. Schwartz, *The Red Ape* (Boston: Houghton Mifflin, 1987).

"In 1978, when they published their article . . ." refers to D. C. Johanson, T. D. White, and Y. Coppens, "A new species of the Genus *Australopithecus* (Primates: Hominidae) from the Pliocene of Eastern Africa," *Kirtlandia,* no. 28: 1–14 (1978).

For a general discussion of *Australopithecus afarensis,* see D. C. Johanson and M. Edey, *Lucy: Beginnings of Humankind* (New York: Warner Books, 1981); for more specific issues and debates, see, for example, D. C. Johanson and T. D. White, "A systematic assessment of early African hominids," *Science* 203: 321–330 (1979); T. D. White, D. C. Johanson, and W. H. Kimbel, "*Australopithecus africanus:* Its phyletic position reconsidered," *South African Journal of Science* 77: 445–470 (1981); W. H. Kimbel, D. C. Johanson, and Y. Rak, "The first skull and other new discoveries of *Australopithecus afarensis* at Hadar, Ethiopia," *Nature* 368: 449–451 (1994); and R. L. Susman, J. T. Stern, Jr., and W. L. Jungers, "Arboreality and bipedality in the Hadar hominids," *Folia Primatologica* 43: 113–156 (1984).

"Thirty-two years ago . . ." is taken from conversations with Alan Walker, as well as from B. Patterson and W. W. Howells, "Hominid humeral fragment from early Pleistocene of northwestern Kenya," *Science* 156: 64–66 (1967); and M. Leakey, C. S. Feibel, I. McDougall, and A. Walker, "New four-million-year-old hominid species from Kanapoi and Allia Bay, Kenya," *Nature* 376: 565–571 (1995).

"Eight years after Patterson and Howells's . . ." refers to H. McHenry and R. Corruccini, "Distal humerus in hominoid evolution," *Folia Primatologica* 23: 227–244 (1975).

"However, Brigitte Senut . . ." refers to B. Senut, "New data on the humerus and its Joints in Plio-Pleistocene hominids," *Colloque d'Anthropologie* 4: 87–94 (1980); and B. Senut and C. Tardieu, "Functional aspects of Plio-Pleistocene hominid limb bones: Implications for Taxonomy and Phylogeny," in E. Delson, ed., *Ancestors: The Hard Evidence* (New York: Alan R. Liss, 1985), 193–201.

"Then came the discovery in 1995. . . ." et seq. cites M. Leakey et al., "New four-million year-old hominid species."

"Peter Andrews . . . wrote the commentary . . ." refers to P. Andrews, "Ecological apes and ancestors," *Nature* 376: 555–556 (1995).

For background to the section "Genetics and Development of the Organism," see, for example, L. Wolpert, "Do we understand development?" *Science* 266: 571–572 (1994); C. Nüsslein-Volhard, "Of flies and fishes," *Science* 266: 572–574 (1994); J. Kimbel, "An ancient molecular mechanism for establishing embryonic polarity?" *Science* 577–578 (1994); N. H. Patel, "Developmental evolution: Insights from studies of insect segmentation," *Science* 266: 581–590 (1994); and D. S. Kessler and D. A. Melton, "Vertebrate embryonic induction: Mesodermal and neural patterning," *Science* 266: 596–604 (1994).

Background information on invertebrates can be found in S. S. Mader, *Inquiry into Life,* 4th ed. (Dubuque, Iowa: W. C. Brown, 1985).

The paragraph that begins "A few years ago" refers to P. Hunt et al., "A distinct *Hox* code for the branchial region of the vertebrae head," *Nature* 353: 861–864 (1991).

"Subsequent to these studies, a team of U.S. and Spanish neurobiologists . . ." cites work reviewed in J. L. R. Rubenstein, S. Martinex, K. Shimamura, and L. Puelles, "The embryonic vertebrate forebrain: The prosomeric model," *Science* 266: 578–580 (1994).

"[T]he fruit-fly homeobox gene called *eyeless* . . ." is discussed in R. Quiring, U. Walldorf, U. Kloter, and W. J. Gehring, "Homology of the *eyeless* gene in *Drosophila* to the *Small eye* gene in mice and *Aniridia* in humans," *Science* 265: 785–789 (1994).

For further reading on insect wing and vertebrate limb development, see, for example, S. Grimm and G. O. Pflugfelder, "Control of the gene *optomotor-blind* in *Drosophila* wing development by *decapentaplegic* and *wingless,*" *Science* 271: 1601–1604 (1996); D. R. Davidson, A. Crawley, R. E. Hill, and C. Tickle, "Position-dependent expression of two related homeobox genes in developing vertebrate limbs," *Nature* 352: 429–431 (1991); and D. Duboule, "How to make a limb?" *Science* 266: 575–576 (1994).

"More recently, another amazing discovery . . ." refers to Y. Muragaki, S. Mundlos, J. Upton, and B. R. Olsen, "Altered growth and branching patterns in synpolydactyly caused by mutations in HOXD13," *Science* 272: 548–551 (1996).

For a discussion of the hypotheses of tetrapod fore-limb and hind-limb development, see comments by M. Coates, P. Thorogood, and P. Ferretti, as well as M. Kessel, "*Hox* genes, fin folds and symmetry," *Nature* 364: 195–197.

For background to the statement "Let us consider the traditional," see, for example, various books by G. C. Williams (e.g., *Adaptation and Natural Selection,* Princeton: Princeton University Press, 1966); and R. Dawkins (e.g., *The Selfish Gene,* New York: Oxford University Press, 1976).

H. B. D. Kettlewell summarized his experiments in "Darwin's missing evidence," *Scientific American* 200: 48–53 (1959).

For "Darwin's idea was that natural selection . . . ," see C. Darwin, *On the Origin of Species by Means of Natural Selection; or, the Preservation of Favored Races in the Struggle for Life* (London: John Murray, 1859).

A summary of early work on fruit-fly genetics ("Among the features . . .") can be found in Warren P. Spencer, "Gene homologies and the mutants of *Drosophila hydei,*" in G. L. Jepsen, G. G. Simpson, and E. Mayr, eds., *Genetics, Paleontology and Evolution* (New York: Atheneum, 1949), 23–44.

For an introduction to Ernst Mayr's voluminous work, including "Mayr's biological species definition," see E. Mayr, *Populations, Species, and Evolution* (Cambridge: Belknap Press of Harvard University Press, 1963).

The full citation to T. Dobzhansky's book is *Mankind Evolving: The Evolution of the Human Species* (New Haven: Yale University Press, 1962).

Chapter 2: How Humans Distinguished Themselves from the Rest of the Animal World

The quote from Empedocles is on page 59 of K. Freeman's *Ancilla to the Pre-Socratic Philosophers* (Cambridge: Harvard University Press, 1956).

The quote from Erasmus Darwin's *The Temple of Nature; or, the Origin of Society: A Poem, with Philosophical Notes* (London: J. Johnson, 1803), is in section 2, additional note 10 (facsimile reprint, Menston, Yorkshire Scolar Press, 1973).

For background on Greco-Roman thought and an overview of Herodotus, Hippocrates, and Aristotle, see *Evolution: Genesis and Revelations with Readings from Empedocles to Wilson,* by C. Leon Harris (Albany: State University of New York Press, 1981); *Ecce Homo: An Annotated Bibliographic History of Physical Anthropology,* by F. Spencer (Westport, Conn.: Greenwood Press, 1986); and *Who's Who in the Ancient World: A Handbook to the Survivors of the Greek and Roman Classics,* by B. Radice (New York: Penguin Books, 1973).

For specific discussions of Aristotle's work, see chapters in *Philosophical Issues in Aristotle's Biology,* ed. A. Gotthelf and J. G. Lennox (Cambridge: Cambridge University Press, 1987).

The quotes from Aristotle are from *Parts of Animals,* with an English translation by A. L. Peck, and a foreword by F. H. A. Marshall (Cambridge: Harvard University Press, 1945). "Two legs like man . . . ," *Parts of Animals,* IV.12 $693b_{2\text{-}13}$. "Now, man, instead of forelegs . . . ," *Parts of Animals,* IV.12 $693a_{25\text{-}31}$.

The Descent of Man, Parts I and II, by Charles Darwin, was first published in 1871 (London: John Murray).

Lucretius's best-known work is *De Rerum Naturae* (On the nature of things), trans. W. H. D. Rouse (Cambridge: Harvard University Press, 1966).

For a summary of the medieval period and the Renaissance, see Harris, *Genesis and Revelations,* and Spencer, *Ecce Homo.*

Isadore of Seville's work is *Etymologiae* (Etymologies), ed. and trans. with annotations by P. K. Marshall (Paris: Les Belles Lettres, 1983).

A discussion of the Great Chain of Being is found in the book of the same name by A. O. Lovejoy (Cambridge: Harvard University Press, 1942).

Copernicus's work is *De Revolutionibus Orbium Coelestium* (On the revolutions of the heavenly spheres), trans. A. M. Duncan (1543; reprint, New York: David & Charles, 1976).

The work by Augustine is *De Civitate Dei* (City of God), abr. and trans. J. W. C. Wand (A.D. 413–426; reprint, London: Oxford University Press, 1963).

A review of Hohenheim's contributions can be found in "The history of anthropology" by T. Bendyshe, in *Memoirs of the Anthropological Society of London* 1: 353–354 (1864).

The two-volume work in question by Bory St. Vincent is *L'Homme* (*Homo*): *Essai Zoologique sur le Genre Humaine* (Paris: Gravier, 1925).

Gesner's four-volume *Historia Animalium* was published in Zürich between 1551 and 1558.

Willughby's classification of primates is on page 158 of "Of Animals," in J. Wilkins, ed., *An Essay towards a Real Character, and a Philosophy of Language* (London: Gellibrand, 1668), 121–168.

"It is an easy matter" is on page 282 of vol. 2, "Show a man" on page 201 of volume 1, and "man does not vary . . ." on page 129 of volume 1 of Leonardo da Vinci's *Notebooks,* ed. and trans. E. MacCurdy (New York: Renal & Hitchcock, 1938).

Jonstonus's book was *Thaumatographia Naturalis* (Amsterdam: Blaeu, 1632).

The first edition of Linnaeus's *Systema Naturae* was published in Leiden.

Ray's most important work was *Synopsis Methodica Animalium Quadrupedum et Serpentini Generis* (London, 1693).

The tenth edition of Linnaeus's *Systema* was published in 1758, in Stockholm.

"No one has any right to be angry . . ." and "[A]nd indeed, to speak the truth . . ." are from pages 12–13 of Linnaeus's *Reflections on the Study of Nature* 1754; trans. J. E. Smith (Dublin, 1786).

Linnaeus's letter is on page 55 of Gmelin's *Reliquias,* trans. E. L. Greene (Washington, D.C., 1909).

For a review of Cuvier and Lamarck, see Harris, *Genesis and Revelations;* and Spencer, *Ecce Homo.* See also Loren Eiseley, *Darwin's Century: Evolution and the Men Who Discovered It* (Garden City, N.Y.: Anchor Books, 1961); and E. Mayr, *The Growth of Biological Thought: Diversity, Evolution, and Inheritance* (Cambridge: Belknap Press of Harvard University Press, 1982).

Also for Lamarck, see Richard W. Burkhardt Jr., *The Spirit of System: Lamarck and Evolutionary Biology* (Cambridge: Harvard University Press, 1977).

The human appendix and other examples of use and disuse by Charles Darwin are in *On the Origin of Species by Means of Natural Selection; or, the Preservation of Favored Races in the Struggle for Life* (1859; facsimile of 1st ed., Cambridge: Harvard University Press, 1964).

For a translation of Buffon, see, for example, *Natural History, General and Particular* (Histoire naturelle), 3d ed., trans. Wm. Smellie (London: A. Strahan, 1791).

The quote from Buffon ("quadrupeds, cetaceous animals . . .") is in volume 10 on page 50, "a brute of a kind so singular . . ." is on page 42, "Can there be more evident proof . . ." is on page 61, and "The passage is sudden . . ." in volume 5 is on page 4.

"I reject his first division . . ." is on pages iii–iv of Pennant's *History of Quadrupeds* (London: B. White, 1781).

Blumenbach's 1775 and 1795 versions of *On the Natural Varieties of Mankind* (De generis humani varietate nativa) were translated by T. Bendyshe, first published in 1865 by the Anthropological Society of London, and reprinted in 1969 by Bergman Publishers, New York.

For Camper, see *Works of the Late Professor Camper in the Connexion between the Science of Anatomy and the Arts of Drawing, Painting, and Statuary, &c.* (1791; trans. T. Cogan, London: Dilly, 1821).

Blumenbach quotes Linnaeus on page 163 of the 1795 edition of *On the Natural Varieties of Mankind.*

For the quotes after "Blumenbach's goal . . . ," see page 164 (ibid.).

The quote from Lucretius is on page 169, and Blumenbach's comment "fleshy, useful, and semicircular amplitude" is on page 168 (ibid.).

Blumenbach's classification of Bimana and Quadrumana was in the 1791 (4th) edition of *Handbuch der Naturgeschicte* (Göttingen: Dietrich).

Goethe's "Letter to J. G. von Herder" is in *Sämtlichen Werke,* vol. 23, 401–432 (Berlin, 1820).

The quote from Blumenbach ("that prerogative . . .") is on page 183, and "arbitrary variety . . ." is on page 184 of the 1795 edition.

My books *The Red Ape* and *What the Bones Tell Us* were published by Houghton Mifflin (1987) and H. Holt (1993), respectively.

Chapter 3: Coming to Grips with the Past

The quote from E. Pidgeon comes from page 22 of "The fossil remains of the animal kingdom," in the *Supplementary Volume on Fossils of the Animal Kingdom Arranged in Conformity with Its Organization by the Baron Cuvier, Member of the Institute of France, Etc. with Additional Descriptions of All of the Species Hitherto Named, and Many Not Noticed* (London: Whittaker & Co., 1830).

For Buffon's Asian theory of human origins, see, for example, *Natural History, General and Particular* (Histoire naturelle), 3d ed., trans. Wm. Smellie (London: A. Strahan, 1791).

Saxo Grammaticus's work is *Gesta Danorum* (*Historia Danica*), published in 1208 and translated by O. Elton in 1905 (London: Norraena Society).

An easy-to-follow review of Paleolithic-tool technologies can be found in R. G. Klein, *The Human Career* (Chicago: University of Chicago Press, 1989); for Paleolithic and Neolithic technologies, see *Encyclopedia of Human Evolution,* ed. I. Tattersall, E. Delson, and J. Van Courvering (New York: Garland Publishing, 1988).

For a review of the significance of ceraunia, see D. K. Grayson, *The Establishment of Human Antiquity* (New York: Academic Press, 1983); and for reference to "elf arrows," see J. Evans, *The Ancient Stone Implements, Weapons, and Ornaments of Great Britain* (London: Longmans, Green Reader, & Dyer, 1872). Gesner's work is *De Rerum Fossilium, Lapidium et Gemmarum Maxime* (Zürich: Tiguri, 1565); Aldrovandi's is *Musaeum Metallicum* (Bonn: Ferronii, 1648).

Mercati's manuscript, *Metallotheca Vaticana,* though written in 1574, was published in 1717, in Rome. The quote from Lucretius (*De Rerum Naturae*) is in book 5, line 925.

Steno's work was published in 1669; the translation, *The Prodromus of Nicolaus Steno's Dissertation Concerning a Solid Body Enclosed by Process of Nature within a Solid,* by J. G. Winter, was published in 1968 (New York: Hafner).

The Natural History of Stafford-shire, by R. Plot, was published in Oxford, in 1686.

Woodward's last work was *Fossils of All Kinds, Digested into a Method, Suitable to Their Mutual Relations and Affinity* (London: Wilkin, 1728).

Spencer proposed an early version of what he would later call "social Darwinism" in 1851. See *Social Statics: The Conditions Essential to Human Happiness Specified, and the First of Them Developed* (New York: Robert Schalkenbach Foundation, 1970).

Darwin's *The Descent of Man and Selection in Relation to Sex* (hereafter abbreviated in this chapter as TDM) was first published in 1871 (London: John Murray), and the revised edition (from which quotes are taken) in 1874 (New York: Merrill and Baker). Quotes from Darwin's *On the Origin of Species by Means of Natural Selection; or, the Preservation of Favored Races in the Struggle for Life* (London: John Murray, 1859) (hereafter abbreviated in this chapter as OTO) are from a facsimile of the first edition (Cambridge: Harvard University Press, 1964).

For "In the eighteenth century, Buffon . . ." see Buffon, *Natural History.*

Huxley's essay, "On the Relation of Man to the Lower Animals," was originally published with two other essays as a collection entitled *Man's Place in Nature* (New York: D. Appleton, 1863).

In the 1896 edition of *Man's Place in Nature* (New York: D. Appleton) (hereafter abbreviated in this chapter as MPN), the original three essays are republished, along with three additional pieces.

"It is quite certain . . ." is on page 97 of the 1863 edition of MPN, as are all other quotes, unless they are specified as being from the 1896 edition.

"The structural differences . . ." comes from page 145 of MPN.

The quote "tenability, or untenability . . ." is on page 147 of MPN.

The comment and illustration of the fetal orangutan is in TDM on page 17.

From among Haeckel's many publications, see *The History of Creation, or, The Development of the Earth and Its Inhabitants by the Action of Natural Causes: A Popular Exposition of the Doctrine of Evolution in General, and of That of Darwin, Goethe, and Lamarck in Particular,* trans. E. R. Lankester (New York: D. Appleton, 1868); and *The Evolution of Man: A Popular Exposition of the Principal Points of Human Ontogeny and Phylogeny,* 3d ed., New York: H. L. Fowle, 1876). The evolutionary tree with orangutans closest to humans is presented in plate 15 of the latter work.

"[I]nasmuch as Development and Vertebrate Anatomy . . ." is from page ix of MPN.

"Light will be thrown on the origin of man and his history" comes from page 488 of OTO.

"In fact . . ." is also from page ix of MPN.

For a review of the German anatomical literature as well as a comparison of apes and humans in the anatomy of the arm, wrist, and hand, see A. H. Schultz, "Characters common to higher primates and characters specific for man," *Quarterly Review of Biology* 11: 259–283, 425–455 (1936).

The quote that begins "In each great region" is on page 151 of TDM.

The quotes in the paragraph that begins "Elsewhere in *The Descent*" are on page 61 of TDM.

The quotes in the paragraph that begins "As Blumenbach" are on page 50 of TDM.

The quote "the reproductive power . . ." is on page 43 of TDM; ". . . ridiculous mockers" and the comment about "savages" and imitation are on page 70; and "The strong tendency . . ." is on page 85.

The quote "a civilized man would perhaps . . ." is on page 74 of TDM.

The quote "the greater differences . . ." is on page 26 of TDM, and the remaining quotes in the same paragraph are on page 52.

The article by Broca cited by Darwin in the revised edition of TDM is "Les Selections," which was published in *Revue d'Anthropologies* (1873); the quote "the seat of intellectual faculties . . ." is on page 53.

The quote "intellect, instincts, senses and voluntary movements . . ." is also on page 53 of TDM.

The conclusions referred to in "But it is not until nearly the end of part 1 . . ." can be found on page 141 of TDM.

The quote "[C]ivilized nations are the descendants of barbarians" is on page 140, as is the quote that follows. The last quote of the paragraph, which begins "It is apparently[,]" is on page 141 of TDM.

A sense of the background to the sentence "One of Darwin's worst enemies was himself" can be gained from L. Eiseley's *Darwin's Century: Evolution and the Men Who Discovered It* (Garden City, N.Y.: Anchor Books, 1961).

For a review of the expansion of paleontology in the latter part of the nineteenth century, see E. H. Colbert, *The Great Dinosaur Hunters and Their Discoveries* (New York: Dover, 1984).

See earlier citations to E. Haeckel.

The long quote that begins "I have applied this name" can be found on pages 181–182 of vol. 2 of Haeckel's *The Evolution of Man.*

The information on the oral presentations by Schaaffhausen and Fuhlrott appears in Schaffhausen's published article of 1857, "On the crania of the most ancient races of man," which was translated in its entirety by G. Busk and published in *Natural History Review,* 158–162 (1861).

Fraipont and Lohest's first publication was "La race de Neanderthal ou de Cannstadt, en Belgique: Recherches ethnographiques sur des ossements humains décourts dans des dépots quaternaires d'une grotte à Spy et détermination de leur age géologique—Note Préliminaire," in *Bulletin des Academie Royale des Sciences Lettres de Beaux-Arts* (Belgium) 12: 741–784 (1886).

The information about von Mayer and dendrites comes from Schaaffhausen's article "The Most Ancient Races."

Huxley's essay "On some fossil remains of man" was one of the three original essays in MPN.

The quote that begins "A small additional amount" is on page 203 of MPN.

The quote following "as the former scholar put it" is from page 206 of MPN.

W. King's article is "The reputed fossil man of the Neanderthal," in the *Quarterly Journal of Science* 1: 88–97 (1864).

The quote that begins "a closer resemblance" is from page 92 of King's article.

"It is . . . to be apprehended" et seq. is also from page 92 of King's article, as is the last quote of the same paragraph.

Chapter 4: Filling in the Gaps of Human Evolution

The quote from Lecomte de Noüy is on page 104 of his *Human Destiny* (New York: Longmans, Green & Co., 1947).

The background on Henry Fairfield Osborn and the importance of central Asia comes in large part from E. H. Colbert, *The Great Dinosaur Hunters and Their Discoveries* (New York: Dover, 1984).

Dubois announced his first discoveries in the article "Palaeontologische onderzoekingen op Java" (Paleontological investigations on Java), which appeared in *Verslag van het Mijnwesen,* third quarter, 10–14 (1892).

The quote that begins "And thus the factual proof" is on the last page (p. 14) of Dubois's 1892 article.

Dubois's 1894 publication is Pithecanthropus erectus, *eine Menschenähnliche Übergangsform aus Java* (Batavia: Landesdruckerei).

The article in which Dubois cited support for his ideas by other scholars is "*Pithecanthropus erectus*—A form from the ancestral stock of mankind," which was published in 1886 in Dutch; the English translation appeared in 1900, in *Smithsonian Report for 1898,* 445–459.

Schoetensack described and named the Mauer jaw in *Der Unterkiefer des* Homo heidelbergensis *aus den Sanden von Mauer bei Heidelberg* (Leipzig: Wilhelm Engelmann, 1908).

Dawson gives his account of discovering the site at Piltdown and its treasures, as well his version of how Smith Woodward had unquestioningly received him and his ideas, in "On the discovery of a Palaeolithic human skull and mandible in flint-bearing gravel overlying the Wealden (Hastings Beds) as Piltdown (Fletching), Sussex," co-authored with Arthur Smith Woodward, *Quarterly Journal of the Geological Society London* 69: 125–151 (1913).

More accurate documentation of Dawson's attempts at getting Smith Woodward interested in the Piltdown site and remains is provided by Ronald Millar in *The Piltdown Men* (New York: Ballantine Books, 1972).

Elliot Smith's discussion of his reconstruction of the Piltdown skull's brain was published as an appendix to Dawson and Smith Woodward's 1913 report entitled "Preliminary report on the cranial cast," 145–147.

Dawson and Smith Woodward announced the discovery of the lower canine of *Eoanthropus* in "Supplementary note on the discovery of a Palaeolithic human skull and mandible at Piltdown (Sussex), *Quarterly Journal of the Geological Society of London* 70: 82–99 (1914).

The publication that exposed the Piltdown forgery was by J. S. Weiner, K. P. Oakley, and W. Le Gros Clark, "The solution of the Piltdown problem," *Bulletin of the British Museum (Natural History)*, Geological Series 2, no. 3 (1953).

For the latest on the Piltdown affair, see, for example, *Unraveling Piltdown: The Science Fraud of the Century and Its Solution,* by John Evangelist Walsh (New York: Random House, 1996); H. Gee, "Box of bones 'clinches' identity of Piltdown palaeontology hoaxer," *Nature* 381: 261–262 (1996); and the comments of E. T. Hall and A. O. Lutes, "Riddle of the tenth man," *Nature* 381: 728 (1996).

The biographical information on Dart and the discoveries on Taung come from Dart, "Association with and impressions of Sir Grafton Elliot Smith," *Mankind* 8: 171–175 (1972), and "Recollections of a reluctant anthropologist," *Journal of Human Evolution* 2: 417–427 (1973); Robert Broom, *Finding the Missing Link,* 2d ed. (London: Watts & Co. 1951); and P. V. Tobias, *Dart, Taung and the "Missing Link"* (Johannesburg: Witwatersrand University Press, 1984).

Dart's article announcing and describing the Taung specimen was "*Australopithecus africanus:* The Man-Ape of South Africa," *Nature* 115: 195–199 (1925).

The quotes from Dart's article ("an extinct race . . ." and "delicate and humanoid character . . .") are on page 196.

For reviews of the different theories based on different primate models on the origin of human bipedalism, see, for example, M. Cartmill, "Basic Primatology and Prosimian Evolution," in F. Spencer, ed., *A History of American Physical Anthropology 1930–1980* (New York: Academic Press, 1982, 147–186); and J. G. Fleagle and W. L. Jungers, "Fifty Years of Higher Primate Phylogeny," in the same volume, 187–230.

The quote that begins "The improved poise of the head" is on page 197, and the quotes in the paragraph that begin "Dart was also struck" are on page 198 of Dart's 1925 article on the Taung specimen.

The quotes that begin "Unlike Pithecanthropus," "in commemoration," and "In anticipating" are on the last page of Dart's 1925 article.

Broom announced *Australopithecus transvaalensis* in "A new fossil anthropoid skull from South Africa," *Nature* 138: 486–488 (1936).

Chapter 5: Humans as Embryos

The quote from Keith is on page 65 of *Concerning Man's Origin* (New York: G. P. Putnam's Sons, 1928).

The interplay between rates of physical versus sexual development is explained simply in Sir Gavin R. de Beer's *Embryology and Evolution* (Oxford at the Clarendon Press, 1930) and in more depth in Stephen Jay Gould's *Ontogeny and Phylogeny* (Cambridge: Belknap Press of Harvard University Press, 1977).

The two fundamental Garstang articles are "Preliminary note on a new theory of the phylogeny of the Chordata," *Zoologischer Anzeiger* 17: 122–125 (1894); and "The morphology of the Tunicate and its bearings on the phylogeny of the Chordata," *Quarterly Journal of Microscopical Science* 72: 51–187 (1928).

L. L. Bolk's influential publications were "Über Langerung, Vershiebung und Neigung des Foramen magnum am Schädel der Primaten," *Zeitschrift für Mophologie und Anthropologie* 7: 611–692 (1915); "The problem of orthognathism," *Proceedings of the Section of Sciences Kon. Akademie Wetens. Amsterdam B* 25: 371–380 (1923); "The chin problem," *Proceedings of the Section of Sciences Kon. Akademie Wetens. Amsterdam* 27: 329–344 (1924); "On the problem of anthropogenesis," *Proceedings of the Section of Sciences Kon. Akademie Wetens. Amsterdam* 29: 465–475 (1926); "La récapitulation ontogenetique comme phénomène harmonique," *Archives of Anatomy, Histology, and Embryology* 5: 85–98 (1926); and *Das Probleme der Menschwerdung* (Jena: Gustav Fischer, 1926).

See Buffon's discussion of the orangutan in *Oeuvres Complètes de Buffon, avec des Extraits de Daubenton, et la Classification de Cuvier, Tome Quatrième, Mammiféres. II* (Paris: Chex Furne et Compe, 1837) and Geoffroy Saint-Hilaire's in "Considerations sur les singes les plus voisins de l'homme," *Comptes Rendue des Academie de Sciences* 2: 92–95 (1936).

De Beer's classic work on comparative and developmental cranial morphology is *The Development of the Vertebrate Skull* (1937; reprint, Chicago: University of Chicago Press, 1985).

For a recent discussion of craniofacial growth changes, especially of the orbital region, in anthropoid primates, see this author's article "*Lufengpithecus* and hominoid phylogeny," in D. R. Begun, C. V. Ward, and M. D. Rose, eds., *Function, Phylogeny, and Fossils: Miocene Hominoid Evolution and Adaptations* (New York: Plenum Press, 1997), 363–388.

In the paragraph that discusses fetalization, the background for the ideas of Bolk, de Beer, and Gould can be found in the works cited above.

D'Arcy W. Thompson's influential monograph *On Growth and Form* was first published in 1917 (Cambridge: Cambridge University Press).

For more recent scenarios linking brow-ridge development to stress-related function, see, for example, B. Endo, "Experimental studies on the mechanical significance of the form of the human facial skeleton," *Journal of the Faculty of Sciences,* vol. 3, pt. 1, sec. 5 (Tokyo: University of Tokyo, 1966), 1–106 and "Analysis of stresses around the orbit due to masseter and temporalis muscles respectively," *Journal of the Anthropological Society of Nip-*

pon 78: 251–266 (1970); M. D. Russell, 1983 University of Michigan doctoral thesis *The Functional and Adaptive Significance of the Supraorbital Torus;* O. J. Oyen, "Bone strain in the orbital region of growing vervet monkeys," *American Journal of Physical Anthropology* 72: 39–40 (1987); and O. J. Oyen, R. W. Rice, and M. S. Cannon, "Browridge structure and function in extant primates and Neanderthals," *American Journal of Physical Anthropology* 51: 83–96 (1979). Studies debunking the form-function argument for brow ridges include M. J. Ravosa, "Browridge development in Cercopithecidae: A test of two models," *American Journal of Physical Anthropology* 78: 535–555 (1988), and "Ontogenetic perspective on mechanical and nonmechanical models of primate circumorbital morphology," *American Journal of Physical Anthropology* 85: 95–112 (1991); P. G. Picq and W. L. Hylander, "Endo's stress analysis of the primate skull and the functional significance of the supraorbital region," *American Journal of Physical Anthropology* 79: 393–398 (1989); and W. L. Hylander, P. G. Picq, and K. R. Johnson, "Masticatory-stress hypotheses and the supraorbital region of primates," *American Journal of Physical Anthropology* 86: 1–36 (1991). C. Loring Brace, for example, presents a discussion of anterior tooth use in Neanderthals in "Cultural factors in the evolution of the human dentition," in M. F. Ashley Montagu, ed., *Culture and the Evolution of Man* (New York: Oxford University Press, 1962), 343–354. The details on Neanderthal masticatory muscle scars are derived from the author's own research.

The quote from de Beer about the browless Piltdown skull is from page 62 of *Embryology and Evolution,* and Elliot Smith's musings about the occipital region of the Piltdown skull are from *Essays on the Evolution of Man* (Oxford: Oxford University Press, 1927).

G. Elliot Smith's comments on the comparability of the Piltdown skull with juveniles apes are in his *Essays on the Evolution of Man.*

The quote from Dart that begins "Unlike Pithecanthropus" is on page 199 of his article "*Australopithecus africanus:* The man-ape of South Africa," *Nature* 115: 195–199 (1925).

The quotes from Keith that begin "In some of the higher primates" and "[i]n the Taungs skull" are both on page 58 of *New Discoveries Relating to the Antiquity of Man* (New York: W. W. Norton, 1931). The quote that begins "the recognizable markings of the Taungs brain" is on pages 82–83, and the quotes "mental status of Australopithecus" and "[W]e may infer" are on page 86.

The quote from Keith's *New Discoveries* that begins "supporting their bodies" is on pages 113–114.

The large quote that begins "The growth transmutations" is on page 58 of Keith's *New Discoveries.* The quotes that begin "[in] the Taungs skull" and "the discovery at Taungs" and the quotes that follow in the same paragraph are on page 116.

One of the best histories of early fossil-hominid discoveries is Robert Broom's semi-autobiographical work, *Finding the Missing Link,* 2d ed. (London: R. Clay & Co., 1951). The quote that begins "As a paleontologist" is on page 25.

The quote from Sollas's letter to Broom is on page 26 (ibid.).

The articles by Sollas are "The Taungs skull," *Nature* 115: 908–909 (1925); and "A sagittal section of the skull of *Australopithecus africanus,*" *Quarterly Journal of the Geological Society of London* 82: 1–11 (1926).

Broom quotes Cooper's "Come to Sterkfontein . . ." on page 42 of *Finding the Missing Link.*

Broom announced *Australopithecus transvaalensis* in "A new fossil anthropoid skull from South Africa," *Nature* 138: 486–488 (1936).

The quotes from Broom's 1936 *Nature* article—"[t]his newly-found primate . . ." and "I therefore think . . ."—are on page 487.

Broom introduced the genus *Plesianthropus* and the new Kromdraai hominid, *Paranthropus robustus,* in the article "The Pleistocene anthropoid apes of South Africa," *Nature* 142: 377–379 (1938), and, with co-author G. W. H. Schepers, provided much more detail on all South African fossil hominids then known in "The South African Fossil Ape-Men, the Australopithecinae," *Transvaal Museum Memoir,* no. 2: 1–271 (1946).

Broom announced the 1947 skull from Sterkfontein in "Discovery of a new skull of the South African ape-man, *Plesianthropus,*" *Nature* 159: 672 (1947) and, as senior author, published a monograph on the specimen with J. T. Robinson and G. W. H. Schepers, "Sterkfontein Ape-Man, *Plesianthropus,*" *Transvaal Museum Memoir,* no. 4: 1–83 (1950).

Broom introduced *Paranthropus crassidens* in "Another new type of fossil ape-man," *Nature* 163: 57 (1949) and co-authored a volume with J. T. Robinson on this and many more specimens, "Swartkrans Ape-Man, *Paranthropus Crassidens,*" *Transvaal Museum Memoir* no. 6: 1–123 (1952).

Broom quotes de Beer on page 89 of *Finding the Missing Link.*

Abbie's publications include "Headform and human evolution," *Journal of Anatomy* 81: 233–258 (1947); "A new approach to the problem of human evolution," *Transactions of the Royal Society of Southern Australia* 75: 70–88 (1952); and "Timing in human evolution," *Proceedings of the Linneaen Society of New South Wales* 83: 197–213 (1958).

M. F. Ashley Montagu's presentation of Asians as neotenic is in "Time, morphology, and neoteny in the evolution of man," in M. F. Ashley Montagu, ed., *Culture and the Evolution of Man* (New York: Oxford University Press, 1962), 324–342.

Abbie's reference to Schultz's "modern negress" and to his own discovery in a "living Aborigine" is on page 77 of the 1952 article.

The quote that begins "even in his extreme form" is from Abbie's 1952 article, page 79, and the one that begins "[w]ith that possible exception" is on page 83.

"If a common generalized foetal form . . ." is on page 84 of Abbie's 1952 article.

The quotes that begin "Development can be looked upon" and "We have no idea" are on page 212 of Abbie's 1958 article, and "While it is true . . ." is on page 84 of the 1952 article.

D. J. Morton's best-known work is *The Human Foot* (New York: Columbia University Press, 1935).

Chapter 6: Development, Inheritance, and Evolutionary Change

The quote from Castle is on page 6 of his *Heredity* (New York: D. Appleton & Co., 1913).

The quote that begins "Natural selection can act only" is on page 95 of Darwin's *On the Origin of Species by Means of Natural Selection* (London: John Murray, 1859).

Von Baer's classic work is *Über Entwicklungsgeschicte der Thiere: Beobachtung und Reflexion* (Königsberg: Bornträger, 1828).

Haeckel presented his biogenetic law in *Generelle Morphologie der Organismen* (Berlin: Georg Reimer, 1866).

The quotes that begin "the simple fundamental idea," "man produces," "constant preference" and "that the interaction" are on page 99 of Haeckel's *The Evolution of Man: A Popular Exposition of the Principal Points of Human Ontogeny and Phylogeny,* 3d ed. (New York: H. L. Fowle, 1876).

Darwin's arguments for a struggle for existence and natural selection are fully detailed in *On the Origin of Species* and discussed in Loren Eiseley, *Darwin's Century: Evolution and the Men Who Discovered It* (Garden City, N.Y.: Anchor Books, 1961); Ernst Mayr, *The Growth of Biological Thought: Diversity, Evolution, and Inheritance* (Cambridge: Belknap Press of Harvard University Press, 1982); and C. Leon Harris, *Evolution: Genesis and Revelations, with Readings from Empedocles to Wilson* (Albany: State University of New York Press, 1981).

For a review of the history of thought on germ cells, and the discovery of sperm and ova, see Haeckel, *The Evolution of Man* as well as S. F. Mason, *A History of the Sciences* (New York: Collier Books, 1962). Haeckel ascribes Leeuwenhoek's discovery of male spermatozoa to the year 1690, whereas Mason puts it in 1677.

The quote (B47) from Notebook B ("The condition . . .") is on page 182 of *Charles Darwin's Notebooks 1836–1844,* transc. and ed. P. H. Barrett, P. J. Gautrey, S. Berbert, D. Kohn, and S. Smith (Ithaca: Cornell University Press, 1987).

Darwin's distinction between inheriting mutilation and muscle size is in note D18 of Notebook D. (See page 336 of Barrett et al.)

From Notebook D, the quotes that begin "[W]hat has long" and "an animal is <only>" are on page 335 of Barrett et al., and the long quote ("When two animals cross") as well as "whole parent" are on page 336.

The quotes that begin "[w]ith respect to future destinies" and "the tribes become blended" are from Notebook D (D38–39). (See page 343 of Barrett et al.)

The quote from Notebook D (D18) "by a succession of . . ." is on page 336 of Barrett et al.

P. Bowler's discussion of Darwin's rejection of geographic isolation is on pages 212–213 of his revised edition of *Evolution: The History of an Idea* (Berkeley: University of California Press, 1989).

C. Darwin's *The Variation of Animals and Plants under Domestication* was published in 1868 (London: John Murray).

Darwin's Journal was edited and annotated by G. R. de Beer and published as *Bulletin of the British Museum (Natural History) Historical Series,* vol. 2, no. 2 (1959). A complementary chronology of Darwin's activities can be found in vol. 1 of Francis Darwin's edited two-volume collection *More Letters of Charles Darwin* (London: John Murray, 1903).

Darwin's letter to Hooker is on pages 280–281 of vol. 1 of F. Darwin's *More Letters.*

For various and at times differing discussions of Darwin's theory of pangenesis in a historical context, see, for example, Bowler, *History of an Idea;* Eiseley, *Darwin's Century;* and Mayr, *The Growth of Biological Thought.*

The quote that begins "after mature reflection" is on page 438 of part 2 of Darwin's TDM.

Darwin's letter to Huxley is on page 287 of vol. 1 of F. Darwin's *More Letters.*

Darwin's letter to Bentham is on page 363 of vol. 2 of F. Darwin's *More Letters.*

For Hippocrates' works, see *The Medical Works of Hippocrates: A New Translation from the Original Greek Made Especially for English Readers by the Collaboration of John Chadwick and W. N. Mann* (Oxford: Blackwell Scientific Publications, 1950).

Galton's article is "Experiments in pangenesis, by breeding from rabbits of a pure variety, into whose circulation blood taken from other varieties had previously been largely transfused," *Proceedings of the Royal Society (Biology)* 19: 393–404 (1871).

The quote that begins "My results thus far" is on page 403 and the one that begins "if the reproductive elements" is on page 404 of Galton's 1871 article.

Darwin's response to Galton is "Pangenesis," *Nature* 3: 502–503 (1871), and the quotes are from the last page.

Galton's response to Darwin is "Pangenesis," *Nature* 4: 5–6 (1871), and the quote that begins "I feel as if" is from the last page.

The second edition of *The Variation of Animals and Plants under Domestication* was published in 1875 (London: John Murray), as was Galton's "A theory of heredity," *Contempo-*

rary Review 27: 80–95, of which a revised version was published in the *Journal of the Anthropological Institute of Great Britain and Ireland* 5: 329–348 (1876). Discussions, sometimes conflicting, of Darwin and Galton can be found in the previously cited works of Bowler, Eiseley, Harris, and Mayr, and also in R. Olby's *Origins of Mendelism,* 2d ed. (Chicago: University of Chicago Press, 1985).

Darwin's letter of November 7, 1875, to Galton is on pages 360–362 of F. Darwin's *More Letters,* and that of December 18, 1875, is on page 362.

The quote that begins "the personal structure" is on page 331 of Galton's "Theory of Heredity."

The quote from Darwin's November 1875 letter that begins "[u]nless you can make" is on page 360 of F. Darwin's *More Letters.*

Wallace's work is *Darwinism* (London: Macmillan, 1891), and the footnote and the quote from it are on pages 442–443.

R. Virchow, *Cellular Pathology as Based upon Physiological and Pathological History. Twenty Lectures Delivered in the Pathological Institute of Berlin during the Months of February, March and April, 1858,* trans. from the second edition by F. Chance (New York: R. M. DeWitt; reprint, New York: Dover, 1971). The quote is on page 55 of J. A. Moore, ed., *Readings in Heredity and Development* (New York: Oxford University Press, 1972).

Weismann's articles are in *Essays upon Heredity and Kindred Biological Problems,* ed. E. B. Poulton, S. Schonland, and A. E. Shipley (translator) (Oxford: Clarendon Press, 1891–1892).

R. Olby, *Origins of Mendelism.*

The quotes that begin "a general law governing the" and "possible to determine") are on pages 1–2 of Mendel's *Experiments in Plant Hybridisation* (1866, Bateson's English translation with footnotes; reprint, Cambridge: Harvard University Press, 1965).

The quote that begins "as many separate experiments" is on page 4 of Mendel's *Plant Hybridisation.*

The quotes that deal with dominant and recessive characters are on page 8 of Mendel's *Plant Hybridisation.*

See R. Olby, "Mendel no Mendelian?" *History of Science* 7: 53–72 (1979).

A discussion of "A/A + . . ." is on pages 25–26 of Mendel's *Plant Hybridisation.*

The quote that begins "There is therefore" is on page 19 of Mendel's *Plant Hybridisation.*

Chapter 7: Genetics and the Demise of Darwinism

The quote from de Vries is on page 7 of *Species and Varieties: Their Origin by Mutation,* 2d ed. (Chicago: Open Court, 1906).

Slightly different overviews of the discovery of Mendelism can be found in P. J. Bowler, *Evolution: The History of an Idea,* rev. ed. (Berkeley: University of California Press, 1989); Loren Eiseley, *Darwin's Century: Evolution and the Men Who Discovered It* (Garden City, N.Y.: Anchor Books, 1961); C. Leon Harris, *Evolution: Genesis and Revelations, with Readings from Empedocles to Wilson* (Albany: State University of New York Press, 1981); and Ernst Mayr, *The Growth of Biological Thought: Diversity, Evolution, and Inheritance* (Cambridge: Belknap Press of Harvard University Press, 1982).

For more detail on de Vries, see P. J. Bowler, "Hugo de Vries and Thomas Hunt Morgan: The mutation theory and the spirit of Darwinism," *Annals of Science* 35: 55–73 (1978); O. G. Meuer, "Hugo de Vries no Mendelian?" *Annals of Science* 42: 189–232 (1985); and W. B. Provine, *The Origins of Theoretical Population Genetics* (Chicago: University of Chicago Press, 1971), which also summarizes the work of Bateson and Morgan discussed later in this chapter.

The English translation of de Vries's 1889 *Intracellular Pangenesis* was by C. S. Gager (Chicago: Open Court, 1910).

The English translation of de Vries's *The Mutation Theory,* 2 vols., was by J. B. Farmer and A. D. Darbishire (Chicago: Open Court, 1910).

In the paragraph with "Bowler made the point," see Bowler's article on de Vries, "Hugo de Vries and Thomas Hunt Morgan."

See Spencer's 1851 *Social Statics: The Conditions Essential to Human Happiness Specified, and the First of Them Developed* for social Darwinism (New York: Robert Schalkenbach Foundation, 1970).

Among Bateson's works that are of note are *Materials for the Study of Variation, Treated with Especial Regard to Discontinuity in the Origin of Species* (New York: Macmillan, 1894); *Mendel's Principles of Heredity* (Cambridge: Cambridge University Press, 1909); and *Problems of Genetics* (New Haven: Yale University Press, 1913). See also B. Bateson, *William Bateson, Naturalist: His Essays & Addresses Together with a Short Account of His Life* (Cambridge: Cambridge University Press, 1928), as well as Provine's *Population Genetics.*

The quote that begins "My brain boils" is on page 38 of B. Bateson, *Naturalist.*

The quote that begins "most of the elements" is on page 5 of Bateson's *Materials for the Study of Variation.*

The quote that begins "It is the best idea" is on pages 42–43 of B. Bateson, *Naturalist.*

Bateson's article in the *Journal of the Royal Society,* vol. 25, 1900 ("Problems of heredity as a subject for horticultural investigation") is reprinted in B. Bateson, *Naturalist,* 171–180.

The quote that begins "These experiments of Mendel's" is on page 177 of B. Bateson, *Naturalist.*

Bateson and Saunders's 1902 *Reports to the Evolution Committee of the Royal Society* is reprinted in large part on pages 83–123 of J. A. Moore, ed., *Readings in Heredity and Development* (New York: Oxford University Press, 1972).

The long quote that begins "The forms of living things are diverse" is on page 2 of Bateson's *Materials for the Study of Variation.*

The quote that begins "On this hypothesis" is on page 7 of Bateson's *Materials for the Study of Variation.*

The quote from Bateson and Saunders that begins "[W]ith the discovery" is on page 90 of Moore, *Heredity and Development.*

The quote that begins "some degree of sterility" is on page 112 of Moore, *Heredity and Development.*

See E. Mayr's review of species concepts and sterility in, for example, *Systematics and the Origin of Species* (New York: Columbia University Press, 1942); *Animal Species and Evolution* (Cambridge: Belknap Press of Harvard University Press, 1963); and *The Growth of Biological Thought: Diversity, Evolution, and Inheritance* (Cambridge: Belknap Press of Harvard University Press, 1982).

The quote from Bateson and Saunders, "divide up the characters among their gametes," is on page 112 of Moore, *Heredity and Development.*

The quote from B. Bateson, *Naturalist,* regarding the Quick Institute is on page 92, and the two longer quotes in which W. Bateson uses the term "genetics" are on page 93.

The quote that begins "How big a disturbance" is on page 102 in B. Bateson, *Naturalist.*

The quotes in the paragraph that begins "The death" are on pages 109–110 of B. Bateson, *Naturalist.*

See *Mendel's Principles of Heredity* (Cambridge: Cambridge University Press, 1909).

The quote that begins "Research was one long delicious" is on page 115 of B. Bateson, *Naturalist.*

W. Sutton's landmark article is "The chromosomes in heredity," *Biological Bulletin* 4: 231–251 (1902).

For recent work on sex determination and the Y chromosome, see, for example, P. Koopman, J. Gubbay, N. Vivian, P. Goodfellow, and R. Lovell-Badge, "Male development of chromosomally female mice transgenic for *Sry*," *Nature* 351: 117–121 (1991).

The quote that begins "[w]e now can see" is on page 288 of Bateson's *Mendel's Principles.*

F. Jenkin's review is "The Origin of Species," *North British Review* 46: 149–171 (1867).

C. Darwin's letters to Hooker are on pages 373–379 of Francis Darwin's collection (*More Letters of Charles Darwin,* London: John Murray, 1903), and the quote is on p. 379.

See page 209 et seq. of L. Eiseley's *Darwin's Century.*

The full quote from C. Darwin is on page 58 of *The Descent of Man and Selection in Relation to Sex* (London: John Murray, 1871).

The quote that begins "If upon the same individual" is on page 274 of Bateson's *Mendel's Principles,* and the one that begins "divisions by which similar parts" is on page 277.

Chapter 8: Rediscovering Darwin

The quote from Newman is on page 63 of *Evolution, Genetics and Eugenics,* 2d ed. (Chicago: University of Chicago Press, 1925).

Varying histories on genetics in the early twentieth century are provided, for example, by E. G. Wilson, "Mendel's Principles of Heredity and the maturation of the germ-cells," *Science* 16: 991–993 (1902), and "The bearing of cytological research on heredity," *Proceedings of the Royal Society* B 88: 333–352 (1914); T. H. Morgan, "On the mechanism of heredity," *Proceedings of the Royal Society* B 94: 162–197 (1922), and "The rise of genetics," *Science* 76: 261–288 (1932); W. B. Provine, *The Origins of Theoretical Population Genetics* (Chicago: University of Chicago Press, 1971); and P. Bowler, *The Mendelian Revolution* (Baltimore: Johns Hopkins University Press, 1989).

See W. S. Sutton's "The chromosomes in heredity," *Biological Bulletin* 4: 231–251 (1902).

See C. McClung, "The accessory chromosome—sex determinant? *Biological Bulletin* 3: 43 (1902).

Among T. H. Morgan's major books were *Evolution and Adaptation* (1903; reprint, New York: Macmillan, 1908); *A Critique of the Theory of Evolution* (Princeton: Princeton University Press, 1916); *Evolution and Genetics* (Princeton: Princeton University Press, 1925); *The Scientific Basis of Evolution* (New York: W. W. Norton, 1935); and, with A. H. Sturtevant, H. J. Muller, and C. B. Bridges, *The Mechanism of Mendelian Heredity* (1915; rev. and repr., New York: Henry Holt, 1926).

In Morgan's article "What are 'factors' in Mendelian explanations?" *Proceedings of the American Breeder's Association* 5: 365–368 (1909), the quote that begins "In the modern interpretation" is on page 365, the one that begins "We assume" is on page 366, and the one that begins "The egg need not contain" is on page 367.

The quote that begins "The theoretical interpretation" is on page 284, and the one that begins "there can remain" is on page 285 of Morgan's *Evolution and Adaptation.*

The quote that begins "[N]ew species comparable in all respects" is on page 103, and the one that begins "If, on the other hand" is on page 405 (ibid.).

See P. Bowler's article "Hugo de Vries and Thomas Hunt Morgan: The mutation theory and the spirit of Darwinism," *Annals of Science* 35: 55–73 (1978).

Morgan's Batesonian ideas are on pages 286–287 of *Evolution and Adaptation.*

The quote that begins "If, however, a species begins" is on pages 286–287 (ibid.).

The quote from Darwin that begins "It is good thus to try" is on page 79 of *On the Origin of Species by Means of Natural Selection for Life* (London: John Murray, 1859).

The quote from Morgan that begins "The kindliness of heart" is on page 116 of *Evolution and Adaptation,* and the one that begins "The destruction of the unfit" is on page 462.

The quote that begins "If we suppose" is on page 464 (ibid.).

Morgan's 1909 presidential address was published as "Chance or purpose in the origin and evolution of adaptation," *Science* 31: 201—210 (1910); and the quote that begins "They *feel* that some" is on page 202.

Morgan's "the dice are loaded" and the quotes that follow are on page 208; and the quote that begins "Is the battle" and the one that follows are on page 209 (ibid.).

See P. Bowler's article "Hugo de Vries and Thomas Hunt Morgan."

The quotes that deal with racial advance up to and including "To them belongs the future" are on page 210 of Morgan's 1910 *Science* article, "Chance or purpose in the origin and evolution of adaptation."

For his doubts, see Morgan's "Chromosomes and heredity," *American Naturalist* 44: 449–496.

See Morgan's *A Critique.*

The quotes in the paragraph that begins "It was in *A Critique*" are on pages 38–39 (ibid.).

See Morgan, Sturtevant, Muller, and Bridges, *Mechanism of Mendelian Heredity.*

W. Johanssen's work is Elemente der Exakten Erblichkeitslehre (Jena: Gustav Fisher, 1909).

See E. B. Wilson, "The bearing of cytological research on heredity," *Proceedings of the Royal Society* B88: 333–352 (1914).

F. A. Janssens' article was "La théorie de la chiasmatypie. Novelle interpétation des cinèse de maturation," *La Cellule* 25: 389–411 (1909).

The quote that begins "It can not too insistently" is on page 117 of Morgan's *A Critique.*

The quote that begins "I do not think such a group" is on page 85 (ibid.).

The case of *Drosophila simulans* is summarized in Morgan's "On the mechanism of Mendelian heredity."

The quote that begins "Evolution of wild species" is on pages 86–87 of *A Critique.*

Morgan discusses selection in chapter 4, and the quotes in this paragraph are on page 154 (ibid.).

The quotes in the paragraph that begins "But just because mutation" are on pages 187–190 (ibid.).

The quote that begins "Does the elimination" is on page 187 (ibid.).

The quote that begins "New and advantageous characters" is on page 88, and the one that begins "Evolution has taken place" is on page 194 (ibid.).

Chapter 9: Genetics Goes Statistical

The quote from Galton is on page 33 of *Inquiries into Human Faculty and Its Development* (London: J. M. Dent & Sons, 1907).

For historical background, see W. B. Provine, *The Origins of Theoretical Population Genetics* (Chicago: University of Chicago Press, 1971); and P. Bowler, *The Mendelian Revolution* (Baltimore: Johns Hopkins University Press, 1989).

See K. Pearson's "On the principle of homotyposis and its relation to heredity, to the variability of the individual, and to that of the race. Part 1. Homotyposis in the Vegetable Kingdom," *Philosophical Transactions of the Royal Society,* Series A, 197: 285–379 (1901); and "On a generalized theory of alternative inheritance, with special reference to Mendel's laws," *Philosophical Transactions of the Royal Society,* Series A, 203: 53–86 (1904).

See G. U. Yule's "Mendel's laws and their probable relations to inter-racial heredity," *New Phytologist* 1: 194–238 (1902).

See G. H. Hardy's "Mendelian proportions in a mixed population," *Science* 28: 49–50 (1908).

A translation of Weingberg's 1908 article, "On the proof of heredity in humans," is in S. H. Boyer, *Papers on Human Genetics* (Englewood Cliffs, N.J.: Prentice-Hall, 1963), 4–15.

See R. C. Punnett, *Mimicry in Butterflies* (Cambridge: Cambridge University Press, 1915); Norton's table is presented and discussed in appendix 1 (pp. 154–156), and its implications are discussed on pages 94–96.

See Fisher's "The correlation between relatives on the supposition of Mendelian inheritance," *Transactions of the Royal Society of Edinburgh* 52: 399–433 (1918); and *The Genetical Theory of Natural Selection* (1st ed., Oxford: Oxford University Press, 1930; 2d ed., from which these quotes are taken, New York: Dover Publications, 1958); the revisions of 1958 are set off from the original text by a different typeface.

The quote that begins "Natural Selection is not Evolution" is on page vii of *The Genetical Theory,* and the one that begins "the great advance" is on page viii.

The quote that begins "The ordinary mathematical" is on p. ix (ibid.).

The quotes in the paragraph that begins "Fisher, a mathematician and a statistician" are on page ix (ibid.).

The quote that begins "the growth of capital" is on page 26, and the one that begins "[a]ny net advantage" is on page 51 (ibid.).

The theorem of natural selection is on page 37, and the quotes that begin "refers only to the variation" and "affords a rational explanation" are on page 49 (ibid.).

The quote that begins "The income derived" is on page 40 (ibid.).

The quotes that begin "the consequences to the organic world" and "An organism is regarded" are on page 41 (ibid.).

The quote that begins "Numerous cases are now known" is on page 56, and the reference to "unsuccessful competitor" is on page 75 (ibid.).

The quote that begins "[e]volution under such human selection" is on pages 69–70 (ibid.).

The quote that begins "It is to be presumed" is on page 85; the one that begins "The slow changes which must" is on page 103; and the one that begins "the difficulty of effecting" is on page 74 (ibid.).

The quote that begins "an evolutionary improvement in any one" is on page 135 (ibid.).

The quote that begins "the differences in colour" is on page 138 (ibid.).

The quote that begins "In many cases without doubt" is on page 139 (ibid.).

The quote that begins "circumstances unfavourable to sexual union" is on page 140 (ibid.).

The quotes that begin "elimination in each extreme" and "[S]election . . . must act gradually" are on page 142 (ibid.).

The quotes in the paragraph that begins "Fisher's concept of speciation" are on page 161 (ibid.).

For the model of species mate recognition, see, for example, Paterson's "The recognition concept of species," in E. Vrba, ed., *Species and Speciation, Transvaal Museum Monograph* no. 4 (Pretoria: Transvaal Museum, 1985), 21–29.

See C. Leon Harris, *Evolution: Genesis and Revelations, with Readings from Empedocles to Wilson* (Albany: State University of New York Press, 1981).

See Morgan, Sturtevant, Muller, and Bridges, *Mechanism of Mendelian Heredity.*

J. B. S. Haldane, *The Causes of Evolution* (New York: Harper & Brothers, 1932).

The quote that begins "I can write of natural selection" is on page 33 (ibid.).

The passage that begins "I have given my reasons" is on pages 158–159 (ibid.).

The quote that begins "If two animals have a common ancestor" is on page 9 (ibid.).

The quotes in the paragraph that begins "True, Haldane admitted" are on pages 22–23; and the quotes in the following paragraph are on pages 213–214 (ibid.).

The quote in the paragraph that begins "Haldane led his audience" is on page 22 and the longer quote that follows is on pages 22–23 (ibid.).

See A. A. Hallam, "The evolution of *Gryphaea*," *Geological Magazine* 99: 571–574; and N. Eldredge and S. J. Gould, "Punctuated equilibria: An alternative to phyletic gradualism, in T. J. M. Schopf, ed., *Models in Paleobiology* (San Francisco: Freeman, Cooper & Co., 1972), 82–115.

The quote "the isolation of a small unrepresentative group of the population" is on page 102 of Haldane's *The Causes of Evolution.* For Mayr's discussion of models of speciation, see his works of 1942 and 1963 (Notes, chap. 8).

Haldane discusses albinism on page 85 of *The Causes of Evolution* (and the quote is on pages 85–85) and a general case of inbreeding on page 104.

The quote in the paragraph that begins "There could also" and the quotes that precede and follow it are on pages 104–105 (ibid.).

The quotes in the paragraph that begins "As for the rapid pace of evolution" are on pages 105–106 (ibid.).

Haldane's definition of *neoteny* is on page 27; the quote that begins "If human evolution is to continue" is on page 150; and the quotes in the paragraph that begins "Haldane's appreciation" are on pages 102–104 ibid.

A brief history of S. Wright is provided in the introductory comments by J. J. Rutledge to Wright's "The relation of livestock breeding to theories of evolution," *Journal of Animal Science* 46: 1192–1200 (1978), as well as by Wright himself in the text of the article.

For W. Castle, see, for example, his *Heredity in Relation to Evolution and Animal Breeding* (New York: D. Appleton, 1913); the quote that begins "From the evidence" is on page 126, and the long quote is on pages 149–151.

On shorthorns, see Wright's "Mendelian analysis of the pure breeds of livestock II: The Duchess family of Shorthorns as bred by Thomas Bates," *Journal of Heredity* 14: 405–422 (1923).

Wright's initial presentations of the shifting balance theory were "Evolution in a Mendelian population," *Anatomical Record* 44: 287 (1929); "Evolution in Mendelian populations," *Genetics* 16: 97–159 (1931); and "The roles of mutation, inbreeding, crossbreeding and selection in evolution," *Proceedings of the Sixth International Congress of Genetics* 1: 356–366 (1932). His well-known diagram of different evolutionary models relative to selective peaks and low points and the intervening saddles first appeared in the latter publication.

The quotes in the paragraph that begins "From Wright's theoretical perspective" are on page 1196 of Wright's "Relation of Livestock Breeding."

Wright's first papers on the species problem are "Breeding structure of populations in relation to speciation," *American Naturalist* 74: 232–48 (1940); and "The statistical consequences of Mendelian heredity in relation to speciation," in J. S. Huxley, ed., *The New Systematics* (Oxford: Oxford University Press, 1940), 161–183; the quote that begins "the processes by which" is on page 232 of the former.

The quote that begins "The course of evolution" is on pages 365–366 in Wright's "Roles of Mutation."

Wright's first criticism of Fisher's dominance theory was "Fisher's dominance theory," *American Naturalist* 63: 274–279 (1929) and of his book, "Review of *The Genetical Theory of Natural Selection* by R. A. Fisher," *Journal of Heredity* 21: 349–356 (1930).

The long quote that begins "Each gene is held in equilibrium" is on pages 205–207 in Wright's "Statistical theory of evolution," *Journal of the American Statistical Association* 26 (supplement): 201–208 (1931).

The quote that begins "Here we undoubtedly have" is on page 362 of Wright's "Roles of Mutation."

See W. Provine's comments on Wright in the former's compiled volume, *Evolution: Selected Papers of Sewall Wright* (Chicago: University of Chicago Press, 1986).

Chapter 10: The Origin of Species Revisited

The quote from Pearl is on pages 1–2 of *Modes of Research in Genetics* (New York: Macmillan, 1915).

See G. de Beer's *Embryology and Evolution* (Oxford at the Clarendon Press, 1930); the quotes in the paragraph that begin "In his small but to the point" are on pages 74–75 (ibid.).

The quotes in the paragraph that begin "De Beer's brief mention" are on page 17 (ibid.).

The quotes in and following the paragraph that begins "If, de Beer mused" start at the bottom of page 19 and continue through page 21 (ibid.).

For biographical information on Goldschmidt, see Gould's introduction to the former's *The Material Basis of Evolution* (1940; facsimile reprint, New Haven: Yale University Press, 1982). For Dobzhansky, see N. Eldredge's *Reinventing Darwin: The Great Debate at the Hight Table of Evolutionary Theory* (New York: John Wiley & Sons, 1995). For Mayr, see his essay "How I Became a Darwinian," in E. Mayr and W. Provine, eds., *The Evolutionary Synthesis* (Cambridge: Harvard University Press, 1980), 413–423.

Goldschmidt published all of his work on the gypsy moth in the volume *Lymantria, Bibliotheca Genetica* 11: 1–180 (1934), which cites his earlier articles on the subject.

Ford and Huxley's study was "Mendelian genes and rates of development in *Gammarus chevreuxi*," *British Journal of Experimental Biology* 5: 112–134 (1927); and Huxley's was "Time relations in amphibian metamorphosis, with some general considerations," *Science Progress* 17: 606–618 (1923).

For a brief discussion of cretinism, see J. Schwartz, *Skeleton Keys: An Introduction to Human Skeletal Morphology, Development, and Analysis* (New York: Oxford University Press, 1995).

The quote that begins "Some animals go on" is on page 26 of de Beer's *Embryology and Evolution.*

The quote that begins "[l]ittle importance" is on page 30 and the one that begins "whatever the evolutionary mechanism" is on page 29 of Wright's "Physiological and evolutionary theories of dominance," *American Naturalist* 68: 25–53 (1934).

The essence of Wright's physiological theory is on page 32, and the quote about Goldschmidt is on page 34 (ibid.).

Wright summarized Goldschmidt's ideas on species in, for example, *Statistical genetics in relation to evolution: Actualités scientifiques et industrielles, 802,* Exposés de Biométrie et de la Statistique Biologique 13 (Paris: Hermann & Cie., 1939).

Dobzhansky published "A critique of the species concept in biology" in 1935, in *Philosophy of Science* 2: 344–355, and the epigram is on page 344.

The two other quotes in the paragraph that begins "In 1935, Dobzhansky took" are on page 347 (ibid.).

The quote that begins "[t]he discontinuity of the living world" is on page 353, and the two quotes that follow are on page 348 (ibid.).

Dobzhansky's species definition is on page 353 (ibid.).

The quote that begins "[i]t is a matter of observation" is on page 348 (ibid.).

The first edition of Dobzhansky's *Genetics and the Origin of Species* was published by Columbia University Press, New York (1937).

Dobzhansky's quote from Bateson is on page 8; the quotes in the sentence that begins "He, however, was interested" are on page 7; and the long quote that begins "Since evolution" is on pages 11–12 (ibid.).

The quote that begins "The theory of evolution asserts" is on page 7 (ibid.).

The quote that begins "The extreme slowness" is on page 179 (ibid.).

The quote that begins "[a] species that would remain" is on page 79 (ibid.).

See Dobzhansky's "The Y chromosome of *Drosophila pseudoobscura*," *Genetics* 20: 366–376 (1935); and "*Drosophila miranda,* a new species," *Genetics* 20: 377–391.

The long quote that begins "Classical genetics conceive" is on pages 114–115 from Dobzhansky's *Genetics and the Origin of Species.*

The quotes in the paragraph that begins "It was in his chapter" are on page 232 (ibid.).

See Goldschmidt, *Material Basis of Evolution.*

The quote that begins "hair in mammals" is on pages 6–7 (ibid.).

The quote that begins "The change from species to species" and the long quote that begins "A systemic mutation" are on page 206 (ibid.).

The quotes in the paragraph that begins "Goldschmidt was unreserved" are on pages 136–137, and those in the paragraph that begins "He blamed the neo-Darwinian's" are on pages 137–138 (ibid.).

The quotes in the paragraphs that begin "Although Goldschmidt had introduced" and "For Goldschmidt, these examples" are on pages 390–392 (ibid.).

Schindewolf's works are *Palaeontologie, Entwicklungslehre und Genetik* (Berlin: Borntraeger, 1936) and *Basic Questions in Paleontology: Geologic Time, Organic Evolution, and Biological Systematics* (German version 1950; English translation by J. Schaefer. Chicago: University of Chicago Press, 1993); and the quote is on page 352 of the latter.

The long quote that begins "I need only quote Schindewolf" is on page 395 of Goldschmidt's *Material Basis of Evolution.*

Wrights' review was "The Material Basis of Evolution (A review of *The Material Basis of Evolution,* by Richard B. Goldschmidt)," *Scientific Monthly* 53: 165–170 (1941); the short quote is on page 168 and the long one is on page 170.

The long quote that begins "According to Goldschmidt" is on pages 52–53 of Dobzhansky's *Genetics and the Origin of Species,* 2d rev. ed. (New York: Columbia University Press, 1941).

The quote that begins "The present writer" is on page 370 (ibid.).

The quote that begins "That speciation is not an abrupt" is on page 159 of Mayr's *Systematics and the Origin of Species* (New York: Columbia University Press, 1942).

The quote that begins "that the accumulation of small" is on page 158 (ibid.).

The quote that begins "First, there is available" is on page 70 (ibid.).

The quote that begins "Characters are usually lost" is on page 296 (ibid.).

Mayr's biological species definition is on page 120, and his illustration of it is on page 160 (ibid.).

The quotes that begin "A new species develops," "definition contains," and "[g]eographical speciation is thinkable" are on page 155 (ibid.).

The quote that begins "The fact that an eminent" is on page 65 (ibid.).

The quotes in the paragraph that begins "In rejecting Goldschmidt's" are on page 162 (ibid.).

The passage in which Mayr accuses Goldschmidt of deception and taxonomic ignorance is on page 137 (ibid.).

The quotes in and following the paragraph that begins "The ad hominem attack" are on page 155 (ibid.).

The quote that begins "[p]aleontologists have too many" is on page 297 (ibid.).

The quote that begins "[m]acromutations are the determining" is on page 351, and the one that begins "that phylogenetic advance" is on page 343 of Schindewolf's *Basic Questions.*

The quote that begins "[W]e may say" is on page 298 of Mayr, 1942, *Systematics and the Origin of Species.*

Jepsen's account of the beginning of the Synthesis is in the foreword to *Genetics, Paleontology, and Evolution,* G. G. Jepsen; G. G. Simpson, and E. Mayr, eds. (Princeton: Princeton University Press, 1949; reprint, New York: Atheneum, 1963), and the quotes from him are on pages v–vi.

Simpson, *Tempo and Mode in Evolution* (New York: Columbia University Press, 1944), and the quote that begins "When breaks or apparent" is on page 105.

The quotes in the paragraph that begins "In his attempt to reconcile" are on page 58 (ibid.).

The quote that begins "The typical pattern" is on page 123 (ibid.).

Wright's review was "Tempo and Mode in Evolution: A critical review (A review of *Tempo and Mode in Evolution,* by George Gaylord Simpson)," Ecology 26: 415–419 (1945), and the quote is on page 417.

Simpson's chapter in *Genetics, Paleontology, and Evolution* is "Rates of evolution in animals," 205–228; the quotes on genetic and morphologic rates are on page 206, and the one on taxonomic rates is on page 208.

Chapter 11: Toward a New Evolution

The quote from Huxley is on page 307 of *Lay Sermons, Addresses, and Reviews* (New York: D. Appleton & Co., 1986).

See Wright's "Character change, speciation, and the higher taxa," *Evolution* 36: 427–443 (1982) and Simpson's *The Major Features of Evolution* (New York: Simon & Schuster, 1953).

The quote that begins "I am not in a position" is from Wright's "Character Change."

The quotes in the paragraph that begins "In the spirit of the Committee" are on pages 113–114 in Simpson et al., eds., *Genetics, Paleontology, and Evolution* (Princeton: Princeton University Press, 1949).

See Mayr's *Animal Species and Evolution* and *Populations, Species, and Evolution* (Cambridge: Belknap Press of Harvard University Press, 1963).

The quote that begins "[s]peciation proceeds" is on page 296 of Mayr's "Speciation and Systematics," 281–298, in Jepsen et al., *Genetics, Paleontology, and Evolution.*

The long quote that begins "Peripheral isolates, no matter how close" is taken from pages 544–546 of Mayr's *Animal Species.*

The quote that begins "Most habitats are saturated" is on page 513, and those in the sentence that begins "Without citing Haldane" are on page 554 (ibid.).

The quote that begins "evolution is a two-stage" is on page 8 (ibid.).

The quote that begins "Whether function precedes" is on pages 295–296 of Mayr's *Systematics and the Origin of Species* (New York: Columbia University Press, 1942).

The quote that begins "Population geneticists, who have worked" is on page 520 of Mayr's *Animal Species.*

See Mayr, "Taxonomic Categories in Fossil Hominids," *Cold Spring Harbor Symposium on Quantitative Biology* 15: 109–118 (1950).

The quote that begins "I now agree" and the one immediately following are on page 631 of Mayr's *Animal Species.*

See Dobzhansky's "Genetic entities in hominid evolution" (pp. 347–362) and Simpson's "The meaning of taxonomic statements" (pp. 1–31), in S. L. Washburn, ed., *Classification and Human Evolution* (Chicago: Aldine, 1963).

Mayr's discussion of human evolution is in chapter 20 of *Animal Species,* and the quote about *Australopithecus africanus* is on page 640.

The quote that begins "since Recent Man" is on page 638, and the others in the same paragraph are on page 644 (ibid.).

The quote that begins "as concerns the fossil hominids" is on page 638 (ibid.).

G. C. Williams, *Adaptation and Natural Selection: A Critique of Some Current Evolutionary Thought* (Princeton: Princeton University Press, 1966).

The quotes in the paragraphs that begin "The essence of Williams's" and "Williams also reduced" are on pages 24 and 25, respectively (ibid.).

The quote that begins "natural selection works only" is on page 32 (ibid.).

The quote that begins "[m]utation is . . . a necessary" is on page 139 (ibid.).

The quote that begins "The important process" is on page 54 (ibid.).

The quotes in the paragraph that begins "Williams was very clear" are on pages 270–271 (ibid.).

The quote that begins "Its first concern" is on page 258, and the quote "the differential survival of alleles" is on page 270 (ibid.).

The quote that begins "The microevolutionary process that adequately" is on page 31 of Williams's *Natural Selection: Domains, Levels, and Challenges* (New York: Oxford University Press, 1992).

Eldredge's article is "The allopatric model and phylogeny in Paleozoic invertebrates," *Evolution* 25: 156–167 (1971).

Eldredge and Gould's article is "Punctuated equilibria: An alternative to phyletic gradualism," in T. J. M. Schopf, ed., *Models in Paleobiology* (San Francisco: Freeman, Cooper & Co., 1972), 82–115.

The quote that begins "We readily admit" is on pages 98–99 (ibid.).

Lerner's book is *Genetic Homeostasis* (New York: John Wiley & Sons, 1954).

The quote that begins "the basic property" is on page 114, and the quotes that follow are on page 115 of Eldredge and Gould's 1971 publication.

See, for example, Gingerich, "Paleontology and phylogeny: Patterns of evolution at the species level in early Tertiary mammals," *American Journal of Science* 276: 1–28 (1976); the quote is on page 26.

Gould and Eldredge's article is "Punctuated equilibria: The tempo and mode of evolution reconsidered," *Paleobiology* 3: 115–151 (1977).

Stanley's article is "A theory of evolution above the species level," *Proceedings of the National Academy of Sciences* 72: 646–650 (1975).

King and Wilson's article is "Evolution at two levels in humans and chimpanzees," *Science* 188: 107–116 (1975), and the quotes are on page 115.

The quote from Zuckerkandl and the one that follows are on page 138 of Gould and Eldredge's 1977 article in *Paleobiology.*

Wright's discussion of punctuated equilibria is in "Character change, speciation" and his quote from Gould and Eldredge (1977) and his paraphrasing of it are on page 440.

The quote that begins "Of special importance" is on page 439 (ibid.).

The quotes in the paragraph that begins "Wright realized" are on page 442 (ibid.).

Charlesworth, Lande, and Slatkin's article is "A neo-Darwinian commentary on macroevolution," *Evolution* 36: 474–498 (1982), and the quote that begins "there is nothing" and the seven quotes that follow are on page 487.

The quote that begins "that small, successive" and the three quotes that follow are on page 488; the quote that begins "punctuationists claim" is on page 493 (ibid.).

Jackson and Cheetham's article is "Phylogeny reconstruction and the tempo of speciation in cheilostome Bryozoa," *Paleobiology* 20: 407–423 (1994), and the background information was reported by R. Kerr in *Science* 267: 1421–1422 (1995).

Oster and Alberch's article is "Evolution and bifurcation of developmental programs," *Evolution* 36: 444–459 (1982), and the first three quotes are on page 444.

Løvtrup's book is *Epigenetics: A Treatise on Theoretical Biology* (New York: John Wiley & Sons, 1974), and the quote is on page 13.

For an introduction to early experimental transplantation of presumptive dental structures, see, for example, J. W. Osborn and A. R. Ten Cate, *Advanced Dental Histology* (Dorchester: John Wright & Sons, 1976) and various chapters in A. A. Dahlberg's edited volume, *Dental Morphology and Evolution* (Chicago: University of Chicago Press, 1971).

The quote that begins "the construction" is on page 452, and those that begin "far removed from" and "emerges as a consequence" are on page 454 of Oster and Alberch's "Evolution and Bifurcation."

For reviews of pattern formation, see Wolpert, "Positional information and pattern formation in limb development," in R. M. Pratt and R. L. Christiansen, eds., *Current Research Trends in Prenatal Craniofacial Development* (North Holland: Elsevier, 1980), 89–101; and D. Summerbell, "Evidence for regulation of growth, size and pattern in the developing chick limb bud," *Journal of Embryology and Experimental Morphology* 65: 129–150 (1981).

See Lumsden's "Pattern formation in the molar dentition of the mouse," *Journale Biologique Buccale* 7: 77–103 (1979).

See Osborn's "Morphogenetic gradients: Fields versus clones," in P. M. Butler and K. A. Joysey, eds., *Development, Function, and Evolution of Teeth* (New York: Academic Press, 1978), 171–201.

For reviews of homeobox genes, see, for example, D. A. Melton, "Pattern formation during animal development," *Science* 252: 234–241 (1991); articles (by Wolpert, Nüsslein-Volhard, Duboule, Kimble, Rubenstein et al., Patel, Cooley and Theurkauf, Kessler and Melton, and Goldberg et al.) in "Frontiers in biology: Development," *Science* 266: 561–614 (1994); and A. Lumsden and R. Krumlauf, "Patterning the vertebrate neuraxis," *Science* 274: 1109–1115 (1996).

For a review of chick-limb homeobox genes, see C. Tickle, "A tool for transgenesis," *Nature* 358: 188–189 (1992).

Shubin and Alberch's article is "A morphogenetic approach to the origin and basic organization of the tetrapod limb," *Evolutionary Biology* 20: 319–387 (1986).

See Sordino et al., "Hox gene expression in teleost fins and the origin of vertebrate digits," *Nature* 375: 678–681 (1995).

See Muragaki et al., "Altered growth and branching patterns in synpolydactyly caused by mutations in HOXD13," *Science* 272: 548–551 (1996).

Zákány and Duboule, "Synpolydactly in mice with a targeted deficiency in the *HoxD* complex," *Nature* 384: 69–71 (1996).

The article by K. Johnston et al. is "A new spontaneous mouse mutation of *Hoxd13* with a polyalanine expansion and phenotype similar to human synpolydactyly," *Human Molecular Genetics* 7: 1033–1038 (1998).

For limb formation, see, for instance, M. J. Cohn et al., "*Hox9* genes and vertebrate limb specification," *Nature* 387: 97–101 (1997).

See N. Shubin, C. Tabin, and S. Carroll, "Fossils, genes and the evolution of animal limbs," *Nature* 388: 639–648 (1997).

M. Averoff and S. M. Cohen, "Evolutionary origin of insect wings from ancestral gills," *Nature* 385: 627–630 (1997).

For the eye, see R. Quiring, U. Walldorf, U. Kloter, and W. Gehring, "Homology of the eyeless gene of *Drosophila* to the *Small eye* gene in mice and *Aniridia* in humans," *Science* 265: 785–789 (1994); G. Halder, P. Callaerts, and W. Gehring, "Induction of ectopic eyes by targeted expression of the *eyeless* gene in *Drosophila,*" *Science* 267: 1788–1792 (1995); and P. Mathers, A. Grinberg, K. Mahon, and M. Jamrich, "The *Rx* homeobox gene is essential for vertebrate eye development," *Nature* 387: 604–607 (1997).

For segmentation of the head region, see J. L. R. Rubenstein, S. Martinez, K. Shimamura, and L. Puelles, "The embryonic vertebrate forebrain: The prosomeric model," *Science* 266: 578–580 (1994); P. Hunt et al., "A distinct *Hox* code for the branchial region of the vertebrate head," *Nature* 353; 861–864 (1991); and M. Figdor and C. Stern, "Segmental organization of embryonic diencephalon," *Nature* 363: 630–634 (1993).

On the appearance of paired fins/limbs, see C. Tabin and E. Laufer, "*Hox* genes and serial homology," *Nature* 361: 692–693 (1993), and the responses by M. Coates, P. Thorogood and P. Ferretti, and M. Kessel, *Nature* 364: 195–197 (1993).

Chapter 12: The New Evolution

Citations previously referenced will not be repeated here.

The quote from Simpson is on page 222 of G. L. Jepsen, G. G. Simpson, and E. Mayr, eds., *Genetics, Paleontology and Evolution* (Princeton: Princeton University Press, 1949).

For body-axis formation, see A. González-Reyes, H. Elliott, and D. St Johnston, "Polarization of both major body axes in *Drosophila* by *gurken-torpedo* signalling," *Nature* 375: 654–658 (1995).

Radial asymmetry is discussed in C. J. Lowe and G. A. Wray, "Radical alterations in the roles of homeobox genes during echinoderm evolution," *Nature* 389: 718–721 (1997).

On the *distal-less* gene, see, for example, S. B. Carroll et al., "Pattern formation and eyespot determination in butterfly wings," *Science* 265: 109–114 (1994).

R. A. Raff, *The Shape of Life: Genes, Development, and the Evolution of Animal Form* (Chicago: University of Chicago Press, 1996).

Gould's essay on animal body shapes is "Kingdoms with wheels," reprinted from *Natural History,* on pages 158–165 of *Hen's Teeth and Horse's Toes: Further Reflections in Natural History* (New York: W. W. Norton, 1984).

Gould's piece on Geoffroy and Gaskell is "As the worm turns," *Natural History* 106: 24–27, 68–73 (1997).

Dawkins's piece on Nilsson and Pelger is "The eye in a twinkling," *Nature* 368: 690–191 (1994), and the quote that begins "[u]nlike human designers" is on page 690.

For a review of baboon hybrids, see C. J. Jolly, "Species, Subspecies, and Baboon Systematics," in W. H. Kimbel and L. Martin, eds., *Species, Species Concepts, and Primate Evolution* (New York: Plenum Press, 1993), 67–107; and on crayfish, W. Roush, "Hybrids consummate species invasion," *Science* 277: 316–317 (1997).

For discussion of the species mate recognition model, see, for example, Hugh Paterson's "The recognition concept of species," in E. Vrba, ed., *Species and Speciation, Transvaal Museum Monograph No. 4* (Pretoria: Transvaal Museum, 1985), 21–29.

On the effects of the mutant *Hsp90* gene in fruit flies see S. L. Rutherford and S. Lindquist, "Hsp90 as a capacitor for morphological evolution," *Nature* 396: 336–342 (1998); for further discussion of this study see J. H. Schwartz, "Homeobox genes, fossils, and the origin of species," *The Anatomical Record/The New Anatomist* (February 1999).

On the *Manx* gene, see B. Swalla and W. Jeffery, "Requirement of the *Manx* gene for expression of chordate features in a tailless ascidian larva," *Science* 274: 1205–1208 (1996).

On feathered dino-birds, see Ji Qiang, P. J. Currie, and M. A. Norrell, "Two feathered dinosaurs from northeastern China," *Nature* 393: 753–761 (1998).

On teeth in birds, see E. Kollar and C. Fisher, "Tooth induction in chick epithelium: Expression of quiescent genes for enamel synthesis," *Science* 207: 993–995 (1980).

On tooth homeobox genes, see D. Stock et al., "The evolution of the vertebrate *Dlx* gene family," *Proceedings of the National Academy of Sciences* 93: 10858–10863 (1996).

H. Grüneberg, *The Genetics of the Mouse* (English translation) (Cambridge: Cambridge University Press, 1943).

Glossary

allele One of a pair of alternative states of a gene.

amino acid A molecule encoded by a codon that contributes to the formation of a protein molecule.

anthropoid A primate of the suborder Anthropoidea, which includes New and Old World monkeys, lesser and great apes, and hominids.

base Two types of nitrogen-based bases are distinguished, purines and pyrimidines. In DNA, these are adenine and guanine (purines) and cytosine and thymine (pyrimidines); in RNA, uracil takes the place of thymine. In order to form a double helix, the bases on one single DNA strand must be matched with the other kind of base on the other single strand (i.e., a purine bonds with a pyrimidine).

bipedalism Locomotion on two feet.

cenogenesis The emergence of a novelty early in development.

chordate A bilaterally symmetrical animal with distinct head-tail, side-to-side, and up-down axes, as well as paired trunk musculature and either a rodlike or a segmented backbone.

chromosome A self-replicating structure within the nucleus of a cell that is composed of DNA and becomes visible at certain phases of cell duplication and division; chromosomes are paired in somatic cells and single in gametes.

codon A sequence of three nucleotide units in a DNA molecule that encodes an amino acid.

crossing-over The overlapping of chromosomes during duplication or division, which can lead to breakage and an exchange of segments between chromosomes.

DNA Deoxyribonucleic acid is located in the chromosomes in the cell nucleus; it regulates protein synthesis.

dominant allele The alternative state of a gene, which is expressed phenotypically in both homozygotes and heterozygotes.

embryogenesis The course of development of an embryo.

epigenesis A theory of development based on the premise that structural differentiation proceeds from the general to the specific.

epithelium The layer of cells that covers the external and internal surfaces of an organism and its organs.

foramen magnum The opening in the occipital bone through which the spinal cord exits the skull.

gamete A sex cell, either sperm or ovum.

gene A self-replicating unit of heredity; a portion of DNA (i.e., a sequence of nucleotide units) that encodes a protein.

genera The plural of the taxonomic term *genus.*

genotype The genetic aspects of an organism.

heterochrony The differential development of an organism's parts; the term is generally used to distinguish between the effects of differential rates of sexual and somatic maturation.

heterozygosis The process of maintaining heterozygosity in a population.

heterozygosity The condition of having two different alleles.

heterozygote An individual or cell bearing two different alleles, one dominant and the other recessive.

homeobox genes A class of highly conserved regulatory, or control, genes.

hominid A member of the family Hominidae, which includes species of the genera *Australopithecus, Paranthropus,* and *Homo.*

hominoid A primate of the superfamily Hominoidea, which includes the lesser and the great apes as well as hominids; the great apes and the hominids are often referred to as the large-bodied hominoids.

homozygosity The condition of having two alleles of the same kind.

homozygote An individual or cell bearing two alleles of the same kind (i.e., dominant or recessive).

keratin A fibrous protein involved in claw, nail, feather, and horn formation.

locus A site or stretch of DNA on a chromosome.

macroevolution Processes that produce changes that result in the origin of species.

macromutation Major change or mutation; a genetic change that involves large stretches of DNA or chromosome segments.

meiosis The two-stage process of cell multiplication and division that results in the formation of sex cells, each of which has half the number of chromosomes of a somatic cell.

mesenchyme Cells, which may derive from the neural crest, that underlie the epithelium and often interact with it early in, and continue to contribute to, the development of an organ or a structure.

microevolution Processes that produce changes in the representation of features within a species.

micromutation Minor change or mutation; a genetic change at the molecular level.

mitochondrial DNA A DNA molecule that is much shorter than nuclear DNA and specific to the mitochondria, which are organelles that lie outside the nucleus and are involved in cell metabolism; thought to be exclusively maternally inherited.

mitosis The process by which a somatic cell reproduces itself by duplicating and dividing into two daughter cells.

morphology The shape of something; a structure, trait, feature.

mutation Generally, a change at the genetic level that provokes a change in the phenotype of the organism.

neoteny Paedomorphosis achieved by prolonging the rate of somatic maturation so that the individual is frozen in a physically juvenile state upon reaching sexual maturity.

neural crest In central-nervous-system development, the opposing edges that emerge during the invagination phase of neural tube formation.

neural crest cells Cells that are released from the neural crest that migrate, particularly into the presumptive head region, and are involved with the epithelium in organ and structure formation.

nucleotide A component of a DNA or RNA molecule that consists of both a sugar and a phosphate group; in concert with a nitrogen-based base, a nucleotide forms a nucleotide unit within the DNA or RNA molecule.

nucleus In organisms that contain DNA, the primary DNA-containing region within a cell, which is set off by its own membrane.

occipital condyles The raised areas (mammals have one on either side of the foramen magnum) that articulate with the first vertebra.

ontogeny The developmental history of an organism.

oocyte The cell that goes through meiosis, through which, in turn, an ovum is produced.

ovum The female gamete.

paedomorphosis The process of retaining the physically childlike state as a reproductively mature individual; this can occur via either progenesis or neoteny.

pentadactyly The state of having five digits on each hand or foot.

phenotype The physical (nongenetic) aspects of an organism.

phylogeny The evolutionary history of an organism.

polar body One of three tiny cell products of meiosis in which the large ovum constitutes the female gamete.

progenesis Paedomorphosis achieved by accelerating the rate of sexual maturation and, consequently, freezing the individual in a physically juvenile state.

punctuation An evolutionary model of large-scale morphological change that occurs rapidly relative to geologic time.

quadrupedalism Locomotion on four feet.

recessive allele The alternative state of a gene that is expressed phenotypically only in homozygotes; in heterozygotes, it is masked by its dominant counterpart.

regulatory gene A gene that controls the expression or activity of other genes.

retina The surface at the back of the vertebrate eye that bears light-sensing structures (rods and cones) and upon which an image is focused.

RNA Ribonucleic acid; messenger RNA transcribes or copies the stretch of DNA from which a protein will be synthesized and carries this information to the ribosomes outside the nucleus; transfer RNA assembles the amino acids in the sequence encoded in the messenger RNA, following the translation of this sequence by ribosomal RNA.

saltationism An evolutionary model of large-scale morphological change that occurs within a very short period of time because of major organismal reorganization.

somatic Referring to the body.

sperm The male gamete.

spindle During chromosome duplication and division, a nexus of fine fibers that appear to tether chromosomes aligned in the middle of the nucleus to the opposite poles of the nucleus and along which it appears that the chromosomes travel as they migrate to opposite poles of the nucleus.

structural gene A gene that contributes directly to the actual formation of a structure.

tetrapod A vertebrate that has four limbs or is a descendant of a four-limbed vertebrate.

transcription The process by which messenger RNA copies a sequence of DNA that will be used to form a protein.

translation The process by which ribosomal RNA takes information from messenger RNA and uses it to orchestrate transfer RNA to put certain amino acids together to form a particular protein.

tripartite brain A brain subdivided into three lobes (forebrain, midbrain, and hindbrain).

tunicate A sea squirt, which is member of a group of animals called Hemichordata, which, as adults, are sessile; the larva of some species are free-swimming and possess features that are reminiscent of chordates.

vertebrate A chordate that is distinguished by having a segmented backbone.

zygote The cell that results from the fertilization of an ovum by a sperm.

Index